Lectures on
Differential
Geometry

SERIES ON UNIVERSITY MATHEMATICS

Editors:

W Y Hsiang : Department of Mathematics, University of California, Berkeley, CA 94720, USA
T T Moh : Department of Mathematics, Purdue University, W. Lafayette IN 47907, USA
e-mail: ttm@math.purdue.edu   Fax: 317-494-6318
S S Ding : Department of Mathematics, Peking University, Beijing, China
M C Kang : Department of Mathematics, National Taiwan University, Taiwan, China
M Miyanishi : Department of Mathematics, Osaka University, Toyonaka, Osaka 560, Japan

SERIES ON UNIVERSITY MATHEMATICS – VOL. 1

Lectures on
Differential
Geometry

S. S. Chern
University of California, USA

W. H. Chen
Peking University, China

K. S. Lam
California State Polytechnic University, Pomona, USA

World Scientific
Singapore • New Jersey • London • Hong Kong

Published by

World Scientific Publishing Co. Pte. Ltd.

P O Box 128, Farrer Road, Singapore 912805

USA office: Suite 1B, 1060 Main Street, River Edge, NJ 07661

UK office: 57 Shelton Street, Covent Garden, London WC2H 9HE

Library of Congress Cataloging-in-Publication Data
Chern, Shiing-Shen, 1911–
   [Wei Fen chi ho chiang 1. English]
   Lectures on differential geometry / S.S. Chern, W.H. Chen, K.S.
Lam.
     p. cm.
   ISBN 9810234945 ISBN 9810241828 (pbk)
   1. Geometry, Differential. I. Chen, Wei-huan. II. Lam, K. S.
(Kai Shue), 1949– . III. Title.
QA641.C4913 1998
516.3'6--dc21
                                   98-22031
                                       CIP

British Library Cataloguing-in-Publication Data
A catalogue record for this book is available from the British Library.

First published 1999
Reprinted 2000

This book is printed on acid and chlorine free paper.

Printed in Singapore by World Scientific Printers

Preface

The present book is a translation and an expansion of an introductory text based on a lecture series delivered in Peking University (the People's Republic of China) in 1980 by a renowned leader in differential geometry, S.S. Chern. The original Chinese text resulted from the efforts of several colleagues[a], and, in its final form[b], was compiled by Wei-Huan Chen of Peking University. The final rendition of this English translation is carried out by the undersigned, under the guidance of S.S. Chern. It has been revised from a preliminary draft by Hung-Chieh Chang, which was prepared under the supervision of Professor T.T. Moh of Purdue University.

This translation aims at preserving, as far as possible, both the contents and style of Professor Chern's lectures, the hallmarks of which are simplicity, directness, and economy of approach together with in-depth treatments of fundamental topics. It should be suitable as a text or a work of reference for a wide audience, including (but not limited to) advanced undergraduate and beginning graduate students in mathematics, as well as physicists interested in the diverse applications of differential geometry to physics. Our hope is that the material in this text will provide a solid and comprehensive background for more advanced and specialized studies.

It is Professor Chern's opinion that the time is ripe for the subject of Finsler geometry to occupy a more prominent position within university curricula in basic differential geometry. It was already alluded to by Riemann in his famous Habilitation speech of 1854; and its relevance to the calculus of variations was stressed by Hilbert in his 1900 Paris Lecture. Since Finsler's thesis work on it in 1918, the subject has seen many important developments, but has lacked the kind of coherence that characterizes Riemannian geometry. Some remarkable recent work has shown, however, that the more natural starting point of Riemannian geometry is the more general Finsler setting, and that many of the beautiful and deep results in the former have Finslerian counterparts. Professor Chern himself, beginning with his early work in the 1940's and in recent collaborations with David Bao, has initiated crucial steps and paved the way in this research.

In view of these developments, a new and rather lengthy chapter prepared by Kai S. Lam and S.S. Chern on Finsler geometry (Chapter Eight) has been added. The last section in Chapter Five of the original Chinese text on completeness in Riemannian geometry has also been revised and reincorporated as section 7 of the new chapter, which treats completeness in Finsler geometry.

To bring the entire subject of differential geometry into perspective, Professor Chern has written a valuable piece, "Historical Notes", specially for the present English edition. This appears as the new Appendix A. Appendix B, entitled "Differential Geometry and Theoretical Physics", was originally authored by Professor Chern in Chinese and included in the Chinese text. The English translation, by the undersigned, appears for the first time in this book. It is hoped that this essay will stimulate a degree of fruitful discussion between mathematicians and physicists.

Professor Chern is well known for his masterful synthesis of deep geometrical insights and skillful calculations. The present text will bear witness to this immensely fruitful mathematical style. A central theme of the text is that global and local problems of differential geometry are equally interesting and important. Even though local objects such as coordinates in a manifold are devoid of intrinsic meaning, local tools, such as Cartan's exterior differential calculus and Ricci's tensor analysis, are extremely useful in the study of manifolds. Hence these tools have been developed and used extensively; but at the same time, the importance of intrinsic objects with invariant properties under a change of coordinates, such as tangent vector fields, differential forms, etc., is also stressed. Throughout the text, the relationship between local and global properties of a manifold, as exemplified, for example, by the Gauss–Bonnet Theorem (Chapter Five) and the Chern classes (Chapter Seven), is emphasized.

As a physicist with relatively little formal training in mathematics, I take great pleasure in expressing my sincere gratitude to Professor Chern. Not only has he graciously put up with the plodding attempts of a novice, but has also, over the course of many months, provided me with generous support and guidance. In addition, he took great pains to introduce me to the beautiful and fascinating developments in Finsler geometry. I have deeply benefitted from this unique opportunity to collaborate with and learn from a great mathematician. The new chapter on Finsler geometry has relied heavily on the joint work of Bao and Chern [Houston Journal of Mathematics, Vol. 19, No. 1, 135–180 (1993)], and preliminary drafts of an upcoming comprehensive treatise, "An Introduction to Riemann–Finsler Geometry," by Bao, Chern and Shen, to be published by Springer-Verlag. I am deeply grateful to Professor David Bao of the University of Houston, to Professor Zhongmin Shen of Indiana University–Purdue University at Indianapolis, and again to Professor Chern, for allowing me to draw from these materials. A special note of heartfelt thanks is owed to David Bao, who has rendered inestimable help in the preparation of the new chapter, by carefully going over the drafts, offering freely expert advice, and generously providing much needed reference material. In many ways, our chapter on Finsler geometry may be viewed as providing an introduction to the Bao–Chern–Shen treatise mentioned above, and the serious reader who wishes to explore the subject at greater depth is well-advised to pursue that definitive work.

I would also like to thank my colleagues in both the Physics and Mathematics Departments of my home institution for their various kind acts of assistance, support, and encouragement, especially physicists John Fang and Soumya Chakravarti, mathematical physicist Martin Nakashima, and mathematicians Bernard Banks and Charles Amelin. To physicist Dr. Barbara Hoeling I owe a special word of gratitude for helping me translate an early paper by L. Berwald from German to English. In addition, I am grateful to the Faculty Sabbatical Program of the California State University for providing the necessary freedom and time for this project. Last but not least, I wish to thank my wife, Dr. Bonnie J. Buratti, and our three boys, Nathan, Reuben, and Aaron, for their simply being part of a wonderful and supportive family.

The authors are indebted to Hung-Chieh Chang and T.T. Moh for their efforts in the initial translation; and Hu Sen, Chen Wei, A.N. Kobayashi, S.H. Gan and Jitan Lu of World Scientific for their marvelous expediency and professionalism in bringing this book into print. Finally, a big word of thanks to Messrs. Geoff Simms and Andres Cardenas, who endured with good humor the incessant alterations in the draft, and who, with their insuperable skills in LaTeX, rendered the manuscript into its present form.

Kai S. Lam
California State Polytechnic University, Pomona

---

[a]The lecture series was made possible by the kind concern and support of the following colleagues in Peking University: Professors Duan Xue-Fu (段学复), Jiang Ze-Han (江泽涵), Wu Jiang-Lei (吴光磊) and Wu Da-Ren (吴大任). The latter two, as well as Tien Chow (田畴), also read the Chinese manuscript and contributed valuable suggestions. In addition, the following individuals all expended considerable efforts in note-taking and teaching assistance during the lectures: Jiang Xue-Cheng (章学诚), You Cheng-Ye (尤承业), Liu Wang-Jin (刘旺金), Han Nien-Guo (韩念国), Zhou Zuo-Ling (周作领), Liu Ying-Ming (刘应明), Sun Zhen-Zu (孙振祖), and Li An-Min (李安民).

[b]*Lectures on Differential Geometry* (微分几何讲义) by S.S. Chern and Wei-Huan Chen, Peking University Press, 1983. The present translation is based on this edition. A second Chinese edition, under the same title and with no textual changes from the first, was published in 1990 by the Lien Ching Ch'u Pan Shih Yeh Co., Taipei, Republic of China.

Contents

Chapter 1

Differentiable Manifolds

§1–1 Definition of Differentiable Manifolds

Differentiable manifolds are generalizations of Euclidean spaces. Roughly speaking, any given point in a manifold has a neighborhood which is homeomorphic to an open set of a Euclidean space. Hence we can establish local coordinates in a neighborhood of every point. A manifold is then the result of pasting together pieces of a Euclidean space.

We will use $\mathbb{R}$ to represent the field of real numbers. Let

$$\mathbb{R}^m = \left\{ x = (x^1, \ldots, x^m) \,\middle|\, x^i \in \mathbb{R}, \qquad 1 \leq i \leq m \right\}, \tag{1.1}$$

that is, the set of all ordered m-tuples of real numbers. The number x^i is called the i-th **coordinate** of the point $x \in \mathbb{R}^m$. For any $x, y \in \mathbb{R}^m, a \in \mathbb{R}$, let

$$\begin{cases} (x+y)^i = x^i + y^i, \\ (ax)^i = ax^i. \end{cases} \tag{1.2}$$

This defines addition and scalar multiplication in $\mathbb{R}^m$, making $\mathbb{R}^m$ an m-dimensional vector space over $\mathbb{R}$.

Besides this linear structure, $\mathbb{R}^m$ also has a standard topological structure. For $x, y \in \mathbb{R}^m$, define

$$d(x, y) = \sqrt{\sum_{i=1}^{m} (x^i - y^i)^2}. \tag{1.3}$$

It is easy to verify that the function $d(x, y)$ satisfies the following three conditions:

1) $d(x, y) \geq 0$, the equality holds if and only if $x = y$;

1

2) $d(x,y) = d(y,x)$;

3) for any $x, y, z \in \mathbb{R}^m$, we have the inequality $d(x,y) + d(y,z) \geq d(x,z)$.

Hence $d(x,y)$ is a metric on $\mathbb{R}^m$, which makes $\mathbb{R}^m$ a metric space. As such, $\mathbb{R}^m$ has the natural topological structure[a]: the unions of open balls $B_{x,r} = \{y \in \mathbb{R}^m \mid d(x,y) < r\}$ ($x \in \mathbb{R}^m, r > 0$) are the open sets. The m-dimensional vector space $\mathbb{R}^m$ with the metric (1.3) is called the m-dimensional **Euclidean space**.

Suppose f is a real-valued function defined on an open set $U \subset \mathbb{R}^m$. If all the k-th order partial derivatives of f exist and are continuous for $k \leq r$, then we say $f \in C^r(U)$. Here r is some positive integer. If $f \in C^r(U)$ for every positive integer r, then we say $f \in C^\infty(U)$. If f is analytic, i.e., if f can be expressed as a convergent series in a neighborhood of any point of U, then we say $f \in C^\omega(U)$.

Definition 1.1. Suppose M is a Hausdorff space. If for any $x \in M$, there exists a neighborhood U of x such that U is homeomorphic to an open set in $\mathbb{R}^m$, then M is called an m-dimensional **manifold** (or m-dimensional **topological manifold**).

If the homeomorphism in Definition 1.1 is $\varphi_U : U \longrightarrow \varphi_U(U)$, where $\varphi_U(U)$ is an open set in $\mathbb{R}^m$, we call (U, φ_U) a **coordinate chart** of M. Since φ_U is a homeomorphism, for any $y \in U$, we can define the coordinates of y to be the coordinates of $u = \varphi_U(y) \in \mathbb{R}^m$, i.e.

$$u^i = (\varphi_U(y))^i, \quad i = 1, \ldots, m. \tag{1.4}$$

The $u^i, i = 1, \ldots, m$, are called the **local coordinates** of the point $y \in U$.

Suppose (U, φ_U) and (V, φ_V) are two coordinate charts of M. If $U \cap V \neq \varnothing$, then $\varphi_U(U \cap V)$ and $\varphi_V(U \cap V)$ are two nonempty open sets in $\mathbb{R}^m$, and the map

$$\varphi_V \circ \varphi_U^{-1}\big|_{\varphi_U(U \cap V)} : \varphi_U(U \cap V) \longrightarrow \varphi_V(U \cap V)$$

defines a homeomorphism between these two open sets, with inverse given by

$$\varphi_U \circ \varphi_V^{-1}\big|_{\varphi_V(U \cap V)}.$$

These are both maps between open sets in a Euclidean space. Expressed in coordinates, $\varphi_V \circ \varphi_U^{-1}$ and $\varphi_U \circ \varphi_V^{-1}$ each represents m real-valued functions

---

[a]For fundamental topological concepts, see for instance Munkres 1975.

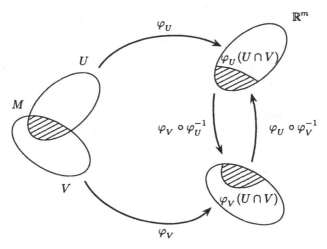

on an open set of a Euclidean space (see Figure 1). We may write

$$y^i = f^i\left(x^1,\ldots,x^m\right) = \left(\varphi_V \circ \varphi_U^{-1}\left(x^1,\ldots,x^m\right)\right)^i, \qquad (1.5)$$
$$\left(x^1,\ldots,x^m\right) \in \varphi_U(U \cap V);$$
$$x^i = g^i\left(y^1,\ldots,y^m\right) = \left(\varphi_U \circ \varphi_V^{-1}\left(y^1,\ldots,y^m\right)\right)^i, \qquad (1.6)$$
$$\left(y^1,\ldots,y^m\right) \in \varphi_V(U \cap V).$$

Since $\varphi_V \circ \varphi_U^{-1}$ and $\varphi_U \circ \varphi_V^{-1}$ are homeomorphisms inverse to each other, f^i and g^i are continuous functions, and

$$\begin{cases} f^i\left(g^1\left(y^1,\ldots,y^m\right),\cdots,g^m\left(y^1,\ldots,y^m\right)\right) = y^i, \\ g^i\left(f^1\left(x^1,\ldots,x^m\right),\cdots,f^m\left(x^1,\ldots,x^m\right)\right) = x^i. \end{cases} \qquad (1.7)$$

We say that the coordinate charts (U,φ_U) and (V,φ_V) are C^r-**compatible** if $U \cap V = \varnothing$, and if $f^i\left(x^1,\ldots,x^m\right)$ and $g^i\left(y^1,\ldots,y^m\right)$ are C^r when $U \cap V \neq \varnothing$.

Definition 1.2. Suppose M is an m-dimensional manifold. If a given set of coordinate charts $\mathcal{A} = \left\{(U,\varphi_U),(V,\varphi_V),(W,\varphi_W),\cdots\right\}$ on M satisfies the following conditions, then we call $\mathcal{A}$ a C^r-**differentiable structure** on M:

1) $\{U,V,W,\ldots\}$ is an open covering of M;

2) any two coordinate charts in $\mathcal{A}$ are C^r-compatible;

3) $\mathcal{A}$ is **maximal**, i.e., if a coordinate chart $(\tilde{U}, \varphi_{\tilde{U}})$ is C^r-compatible with all coordinate charts in $\mathcal{A}$, then $(\tilde{U}, \varphi_{\tilde{U}}) \in \mathcal{A}$.

If a C^r-differentiable structure is given on M, then M is called a C^r-**differentiable manifold**. A coordinate chart in a given differentiable structure is called a **compatible (admissible) coordinate chart** of M. From now on, a **local coordinate system** of a point p on a differentiable manifold M refers to a coordinate system obtained from an admissible coordinate chart containing p.

Remark 1. Conditions 1) and 2) in Definition 1.2 are primary. It is not hard to show that if a set $\mathcal{A}'$ of coordinate charts satisfies 1) and 2), then for any positive integer s, $0 < s \le r$, there exists a unique C^s-differentiable structure $\mathcal{A}$ such that $\mathcal{A}' \subset \mathcal{A}$. In fact, suppose $\mathcal{A}$ represents the set of all coordinate charts which are C^s-compatible with every coordinate chart in $\mathcal{A}'$, then $\mathcal{A}$ is a C^s-differentiable structure uniquely determined by $\mathcal{A}'$. Hence, to construct a differentiable manifold, we need only choose a covering by compatible charts.

Remark 2. In this book, we also assume that any manifold M is a second countable topological space, i.e., M has a countable topological basis (see footnote on page 2).

Remark 3. If a C^∞-differentiable structure is given on M, then M is called a **smooth manifold**. If M has a C^ω-differentiable structure, then M is called an **analytic manifold**. In this book, we are mostly interested in smooth manifolds. When there is no confusion, the term manifold will mean smooth manifold.

Example 1. For $M = \mathbb{R}^m$, let $U = M$ and φ_U be the identity map. Then $\{(U, \varphi_U)\}$ is a coordinate covering of $\mathbb{R}^m$. This provides a smooth differentiable structure on $\mathbb{R}^m$, called the **standard differentiable structure** of $\mathbb{R}^m$.

Example 2. Consider the m-dimensional unit sphere

$$S^m = \left\{ x \in \mathbb{R}^{m+1} \, \middle| \, \left(x^1\right)^2 + \cdots + \left(x^{m+1}\right)^2 = 1 \right\}.$$

For $m = 1$, take the following four coordinate charts:

$$\begin{cases} U_1 \left\{ x \in S^1 \, \middle| \, x^2 > 0 \right\}, \varphi_{U_1}(x) = x^1, \\ U_2 \left\{ x \in S^1 \, \middle| \, x^2 < 0 \right\}, \varphi_{U_2}(x) = x^1, \\ V_1 \left\{ x \in S^1 \, \middle| \, x^1 > 0 \right\}, \varphi_{V_1}(x) = x^2, \\ V_2 \left\{ x \in S^1 \, \middle| \, x^1 < 0 \right\}, \varphi_{V_2}(x) = x^2. \end{cases} \qquad (1.8)$$

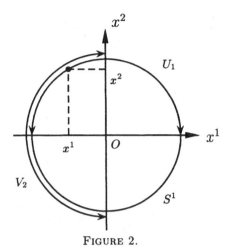

Obviously, $\{U_1, U_2, V_1, V_2\}$ is an open covering of S^1. In the intersection $U_1 \cap V_2$, we have (see Figure 2)

$$\begin{cases} x^2 = \sqrt{1 - (x^1)^2} \; > 0, \\ x^1 = -\sqrt{1 - (x^2)^2} < 0. \end{cases} \tag{1.9}$$

These are both C^∞ functions, thus $\left(U_1, \varphi_{U_1}\right)$ and $\left(V_2, \varphi_{V_2}\right)$ are C^∞-compatible. Similarly, any other pair of the given coordinate charts are C^∞-compatible. Hence these coordinate charts suffice to make S^1 a 1-dimensional smooth manifold. For $m > 1$, the smooth structure on S^m can be defined similarly.

Example 3. The m-dimensional projective space P^m. Define a relation $\sim$ in $\mathbb{R}^{m+1} - \{0\}$ as follows: for $x, y \in \mathbb{R}^{m+1} - \{0\}$, $x \sim y$ if and only if there exists a real number a such that $x = ay$. Obviously, $\sim$ is an equivalence relation. For $x \in \mathbb{R}^{m+1} - \{0\}$, denote the equivalence class of x by

$$[x] = \left[x^1, \dots, x^{m+1}\right].$$

The m-dimensional projective space is the quotient space

$$\begin{aligned} P^m &= \left(\mathbb{R}^{m+1} - \{0\}\right)/\sim \\ &= \left\{[x] \,|\, x \in \mathbb{R}^{m+1} - \{0\}\right\}. \end{aligned} \tag{1.10}$$

The numbers of the $(m+1)$-tuple $(x^1, \ldots, x^{m+1})$ are called the **homogeneous coordinates** of $[x]$. They are determined by $[x]$ up to a a nonzero factor. P^m is thus the space of all straight lines in $\mathbb{R}^{m+1}$ which pass through the origin.

Let

$$
\begin{cases}
U_i = \left\{ [x^1, \ldots, x^{m+1}] \,\middle|\, x^i \neq 0 \right\}, \\
\varphi_i([x]) = (\,_i\xi_1, \ldots, \,_i\xi_{i-1}, \,_i\xi_{i+1}, \ldots, \,_i\xi_{m+1}),
\end{cases}
\tag{1.11}
$$

where $1 \leq i \leq m+1$, $_i\xi_h = x^h/x^i$ $(h \neq i)$. Obviously, $\{U_i, 1 \leq i \leq m+1\}$ forms an open covering of P^m. On $U_i \cap U_j$, $i \neq j$, the change of coordinates is given by

$$
\begin{cases}
_j\xi_h = \dfrac{_i\xi_h}{_i\xi_j}, \qquad h \neq i, j; \\[2mm]
_j\xi_i = \dfrac{1}{_i\xi_j}.
\end{cases}
\tag{1.12}
$$

Hence $\{(U_i, \varphi_i)\}_{1 \leq i \leq m+1}$ suffices to generate a smooth structure on P^m.

Remark. In each of the above examples, the respective coordinate charts given are in fact C^ω-compatible also, and so provide the structures for $\mathbb{R}^m$, S^m, and P^m as analytic manifolds.

Example 4 (Milnor's Exotic Sphere). There may exist distinct differentiable structures on a single topological manifold. J. Milnor gave a famous example (Milnor 1956), which shows that there exist nonisomorphic smooth structures on homeomorphic topological manifolds (see the discussion following the remark to definition 1.3 below). Hence a differentiable structure is more than a topological structure. A complete understanding of the Milnor sphere is outside the scope of this text. Here we will give only a brief description of the main ideas. [A more recent example is the existence of distinct smooth structures on $\mathbb{R}^4$ discovered by S. K. Donaldson (see Donaldson and Kronheimer 1991)].

Choose two antipodal points A and B in S^4. Let

$$
U_1 = S^4 - \{A\}, \qquad U_2 = S^4 - \{B\}.
\tag{1.13}
$$

Then U_1 and U_2 form an open covering of S^4. We wish to paste the trivial sphere bundles $U_1 \times S^3$ and $U_2 \times S^3$ together to get the 3-sphere bundle Σ^7 over S^4.

Under the stereographic projection, U_1 and U_2 are both homeomorphic to $\mathbb{R}^4$, and $U_1 \cap U_2$ is homeomorphic to $\mathbb{R}^4 - \{0\}$. Identify the elements of $\mathbb{R}^4 - \{0\}$ as quaternions, and choose an odd number κ, where $\kappa^2 - 1 \not\equiv 0 \bmod 7$.

Consider the map $\tau : \left(\mathbb{R}^4 - \{0\}\right) \times S^3 \longrightarrow \left(\mathbb{R}^4 - \{0\}\right) \times S^3$, such that for every $(u, v) \in \left(\mathbb{R}^4 - \{0\}\right) \times S^3$, we have

$$\tau(u, v) = \left(\frac{u}{\|u\|^2}, \frac{u^h v u^j}{\|u\|} \right), \tag{1.14}$$

where

$$h = \frac{\kappa + 1}{2}, \qquad j = \frac{1 - \kappa}{2}, \tag{1.15}$$

and in (1.14) the multiplication and the norm $\|\ \|$ are in the sense of quaternions. Obviously τ is a smooth map. We can thus paste $U_1 \times S^3$ and $U_2 \times S^3$ together using τ. It can be proved that the Σ^7 constructed in this way is homeomorphic to the 7-dimensional unit sphere S^7, but its differentiable structure is different from the standard differentiable structure of S^7 (Example 2).

On a smooth manifold, the concept of a smooth function is well-defined. Let f be a real-valued function defined on an m-dimensional smooth manifold M. If $p \in M$, and (U, φ_U) is a compatible coordinate chart containing p, then $f \circ \varphi_U^{-1}$ is a real-valued function defined on the open subset $\varphi_U(U)$ of the Euclidean space $\mathbb{R}^m$. If $f \circ \varphi_U^{-1}$ is C^∞ at the point $\varphi_U(p) \in \mathbb{R}^m$, we say that the function f is C^∞ at $p \in M$.

The differentiability of the function f at the point p is independent of the choice of the compatible coordinate chart containing p. In fact, for another compatible coordinate chart (V, φ_V) containing p such that $U \cap V \neq \varnothing$, we have

$$f \circ \varphi_V^{-1} = \left(f \circ \varphi_U^{-1} \right) \circ \left(\varphi_U \circ \varphi_V^{-1} \right).$$

Since $\varphi_U \circ \varphi_V^{-1}$ is smooth, we see that $f \circ \varphi_V^{-1}$ and $f \circ \varphi_U^{-1}$ are differentiable at the same point p.

If the real-valued function f is C^∞ at every point in M, then we call f a C^∞, or **smooth**, function on M. We shall denote the set of all smooth functions on M by $C^\infty(M)$.

Smooth real-valued functions are just important special cases of smooth maps between smooth manifolds.

Definition 1.3. Suppose $f : M \longrightarrow N$ is a continuous map from one smooth manifold M to another, N, where $\dim M = m$ and $\dim N = n$. If there exist compatible[b] coordinate charts (U, φ_U) at the point $p \in M$ and (V, ψ_V) at $f(p) \in N$ such that the map

$$\psi_V \circ f \circ \varphi_U^{-1} : \varphi_U(U) \longrightarrow \psi_V(V)$$

---

[b]That is, contained in the smooth structures of the respective manifolds.

is C^∞ at the point $\varphi_U(p)$, then the map f is called C^∞ at p. If the map f is C^∞ at every point p in M, then we say that f is a **smooth map** from M to N.

Remark. Since $\psi_V \circ f \circ \varphi_U^{-1}$ is a continuous map from an open set $\varphi_U(U) \subset \mathbb{R}^m$ to another open set $\psi_V(V) \subset \mathbb{R}^n$, its differentiability at the point $\varphi_U(p)$ is defined. Obviously the differentiability of f at p is independent of the choice of compatible coordinate charts (U, φ_U) and (V, φ_V).

In the case $\dim M = \dim N$, if $f : M \longrightarrow N$ is a homeomorphism and f, f^{-1} are both smooth maps, then we call $f : M \longrightarrow N$ a **diffeomorphism**. If the smooth manifolds M and N are diffeomorphic, then we say that the corresponding smooth structures of the manifolds are **isomorphic**. In the above example, the Milnor sphere Σ^7 is homeomorphic but not diffeomorphic to S^7. Hence their smooth structures are not isomorphic.

Another important special case of smooth maps between smooth manifolds is that of **parametrized curves** on manifolds, in which M is an open interval $(a, b) \subset \mathbb{R}^1$. A smooth map $f : (a, b) \longrightarrow N$ from M to the manifold N is a parametrized curve in the manifold N.

Now suppose M and N are m-dimensional and n-dimensional manifolds with differentiable structures $\{(U_\alpha, \varphi_\alpha)\}_{\alpha \in A}$ and $\{(V_\beta, \psi_\beta)\}_{\beta \in \mathcal{B}}$, respectively. We can construct a new $(m + n)$-dimensional smooth manifold $M \times N$ by the following method. First, we see that $\{U_\alpha \times V_\beta\}_{\alpha \in A, \beta \in \mathcal{B}}$ forms an open covering of the topological product space $M \times N$. Then we define maps $\varphi_\alpha \times \psi_\beta : U_\alpha \times V_\beta \longrightarrow \mathbb{R}^{m+n}$ such that

$$\varphi_\alpha \times \psi_\beta(p, q) = (\varphi_\alpha(p), \psi_\beta(q)), \qquad (1.16)$$
$$(p, q) \in U_\alpha \times V_\beta.$$

Thus $(U_\alpha \times V_\beta, \varphi_\alpha \times \psi_\beta)$ is a coordinate chart of $M \times N$. It is easy to prove that all the coordinate charts obtained in this way are C^∞-compatible, and hence they determine a smooth differentiable structure on $M \times N$.

Definition 1.4. The smooth differentiable structure determined by the C^∞-compatible coordinate covering $\{(U_\alpha \times V_\beta, \varphi_\alpha \times \psi_\beta)\}_{\alpha \in A, \beta \in \mathcal{B}}$ of the topological product space $M \times N$ makes $M \times N$ an $(m + n)$-dimensional smooth manifold, called the **product manifold** of M and N.

The natural projections of the product manifold $M \times N$ onto its factors are denoted by

$$\pi_1 : M \times N \longrightarrow M, \qquad \pi_2 : M \times N \longrightarrow N,$$

where, for any $(x, y) \in M \times N$,

$$\pi_1(x, y) = x, \qquad \pi_2(x, y) = y.$$

Obviously these are both smooth maps.

§1–2   Tangent Spaces

At every point on a regular curve (or surface), we have the notion of the tangent line (or tangent plane). Similarly, given a differentiable structure on a topological manifold, we can approximate a neighborhood of any point by a linear space. More precisely, the concepts of the tangent space and the cotangent space can be introduced. We begin with the cotangent space.

Suppose M is an m-dimensional smooth manifold. Fix a point $p \in M$, and let f be a C^∞ function[c] defined in a neighborhood of p. Denote the set of all these functions by C_p^∞. Naturally, the domains of two different functions in C_p^∞ may be different, but addition and multiplication in the function space C_p^∞ are still well-defined. Suppose $f, g \in C_p^\infty$ with domains U and V respectively. Then $U \cap V$ is also a neighborhood containing p. Thus $f + g$ and $f \cdot g$ can be defined as functions on $U \cap V$, that is, $f + g$ and $f \cdot g \in C_p^\infty$.

Define a relation $\sim$ in C_p^∞ as follows. Suppose $f, g \in C_p^\infty$. Then $f \sim g$ if and only if there exists an open neighborhood H of the point p such that $f|_H = g|_H$. Obviously $\sim$ is an equivalence relation in C_p^∞. We will denote the equivalence class of f by $[f]$, which is called a C^∞-**germ** at p on M. Let

$$\mathcal{F}_p = C_p^\infty / \sim = \left\{ [f] \,\middle|\, f \in C_p^\infty \right\}.$$

Then, by defining addition and scalar multiplication, $\mathcal{F}_p$ becomes a linear space over $\mathbb{R}$: for $[f], [g] \in \mathcal{F}_p, a \in \mathbb{R}$, define

$$\begin{cases} [f] + [g] = [f + g], \\ \quad a[f] = [af]. \end{cases} \tag{2.2}$$

In this definition, the right hand sides of (2.2) are independent of the choices of $f \in [f]$ and $g \in [g]$. The reader should verify that $\mathcal{F}_p$ is an infinite-dimensional real linear space.

Suppose γ is a parametrized curve in M through a point p. Then there exists a positive number δ such that $\gamma : (-\delta, \delta) \longrightarrow M$ is a C^∞ map and $\gamma(0) = p$. Denote the set of all these parametrized curves by Γ_p.

For $\gamma \in \Gamma_p, [f] \in \mathcal{F}_p$, let (see Figure 3)

$$\ll \gamma, [f] \gg = \left. \frac{d(f \circ \gamma)}{dt} \right|_{t=0}, \qquad -\delta < t < \delta. \tag{2.3}$$

---

[c]Suppose f is a function defined on an open set $V \subset M$. If the function $f \circ \varphi_U^{-1}$ is C^∞ on the open set $\varphi_U(U \cap V) \subset \mathbb{R}^m$ for any admissible coordinate chart (U, φ_U), where $U \cap V \neq \varnothing$, then we say f is a C^∞ function defined on V. In fact, V has a differentiable structure induced from M (see section §1–3). Thus f is a C^∞ function on the differentiable manifold V.

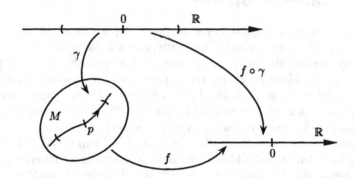

FIGURE 3.

Obviously, for a fixed γ, the value on the right hand side above is determined by $[f]$ and independent of the choice of $f \in [f]$. Also, $\ll \, , \gg$ is linear in the second variable, i.e., for arbitrary $\gamma \in \Gamma_p$, $[f], [g] \in \mathcal{F}_p$, $a \in \mathbb{R}$, we have

$$\ll \gamma, [f] + [g] \gg \, = \, \ll \gamma, [f] \gg + \ll \gamma, [g] \gg,$$
$$\ll \gamma, a[f] \gg \, = \, a \ll \gamma, [f] \gg . \tag{2.4}$$

Let

$$\mathcal{H}_p = \{[f] \in \mathcal{F}_p \mid \, \ll \gamma, [f] \gg \, = 0, \quad \forall \gamma \in \Gamma_p \}. \tag{2.5}$$

Then $\mathcal{H}_p$ is a linear subspace of $\mathcal{F}_p$.

Theorem 2.1. *Suppose $[f] \in \mathcal{F}_p$. For an admissible coordinate chart (U, φ_U), let*

$$F\left(x^1, \ldots, x^m\right) = f \circ \varphi_U^{-1}\left(x^1, \ldots, x^m\right). \tag{2.6}$$

Then $[f] \in \mathcal{H}_p$ if and only if

$$\left. \frac{\partial F}{\partial x^i} \right|_{\varphi_U(p)} = 0, \qquad 1 \leq i \leq m.$$

Proof. Suppose $\gamma \in \Gamma_p$, with coordinate representation

$$\left(\varphi_U \circ \gamma(t)\right)^i = x^i(t), \qquad -\delta < t < \delta. \tag{2.7}$$

$$\ll \gamma, [f] \gg = \frac{d}{dt}(f \circ \gamma)\Big|_{t=0}.$$

$$= \frac{d}{dt} F\left(x^1(t), \cdots, x^m(t)\right)\Big|_{t=0} \qquad (2.8)$$

$$= \sum_{i=1}^{m} \left(\frac{\partial F}{\partial x^i}\Big|_{\varphi_U(p)} \cdot \frac{dx^i(t)}{dt}\Big|_{t=0}\right).$$

Since we may choose the appropriate γ to get any real value for $\frac{dx^i(t)}{dt}\Big|_{t=0}$, a necessary and sufficient condition for $\ll \gamma, [f] \gg = 0$ for arbitrary $\gamma \in \Gamma_p$ is

$$\frac{\partial F}{\partial x^i}\Big|_{\varphi_U(p)} = 0, \qquad 1 \le i \le m.$$

$\square$

We can summarize Theorem 2.1 as follows. The subspace $\mathcal{H}_p$ is exactly the linear space of germs of smooth functions whose partial derivatives with respect to local coordinates all vanish at p.

Definition 2.1. The quotient space $\mathcal{F}_p/\mathcal{H}_p$ is called the **cotangent space** of M at p, denoted by T_p^* (or $T_p^*(M)$). The $\mathcal{H}_p$-equivalence class of the function germ $[f]$ is denoted by $\widetilde{[f]}$ or $(df)_p$, and is called a **cotangent vector** on M at p.

T_p^* is a linear space. It has a linear structure induced from the linear space $\mathcal{F}_p$, i.e. for $[f], [g] \in \mathcal{F}_p$, $a \in \mathbb{R}$ we have

$$\begin{cases} \widetilde{[f]} + \widetilde{[g]} = (\widetilde{[f] + [g]}), \\ a \cdot \widetilde{[f]} = (\widetilde{a[f]}). \end{cases} \qquad (2.9)$$

Theorem 2.2. *Suppose* $f^1, \cdots, f^s \in C_p^\infty$ *and* $F\left(y^1, \cdots, y^s\right)$ *is a smooth function in a neighborhood of* $\left(f^1(p), \cdots, f^s(p)\right) \in \mathbb{R}^s$. *Then* $f = F\left(f^1, \cdots, f^s\right) \in C_p^\infty$ *and*

$$(df)_p = \sum_{k=1}^{s} \left[\left(\frac{\partial F}{\partial f^k}\left(f^1(p), \cdots, f^s(p)\right)\right) \cdot (df^k)_p\right]. \qquad (2.10)$$

Proof. Suppose the domain of f^k containing p is U_k. Then f is defined in $\bigcap_{k=1}^{s} U_k$, and for $q \in \bigcap_{k=1}^{s} U_k$,

$$f(q) = F\left(f^1(q), \cdots, f^s(q)\right).$$

Since F is a smooth function, $f \in C_p^\infty$. Let $a_k = \dfrac{\partial F}{\partial f^k}\left(f^1(p), \cdots, f^s(p)\right)$. Then for any $\gamma \in \Gamma_p$,

$$
\begin{aligned}
\ll \gamma, [f] \gg &= \left.\frac{d}{dt}\right|_{t=0} (f \circ \gamma) \\
&= \left.\frac{d}{dt}\right|_{t=0} F\left(f^1 \circ \gamma(t), \cdots, f^s \circ \gamma(t)\right) \\
&= \sum_{k=1}^{s} a_k \left.\frac{d}{dt}\right|_{t=0} \left(f^k \circ \gamma(t)\right) \\
&= \ll \gamma, \sum_{k=1}^{s} a_k \left[f^k\right] \gg .
\end{aligned}
$$

Thus

$$
[f] - \sum_{k=1}^{s} a_k \left[f^k\right] \in \mathcal{H}_p,
$$

i.e.,

$$
(df)_p = \sum_{k=1}^{s} a_k (df^k)_p.
$$

$\square$

Corollary 1. *For any* $f, g \in C_p^\infty$, $a \in \mathbb{R}$, *we have*

$$
d(f + g)_p = (df)_p + (dg)_p, \tag{2.11}
$$
$$
d(af)_p = a \cdot (df)_p, \tag{2.12}
$$
$$
d(fg)_p = f(p) \cdot (dg)_p + g(p) \cdot (df)_p. \tag{2.13}
$$

We see that (2.11) and (2.12) are the same as (2.9), and (2.13) follows directly from Theorem 2.2. $\square$

Corollary 2. $\dim T_p^* = m$.

Proof. Choose an admissible coordinate chart (U, φ_U), and define local coordinates u^i by

$$
u^i(q) = (\varphi_U(q))^i = x^i \circ \varphi_U(q), \quad q \in U, \tag{2.14}
$$

where x^i is a given coordinate system in $\mathbb{R}^m$. Then $u^i \in C_p^\infty$, $(du^i)_p \in T_p^*$. We will prove that $\{(du^i)_p, 1 \leq i \leq m\}$ is a basis for T_p^*.

Suppose $(df)_p \in T_p^*$. Then $f \circ \varphi_U^{-1}$ is a smooth function defined on an open set of $\mathbb{R}^m$. Let $F(x^1, \ldots, x^m) = f \circ \varphi_U^{-1}(x^1, \ldots, x^m)$. Thus

$$f = F(u^1, \ldots u^m). \tag{2.15}$$

By Theorem 2.2,

$$(df)_p = \sum_{i=1}^m \left[\left(\frac{\partial F}{\partial u^i}(u^1(p), \cdots, u^m(p)) \right) \cdot (du^i)_p \right]. \tag{2.16}$$

Thus $(df)_p$ is a linear combination of the $(du^i)_p$, $1 \leq i \leq m$.

If there exist real numbers $a_i, 1 \leq i \leq m$, such that

$$\sum_{i=1}^m a_i(du^i)_p = 0, \tag{2.17}$$

i.e.

$$\sum_{i=1}^m a_i \left[u^i \right] \subset \mathcal{H}_p,$$

then for any $\gamma \in \Gamma_p$, we have

$$\ll \gamma, \sum_{i=1}^m a_i \left[u^i \right] \gg = \sum_{i=1}^m a_i \left. \frac{d(u^i \circ \gamma(t))}{dt} \right|_{t=0} = 0. \tag{2.18}$$

Choose $\lambda_k \in \Gamma_p, 1 \leq k \leq m$ such that

$$u^i \circ \lambda_k(t) = u^i(p) + \delta_k^i t, \tag{2.19}$$

where

$$\delta_k^i = \begin{cases} 1, & i = k, \\ 0, & i \neq k. \end{cases}$$

Then

$$\left. \frac{d(u^i \circ \lambda_k(t))}{dt} \right|_{t=0} = \delta_k^i.$$

Let $\gamma = \lambda_k$. By (2.18) $a_k = 0, 1 \leq k \leq m$, i.e., $\{(du^i)_p, 1 \leq i \leq m\}$ is linearly independent. Therefore it forms a basis for T_p^*, called the **natural basis** of T_p^* with respect to the local coordinate system u^i. Thus T_p^* is an m-dimensional linear space. □

By definition, $[f] - [g] \in \mathcal{H}_p$ if and only if $\ll \gamma, [f] \gg \, = \, \ll \gamma, [g] \gg$ for all $\gamma \in \Gamma_p$, so we can define

$$\ll \gamma, (df)_p \gg \, = \, \ll \gamma, [f] \gg, \qquad \gamma \in \Gamma_p, \qquad (df)_p \in T_p^*. \qquad (2.20)$$

Now define a relation $\sim$ in Γ_p as follows. Suppose $\gamma, \gamma' \in \Gamma_p$. Then $\gamma \sim \gamma'$ if and only if for any $(df)_p \in T_p^*$,

$$\ll \gamma, (df)_p \gg \, = \, \ll \gamma', (df)_p \gg. \qquad (2.21)$$

Obviously this is an equivalence relation. Denote the equivalence class of γ by $[\gamma]$. Hence we can define

$$\langle [\gamma], (df)_p \rangle \, = \, \ll \gamma, (df)_p \gg. \qquad (2.22)$$

We will prove that the $[\gamma], \gamma \in \Gamma_p$, form the dual space of T_p^*. For this purpose we will use local coordinate systems.

Under the local coordinates u^i, suppose $\gamma \in \Gamma_p$ is given by the functions

$$u^i = u^i(t), \qquad 1 \le i \le m. \qquad (2.23)$$

Then (2.22) can be written as

$$\langle [\gamma], (df)_p \rangle = \sum_{i=1}^{m} a_i \xi^i, \qquad (2.24)$$

where

$$a_i = \left(\frac{\partial (f \circ \varphi_U^{-1})}{\partial u^i} \right)_{\varphi_U(p)}, \qquad \xi^i = \left(\frac{du^i}{dt} \right)_{t=0}. \qquad (2.25)$$

The coefficients a_i are exactly the components of the cotangent vector $(df)_p$ with respect to the natural basis $(du^i)_p$ [see (2.16)]. Obviously, $\langle [\gamma], (df)_p \rangle$ is a linear function on T_p^*, which is determined by the components ξ^i. Choose γ such that

$$u^i(t) = u^i(p) + \xi^i t \qquad (2.26)$$

with ξ^i arbitrary. Thus the $\langle [\gamma], (df)_p \rangle, \gamma \in \Gamma_p$, represent the totality of linear functionals on T_p^* and form its dual space, T_p, called the **tangent space** of M at p. Elements in the tangent space are called **tangent vectors**.

The geometric meaning of tangent vectors is quite simple: if $\gamma' \in \Gamma_p$ is given by functions

$$u^i = u'^i(t), \qquad 1 \le i \le m,$$

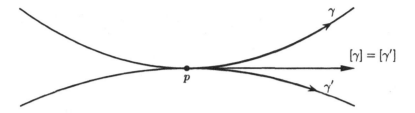

FIGURE 4.

then a necessary and sufficient condition for $[\gamma] = [\gamma']$ is

$$\left(\frac{du^i}{dt}\right)_{t=0} = \left(\frac{du'^i}{dt}\right)_{t=0}.$$

Hence the equivalence of γ and γ' means that these two parametrized curves have the same tangent vector at the point p (see Figure 4). Thus we identify a tangent vector X of M at p with the set of all parametrized curves through p with a common tangent vector.

By the discussion above, the function

$$\langle X, (df)_p \rangle, \qquad X = [\gamma] \in T_p, \ (df)_p \in T_p^*$$

is bilinear, i.e., linear in either variable. Suppose parametrized curves λ_k, $1 \leq k \leq m$, are given as in (2.19). Then

$$\langle [\lambda_k], (du^i)_p \rangle = \delta_k^i. \tag{2.27}$$

Therefore $\{[\lambda_k], 1 \leq k \leq m\}$ is the dual basis of $\{(du^i)_p, 1 \leq i \leq m\}$. (For the definition of dual basis, see section §2–1 of the next chapter.)

There is another meaning of the tangent vectors $[\lambda_k]$. We have

$$\begin{aligned}
\langle [\lambda_k], (df)_p \rangle &= \left\langle [\lambda_k], \sum_{i=1}^{m} \left\{ \left(\frac{\partial f}{\partial u^i}\right)_p \cdot (du^i)_p \right\} \right\rangle \\
&= \left(\frac{\partial f}{\partial u^k}\right)_p,
\end{aligned} \tag{2.28}$$

where $\partial f/\partial u^i$ means $\partial(f \circ \varphi_U{}^{-1})/\partial u^i$. Thus the $[\lambda_k]$ are the partial differential operators $(\partial/\partial u^k)$ on the function germs $[f]$; and (2.27) can be written as

$$\left\langle \frac{\partial}{\partial u^k}\bigg|_p, (du^i)_p \right\rangle = \delta_k^i. \tag{2.29}$$

We call the dual basis of $\{(du^i)_p, 1 \leq i \leq m\}$ in T_p the **natural basis** of the tangent space T_p under the local coordinate system (u^i). From (2.24) we have

$$[\gamma] = \sum_{i=1}^{m} \xi^i \frac{\partial}{\partial u^i}\Big|_p .$$

Thus ξ^i are the components of the tangent vector $[\gamma]$ with respect to the natural basis. If $[\gamma'] \in T_p$ has components ξ'^i, then $[\gamma] + [\gamma']$ is determined by the components $\xi^i + \xi'^i$. Similarly the tangent vector $a \cdot [\gamma]$ $(a \in \mathbb{R})$ has components $a\xi^i$.

For simplicity, we sometimes suppress the lower index p of tangent and cotangent vectors when there is no confusion.

Definition 2.2. Suppose f is a C^∞-function defined near p. Then $(df)_p \in T_p^*$ is also called the **differential** of f at the point p. If $(df)_p = 0$, then p is called a **critical point** of f.

The study of critical points of smooth functions on M is an important topic in differentiable manifolds, called **Morse Theory**. The reader can refer to Milnor, 1963.

Definition 2.3. Suppose $X \in T_p$, $f \in C_p^\infty$. Denote

$$X f = \langle X, (df)_p \rangle. \tag{2.30}$$

$X f$ is called the **directional derivative** of the function f along the vector X.

The following theorem gives some properties of the directional derivative.

Theorem 2.3. *Suppose* $X \in T_p$, $f, g \in C_p^\infty$, $\alpha, \beta \in \mathbb{R}$. *Then*

1) $X(\alpha f + \beta g) = \alpha \cdot X f + \beta \cdot X g$;
2) $X(fg) = f(p) \cdot X g + g(p) \cdot X f$.

Proof. These follow from Corollary 1 of Theorem 2.2 directly. □

Remark 1. Statement 1) of Theorem 2.3 indicates that a tangent vector X can also be viewed as a linear operator on C_p^∞. Using 1) and 2), we see that the result of X operating on any constant function c is 0.

Remark 2. Frequently, in the literature[d], properties 1) and 2) are used to define tangent vectors. In fact, all the operators on C_p^∞ satisfying these two properties form a linear space dual to T_p^*, which must then be identical to T_p.

---

[d]For example, see Chevalley 1946.

Under local coordinates u^i, a tangent vector $X = [\gamma] \in T_p$ and a cotangent vector $a = df \in T_p^*$ have linear representations in terms of natural bases:

$$X = \sum_{i=1}^m \xi^i \frac{\partial}{\partial u^i}, \quad a = \sum_{i=1}^m a_i du^i, \tag{2.31}$$

where

$$\xi^i = \frac{d(u^i \circ \gamma)}{dt}, \quad a_i = \frac{\partial f}{\partial u^i}.$$

Under another local coordinate system u'^i, if the components of X and a with respect to the corresponding natural bases are ξ'^i and a'_i, respectively, then they satisfy the following transformation rules:

$$\xi'^j = \sum_{i=1}^m \xi^i \frac{\partial u'^j}{\partial u^i}, \tag{2.32}$$

$$a_i = \sum_{j=1}^m a'_j \frac{\partial u'^j}{\partial u^i}, \tag{2.33}$$

where

$$\frac{\partial u'^j}{\partial u^i} = \frac{\partial(\varphi'_U \circ \varphi_U^{-1})^j}{\partial u^i}$$

is the Jacobian matrix of the change of coordinates $\varphi'_U \circ \varphi_U^{-1}$. In classical tensor analysis, the vectors satisfying (2.32) are called **contravariant**, and those satisfying (2.33) are called **covariant**, vectors.

Smooth maps between smooth manifolds induce linear maps between tangent spaces and between cotangent spaces. Suppose $F : M \longrightarrow N$ is a smooth map, $p \in M$, and $q = F(p)$. Define the map $F^* : T_q^* \longrightarrow T_p^*$ as follows:

$$F^*(df) = d(f \circ F), \quad df \in T_q^*. \tag{2.34}$$

Obviously this is a linear map, called the **differential** of the map F.

Consider next the adjoint of F^*, namely the map $F_* : T_p \longrightarrow T_q$ defined for $X \in T_p$, $a \in T_q^*$ as follows:

$$\langle F_* X, a \rangle = \langle X, F^* a \rangle. \tag{2.35}$$

F_* is called the **tangent map** induced by F.

Suppose u^i and v^α are local coordinates near p and q, respectively. Then the map F can be expressed near p by the functions

$$v^\alpha = F^\alpha \left(u^1, \ldots, u^m \right), \quad 1 \le \alpha \le n. \tag{2.36}$$

Thus the action of F^* on the natural basis $\{dv^\alpha, 1 \le \alpha \le n\}$ is given by

$$F^*(dv^\alpha) = d(v^\alpha \circ F)$$

$$= \sum_{i=1}^{m} \left(\frac{\partial F^\alpha}{\partial u^i}\right)_p du^i. \tag{2.37}$$

The matrix representation of F^* in the natural bases $\{dv^\alpha\}$ and $\{du^i\}$ is exactly the Jacobian matrix $(\partial F^\alpha / \partial u^i)_p$.

Similarly, the action of F_* on the natural basis $\{\partial/\partial u^i\}$ is given by

$$\left\langle F_*\left(\frac{\partial}{\partial u^i}\right), dv^\alpha \right\rangle = \left\langle \frac{\partial}{\partial u^i}, F^*(dv^\alpha) \right\rangle$$

$$= \sum_{j=1}^{m} \left\langle \frac{\partial}{\partial u^i}, du^j \right\rangle \left(\frac{\partial F^\alpha}{\partial u^j}\right)_p$$

$$= \left\langle \sum_{\beta=1}^{n} \left(\frac{\partial F^\beta}{\partial u^i}\right)_p \frac{\partial}{\partial v^\beta}, dv^\alpha \right\rangle,$$

i.e.,

$$F_*\left(\frac{\partial}{\partial u^i}\right) = \sum_{\beta=1}^{n} \left(\frac{\partial F^\beta}{\partial u^i}\right)_p \frac{\partial}{\partial v^\beta}. \tag{2.38}$$

Hence the matrix representation of the tangent map F_* under the natural bases $\{\partial/\partial u^i\}$ and $\{\partial/\partial v^\alpha\}$ is still the Jacobian matrix $(\partial F^\alpha/\partial u^i)_p$.

§1–3   Submanifolds

Before discussing submanifolds, we will first study tangent maps induced by smooth maps between smooth manifolds. Given a smooth map $\varphi : M \longrightarrow N$, for any point $p \in M$ there exists an induced tangent map between the corresponding tangent spaces, $\varphi_* : T_p(M) \longrightarrow T_q(N)$, where $q = \varphi(p)$. The crucial point is that the properties of the tangent map φ_* at $p \in M$ determine the properties of the map φ in a neighborhood of p. A classical result in this regard is the **inverse function theorem** in calculus.

Theorem 3.1. *Suppose W is an open subset of $\mathbb{R}^n$ and $f : W \longrightarrow \mathbb{R}^n$ is a smooth map. If at a point $x_0 \in W$ the determinant of the Jacobian matrix is nonzero, i.e.,*

$$\det \left(\frac{\partial f^i}{\partial x^j}\right)\bigg|_{x_0} \ne 0,$$

then there exists a neighborhood $U \subset W$ of x_0 in $\mathbb{R}^n$ such that $V = f(U)$ is a neighborhood of the point $f(x_0)$ in $\mathbb{R}^n$, and f has a smooth inverse on V

$$g = f^{-1} : V \longrightarrow U. \tag{3.1}$$

By the discussion in the last part of the preceding section, the Jacobian matrix $(\partial f^i / \partial x^j)$ of f is precisely the matrix representation of f_* under the natural basis. Therefore $(\partial f^i / \partial x^j)|_{x_0} \neq 0$ implies that

$$f_* : T_{x_0}(W)(\simeq \mathbb{R}^n) \longrightarrow T_{f(x_0)}(\mathbb{R}^n)(\simeq \mathbb{R}^n)$$

is an isomorphism. That g is the inverse of f means

$$g \circ f = \mathrm{id} : U \longrightarrow U, \qquad f \circ g = \mathrm{id} : V \longrightarrow V. \tag{3.2}$$

Since f and g are both smooth maps, the restriction of f to U provides a diffeomorphism from U to V. The inverse function theorem says that if the tangent map f_* of f is an isomorphism at a point, then f is a diffeomorphism from a neighborhood of that point to an open set in $\mathbb{R}^n$.

Using local coordinate systems, it is not difficult to generalize Theorem 3.1 to the case of manifolds.

Theorem 3.2. *Suppose M and N are both n-dimensional smooth manifolds, and $f : M \longrightarrow N$ is a smooth map. If at a point $p \in M$, the tangent map $f_* : T_p(M) \longrightarrow T_{f(p)}(N)$ is an isomorphism, then there exists a neighborhood U of p in M such that $V = f(U)$ is a neighborhood of $f(p)$ in N and $f|_U : U \longrightarrow V$ is a diffeomorphism.*

Proof. Since $f : M \longrightarrow N$ is a smooth map, we can choose local coordinates (U_0, φ) at $p \in M$ and (V_0, ψ) at $q = f(p) \in N$ such that $f(U_0) \subset V_0$ and

$$\tilde{f} = \psi \circ f \circ \varphi^{-1} : \varphi(U_0) \longrightarrow \psi(V_0) \subset \mathbb{R}^n \tag{3.3}$$

is a smooth map. Obviously the determinant of the Jacobian of $\tilde{f}$ at point $\varphi(p)$ is nonzero. By Theorem 3.1, there exist neighborhoods $\tilde{U} \subset \varphi(U_0)$ and $\tilde{V} \subset \psi(V_0)$ of $\varphi(p)$ and $\psi(q)$ in $\mathbb{R}^n$, respectively, such that $\tilde{f}|_{\tilde{U}} : \tilde{U} \longrightarrow \tilde{V}$ is a diffeomorphism.

Let $U = \varphi^{-1}(\tilde{U})$, $V = \psi^{-1}(\tilde{V})$. Then U and V are neighborhoods of p and q in M and N, respectively, and

$$f = \psi^{-1} \circ \tilde{f} \circ \varphi : U \longrightarrow V \tag{3.4}$$

is a diffeomorphism. $\square$

Remark. Since the manifolds M and N in Theorem 3.2 have the same dimension, the condition "the tangent map f_* is an isomorphism" is equivalent to "f_* is injective." If M is an m-dimensional and N an n-dimensional manifold, $f : M \longrightarrow N$ is smooth, and the tangent map f_* is injective at a point p, then we say that f_* is **nondegenerate** at p. Obviously, in this case, $m \leq n$, and the rank of the Jacobian matrix of f at p is m.

We have the following theorem as an application of Theorem 3.2.

Theorem 3.3. *Suppose M is an m-dimensional and N an n-dimensional manifold, $m < n$. If $f : M \longrightarrow N$ is a smooth map and the tangent map f_* is nondegenerate at a point $p \in M$, then there exist local coordinate systems $(U; u^i)$ near p and $(V; v^\alpha)$ near $q = f(p)$ such that $f(U) \subset V$, and the map $f|_U$ can be expressed by local coordinates as follows: for any $x \in U$,*

$$\begin{cases} v^i(f(x)) = u^i(x), & 1 \leq i \leq m, \\ v^\gamma(f(x)) = 0, & m+1 \leq \gamma \leq n. \end{cases} \tag{3.5}$$

Proof. Suppose $(U; u^i)$ and $(V; v^\alpha)$ are local coordinate systems at points p and q, respectively, and the representation of f under these systems is

$$v^\alpha = f^\alpha(u^1, \ldots, u^m), \qquad 1 \leq \alpha \leq n.$$

Assume $u^i(p) = 0$, $v^\alpha(q) = 0$. Since f_* is nondegenerate at p, we may assume that

$$\left. \frac{\partial(f^1, \ldots, f^m)}{\partial(u^1, \ldots, u^m)} \right|_{u^i=0} \neq 0. \tag{3.6}$$

Let $I_{n-m} = \{(w^{m+1}, \ldots, w^n) \mid |w^\gamma| \leq \delta, \ m+1 \leq \gamma \leq n\}$, where δ is a positive number. By suitably shrinking the neighborhood U and choosing a sufficiently small δ, we can define a smooth map $\tilde{f} : U \times I_{n-m} \longrightarrow V$ such that

$$\begin{cases} \tilde{f}^i(u^1, \ldots, u^m, w^{m+1}, \ldots, w^n) = f^i(u^1, \ldots, u^m), \\ \tilde{f}^\gamma(u^1, \ldots, u^m, w^{m+1}, \ldots, w^n) = w^\gamma + f^\gamma(u^1, \ldots, u^m), \\ 1 \leq i \leq m, \qquad m+1 \leq \gamma \leq n. \end{cases} \tag{3.7}$$

It is obvious that the Jacobian matrix of $\tilde{f}$ at $(u^i, w^\gamma) = (0, 0)$ is nondegenerate. It follows by Theorem 3.2 that $\tilde{f}$ is a diffeomorphism in a neighborhood of $(0,0)$. We may assume that $\tilde{f} : U \times I_{n-m} \longrightarrow V$ is a diffeomorphism. Then $\{u^i, w^\gamma\}$ can be chosen as the local coordinate system v^α in the neighborhood V of q. In this system, $\tilde{f}$ becomes the identity map, that is,

$$\begin{cases} v^i = u^i, & 1 \leq i \leq m, \\ v^\gamma = w^\gamma, & m+1 \leq \gamma \leq n. \end{cases} \tag{3.8}$$

Clearly, $\tilde{f}\Big|_{U\times\{0\}} = f|_U$. Thus, under the coordinate system above, the map $f|_U$ is given as in (3.5), i.e.,

$$f(u^1,\ldots,u^m) = (u^1,\ldots,u^m,0,\ldots,0).$$

$\square$

We thus see that if the map f_* is injective at the point p, then f is also injective near p.

Definition 3.1. Suppose M and N are smooth manifolds. If there is a smooth map $\varphi : M \longrightarrow N$ such that

1) φ is injective;
2) at any point $p \in M$, the tangent map $\varphi_* : T_p(M) \longrightarrow T_{\varphi(p)}(N)$ is nondegenerate,

then (φ, M) is called a **smooth submanifold**, or **imbedded submanifold**, of N.

If φ satisfies 2) only, then φ is called an **immersion**, and (φ, M) an **immersed submanifold** of N.

An immersion is locally injective, but not necessarily so globally. The difference between immersed and imbedded submanifolds is that in the former the image $\varphi(M)$ may have self-intersection points, whereas in the latter it does not.

Example 1. Open submanifolds. Suppose U is an open subset of N. By restricting the smooth structure of N to U, we obtain a smooth structure on U, which makes U a smooth manifold with the same dimension as N. Let $\varphi = \mathrm{id} : U \longrightarrow N$ be the identity map. Then (φ, U) becomes an imbedded submanifold of N, called an **open submanifold** of N.

Example 2. Closed submanifolds. Suppose (φ, M) is a smooth submanifold of N. If

1) $\varphi(M)$ is a closed subset of N;
2) for any point $q \in \varphi(M)$ there exists a local coordinate system $(U; u^i)$ such that $\varphi(M) \cap U$ is defined by

$$u^{m+1} = u^{m+2} = \cdots = u^n = 0,$$

where $m = \dim M$,

then we call (φ, M) a **closed submanifold** of N.

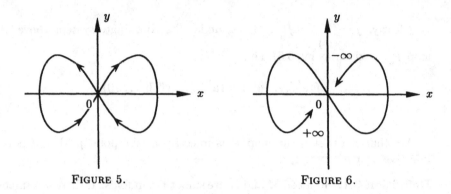

FIGURE 5.   FIGURE 6.

For example, the unit sphere $S^n \subset \mathbb{R}^{n+1}$ and the identity map $\varphi : S^n \longrightarrow \mathbb{R}^{n+1}$ define a closed submanifold of $\mathbb{R}^{n+1}$.

Example 3. Let $F : \mathbb{R} \longrightarrow \mathbb{R}^2$ be defined by

$$F(t) = \left(2\cos\left(t - \frac{1}{2}\pi \right), \sin 2\left(t - \frac{1}{2}\pi \right) \right). \tag{3.9}$$

Then $(F, \mathbb{R})$ is an immersed submanifold of $\mathbb{R}^2$, but not an imbedded submanifold (Figure 5).

Example 4. Let $G : \mathbb{R} \longrightarrow \mathbb{R}^2$ be defined by

$$G(t) = \left(2\cos\left(2\arctan\left(t + \frac{\pi}{2} \right) \right), \sin 2\left(2\arctan\left(t + \frac{\pi}{2} \right) \right) \right). \tag{3.10}$$

Then $(G, \mathbb{R})$ is an imbedded submanifold of $\mathbb{R}^2$ (see Figure 6). At $t = 0$, $G(0) = (0, 0)$, but as $t \to \pm\infty$, $G(t) \to (0, 0)$ also.

Example 5. Let

$$F(t) = \begin{cases} \left(\dfrac{3}{t^2}, \sin \pi t \right), & 1 \le t < +\infty, \\ (0, t + 2), & -\infty < t \le -1, \end{cases} \tag{3.11}$$

and extend F smoothly to include the interval $[-1, 1]$ by connecting the two points $(3, 0)$ and $(0, 1)$ (as shown by the dotted line in Figure 7). As $t \to +\infty$, the curve approaches its own segment $-3 \le t \le -1$ infinitely closely. $(F, \mathbb{R})$ is an imbedded submanifold of $\mathbb{R}^2$.

Example 6. The torus $T^2 = S^1 \times S^1$ (see Figure 8) can be viewed as a 2-dimensional manifold obtained by identifying opposite sides of the unit square

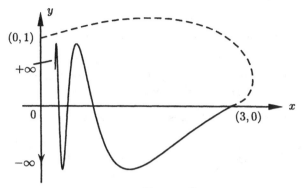

FIGURE 7.

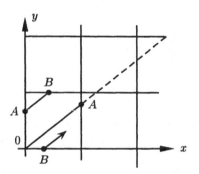

FIGURE 8.

in the plane $\mathbb{R}^2$. Any point on the torus can then be expressed as an ordered pair of real numbers $(x, y) \bmod 1$.

Now take any two real numbers a, b such that $\frac{a}{b}$ is irrational. Consider the map $\varphi : \mathbb{R}^1 \longrightarrow T^2$, where

$$\varphi(t) = (at \bmod 1, bt \bmod 1).$$

Clearly, $(\varphi, \mathbb{R})$ is an imbedded submanifold of T^2; but the image $\varphi(\mathbb{R})$ is dense in T^2. When $\frac{a}{b}$ is rational, $\varphi(\mathbb{R})$ is an immersed submanifold of T^2.

For an imbedded submanifold (φ, M), since φ is injective, the differentiable structure on M can be transported to $\varphi(M)$, making $\varphi : M \longrightarrow \varphi(M)$ a diffeomorphism. On the other hand, being a subset of N, $\varphi(M)$ has an induced topology from N. The last three examples show that the topology on $\varphi(M)$ obtained from M through φ is not necessarily the same as the one induced from

N. Generally, the topology from M through φ is stronger than the topology induced from N. This motivates the following definition.

Definition 3.2. Suppose (φ, M) is a smooth submanifold of N. If $\varphi : M \longrightarrow \varphi(M) \subset N$ is a homeomorphism, then (φ, M) is called a **regular submanifold** of N, and φ is called a **regular imbedding** of M into N.

The following theorem characterizes regular submanifolds.

Theorem 3.4. *Suppose (φ, M) is an m-dimensional submanifold of an n-dimensional smooth manifold N. A necessary and sufficient condition for (φ, M) to be a regular submanifold of N is that it is a closed submanifold of an open submanifold of N.*

Proof. For sufficiency, we only need to show that a closed submanifold (φ, M) of N is a regular submanifold. Choose an arbitrary point $p \in M$. By the definition of closed submanifolds, there exists a local coordinate system $(V; v^\alpha)$ at the point $q = \varphi(p)$ in N such that $\varphi(M) \cap V$ is defined by the equations

$$v^{m+1} = \cdots = v^n = 0. \tag{3.12}$$

Since φ is continuous, there exists a local coordinate system $(U; u^i)$ such that $\varphi(U) \subset V$. We may assume that $u^i(p) = 0$, $v^\alpha(q) = 0$, and $V = \{(v^1, \ldots, v^n) \mid |v^\alpha| < \delta\}$, where δ is some positive number. Thus $\varphi(U) \subset \varphi(M) \cap V$.

The goal is to prove that $\varphi^{-1} : \varphi(M) \subset N \longrightarrow M$ is also continuous. For this, it is sufficient to show that we can choose a sufficiently small $\delta_1 > 0$ such that the preimage of $\varphi(M) \cap V_1$ under φ is contained in U. By (3.12), the map $\varphi|_U$ can be expressed locally by

$$\begin{cases} v^i = \varphi^i(u^1, \ldots, u^m), & 1 \le i \le m, \\ v^\gamma = 0, & m+1 \le \gamma \le n. \end{cases} \tag{3.13}$$

Hence the Jacobian $\left.\frac{\partial(\varphi^i, \cdots, \varphi^m)}{\partial(u^1, \cdots, u^m)}\right|_{u^i=0} \ne 0$. By Theorem 3.1 there exists δ_1 with $0 < \delta_1 < \delta$ such that there is an inverse function set

$$u^i = \psi^i(v^1, \ldots, v^m), \ |v^i| < \delta_1$$

of the function set $(\varphi^1, \cdots, \varphi^m)$. Let

$$V_1 = \left\{ \left(v^1, \ldots, v^m, v^{m+1}, \ldots, v^n\right) \mid \ |v^\alpha| < \delta_1 \right\}.$$

Then the entire preimage of $\varphi(M) \cap V_1$ under φ is contained in U. Hence $\varphi : M \longrightarrow \varphi(M) \subset N$ is a homeomorphism, which implies that (φ, M) is a regular submanifold of N.

Conversely, suppose (φ, M) is a regular submanifold of N. Let $p \in M$. Then for any neighborhood $U \subset M$ of p, there is a neighborhood V of the point $q = \varphi(p)$ such that $\varphi(U) = \varphi(M) \cap V$. By Theorem 3.3, there exist local coordinate systems $(U_1; u^i)$ for p and $(V_1; v^\alpha)$ for q such that $\varphi(U_1) \subset V_1$, and $\varphi|_{U_1}$ can be expressed in local coordinates as

$$\varphi\left(u^1, \dots, u^m\right) = \left(u^1, \dots, u^m, 0, \dots, 0\right). \tag{3.14}$$

We may assume that $U_1 \subset U$. Hence we can choose $V_1 \subset V$ with $\varphi(U_1) = \varphi(M) \cap V_1$. By (3.14), $\varphi(M) \cap V_1$ is defined by

$$v^{m+1} = \cdots = v^n = 0. \tag{3.15}$$

For any point $q \in \varphi(M)$, use V_q to represent the neighborhood of q in N constructed above, where (3.15) holds. Let $W = \bigcup_{q \in \varphi(M)} V_q$. It is obvious that W is an open submanifold of N containing $\varphi(M)$. We need to show that, as a topological subspace of N, $\varphi(M)$ is relatively closed in W, that is, $W \cap \overline{\varphi(M)} = \varphi(M)$, where $\overline{\varphi(M)}$ means the closure of $\varphi(M)$ in N. Choose any point $s \in W \cap \overline{\varphi(M)}$. By the definition of W, there exists $q \in \varphi(M)$ such that $s \in V_q$. By (3.15), $\varphi(M) \cap V_q$ is a coordinate plane in V_q; hence it is a relative closed subset of V_q. Now we have already assumed that $s \in \overline{\varphi(M)} \cap V_q$, that is, s is in the relative closure of $\varphi(M) \cap V_q$ in V_q, so $s \in \varphi(M) \cap V_q$. Therefore $W \cap \overline{\varphi(M)} \subset \varphi(M)$, that is,

$$W \cap \overline{\varphi(M)} = \varphi(M).$$

This proves that (φ, M) is a closed submanifold of the open submanifold W of N. $\qquad\square$

In proving the above theorem, we have also proved the following:

Corollary. *A submanifold (φ, M) of a smooth manifold N is a regular submanifold if and only if for any point $p \in M$, there exists a coordinate chart $(V; v^\alpha), v^\alpha(q) = 0$ at the point $q = \varphi(p)$ in N, such that $\varphi(M) \cap V$ is defined by*

$$v^{m+1} = v^{m+2} = \cdots = v^n = 0.$$

Theorem 3.5. *Suppose (φ, M) is a submanifold of a smooth manifold N. If M is compact, then $\varphi : M \longrightarrow N$ is a regular imbedding.*

Proof. Because $\varphi(M)$, as a topological subspace of N, is a Hausdorff space, and $\varphi : M \longrightarrow \varphi(M) \subset N$ is a $1 - 1$ continuous map from the compact space M to a Hausdorff space, $\varphi : M \longrightarrow \varphi(M) \subset N$ is a homeomorphism, that is, (φ, M) is a regular submanifold of N. $\qquad\square$

H. Whitney proved that any m-dimensional smooth manifold can be imbedded into a $(2m + 1)$-dimensional Euclidean space as a submanifold (see Auslander and Mackenzie 1963). In the following we will prove a weaker version of this theorem (Theorem 3.6) for compact manifolds. To carry out the proof we will need a few lemmas which are useful also in other problems.

Lemma 1. *Suppose D_1 and D_2 are two concentric balls in $\mathbb{R}^m$, with $\overline{D_1} \subset D_2$. Then there exists a smooth, real-valued function f defined on $\mathbb{R}^m$ such that*

1) $0 \leq f \leq 1$;

2) $f(x) = \begin{cases} 1, & x \in D_1 \\ 0, & x \in \mathbb{R}^m - D_2. \end{cases}$

Proof. We may assume that D_1 and D_2 are centered at the origin with radii a and b, $0 < a < b$, respectively. Let

$$g(t) = \begin{cases} \exp\left\{\dfrac{1}{(t-a^2)(t-b^2)}\right\}, & t \in (a^2, b^2), \\ 0, & t \notin (a^2, b^2). \end{cases} \qquad (3.16)$$

Clearly g is a smooth function on $\mathbb{R}^1$. Now let

$$F(t) = \frac{\int\limits_{t}^{\infty} g(s)\, ds}{\int\limits_{-\infty}^{\infty} g(s)\, ds}. \qquad (3.17)$$

F is also a smooth function on $\mathbb{R}^1$, and $0 \leq F \leq 1$. When $t \leq a^2$, $F(t) = 1$; and when $t \geq b^2$, $F(t) = 0$. Let

$$f(x^1, \ldots, x^m) = F\left((x^1)^2 + \cdots + (x^m)^2\right). \qquad (3.18)$$

Then f satisfies the conditions required. $\square$

Lemma 2. *Suppose U and V are two nonempty open sets in $\mathbb{R}^m$ and $\overline{V}$ is compact, with $\overline{V} \subset U$. Then there exists a smooth real-valued function f defined on $\mathbb{R}^m$ such that*

1) $0 \leq f \leq 1$;

2) $f(x) = \begin{cases} 1, & x \in V, \\ 0, & x \in \mathbb{R}^m - U. \end{cases}$

Proof. Since $\overline{V}$ is compact and $\overline{V} \subset U$, there exist finitely many sets of balls $\left\{ D_i^{(1)}, D_i^{(2)} \right\}_{1 \leq i \leq r}$ such that $D_i^{(1)}, D_i^{(2)}$ are pairwise concentric,

$$\overline{D}_i^{(1)} \subset D_i^{(2)} \subset U, \qquad (3.19)$$

and $\left\{ D_i^{(1)} \right\}_{1 \leq i \leq r}$ is an open covering for $\overline{V}$. By Lemma 1, for every $i, 1 \leq i \leq r$, there exists a smooth function $f_i : \mathbb{R}^m \longrightarrow \mathbb{R}^1$ such that $0 \leq f_i \leq 1$, and

$$f_i(x) = \begin{cases} 1, & x \in D_i^{(1)}, \\ 0, & x \notin D_i^{(2)}. \end{cases} \qquad (3.20)$$

Let

$$f = 1 - \prod_{i=1}^{r} (1 - f_i). \qquad (3.21)$$

Then it is easy to verify that f satisfies the requirements of Lemma 2. $\qquad \square$

Lemma 3. *Suppose* (U, φ_U) *is a coordinate chart in a smooth manifold* M, $V \neq \varnothing$ *is an open set in* M *with* $\overline{V}$ *compact, and* $\overline{V} \subset U$. *Then there exists a smooth function* $h : M \longrightarrow \mathbb{R}^1$ *such that*

1) $0 \leq h \leq 1$;

2) $h(p) = \begin{cases} 1, & p \in V, \\ 0, & p \notin U. \end{cases}$

Proof. Since $\overline{V}$ is compact and $\overline{V} \subset U$, we may use the local compactness of M to construct an open set U_1 such that

$$\overline{V} \subset U_1 \subset \overline{U_1} \subset U.$$

Suppose $\dim M = m$. Then $\varphi_U(V)$ and $\varphi_U(U_1)$ are both open sets in $\mathbb{R}^m$. By Lemma 2, there is a smooth function f on $\mathbb{R}^m$ such that $0 \leq f \leq 1$ and

$$f(x) = \begin{cases} 1, & x \in \varphi_U(V), \\ 0, & x \notin \varphi_U(U_1). \end{cases} \qquad (3.22)$$

Let

$$h(p) = \begin{cases} f \circ \varphi_U(p), & p \in U, \\ 0, & p \notin U. \end{cases} \qquad (3.23)$$

Then h is a smooth function on M which satisfies the requirements of Lemma 3. $\qquad \square$

Remark. The conditions in Lemma 3 can be modified as follows: Suppose U and V are two nonempty subsets of a smooth manifold M, where $\overline{V}$ is compact and $\overline{V} \subset U$. Then the smooth, real-valued function h described above still exists. (The reader should verify this.)

Theorem 3.6. *Suppose M is a compact m-dimensional smooth manifold. Then there exists a positive integer n and a smooth map $\varphi : M \longrightarrow \mathbb{R}^n$ such that (φ, M) is a regular submanifold of $\mathbb{R}^n$.*

Proof. Since M is a compact manifold, there exists a finite open covering $\{V_j\}_{1 \leq j \leq r}$ such that every $\overline{V}_j$ is compact and contained in some coordinate neighborhood U_j. Suppose the local coordinate functions in U_j are $u^i_{(j)}$, $1 \leq i \leq m$. Obviously, there exist open sets W_j such that

$$\overline{V}_j \subset W_j \subset \overline{W}_j \subset U_j.$$

By Lemma 3, for each j, $1 \leq j \leq r$, there exists a smooth function f_j on M such that $1 \leq f_j \leq 1$ and

$$f_j(p) = \begin{cases} 1, & p \in V_j, \\ 0, & p \notin W_j. \end{cases} \tag{3.24}$$

Define $n = r(m+1)$ smooth functions on M:

$$x^0_j = f_j$$

$$x^i_j(p) = \begin{cases} u^i_{(j)}(p) \cdot f_j(p), & p \in U_j, \\ 0, & p \notin U_j, \end{cases} \tag{3.25}$$

where $1 \leq i \leq m$, $1 \leq j \leq r$. View (x^0_j, x^i_j) as a point in $\mathbb{R}^n$. Then (3.25) gives a map $\varphi : M \longrightarrow \mathbb{R}^n$. First we prove that (φ, M) is a submanifold of $\mathbb{R}^n$.

If $p, q \in M$, and $\varphi(p) = \varphi(q)$, then

$$x^0_j(p) = x^0_j(q), \qquad x^i_j(p) = x^i_j(q), \tag{3.26}$$

$$1 \leq i \leq m, \qquad 1 \leq j \leq r.$$

Since $\{V_j\}_{1 \leq j \leq r}$ is a covering of M, there exists a k, $1 \leq k \leq r$, such that $p \in V_k$. Hence

$$f_k(q) = x^0_k(q) = x^0_k(p) = f_k(p) = 1, \tag{3.27}$$

$$u^i_{(k)}(q) = u^i_{(k)}(p), \quad 1 \leq i \leq m. \tag{3.28}$$

This shows that q belongs to U_k and both p and q have the same local coordinates in U_k. Hence $q = p$, that is, φ is injective.

Suppose $p \in M$. Then there must be a $k, 1 \leq k \leq r$, such that $p \in V_k$. Then $f_k(p) = 1$ and hence

$$x^i_k\big|_{V_k} = u^i_{(k)}.$$

Thus

$$\frac{\partial(x^1_k, \ldots, x^m_k)}{\partial(u^1_{(k)}, \ldots, u^m_{(k)})}\bigg|_p = 1.$$

This shows that the tangent map φ_* is nondegenerate at p, which means (φ, M) is an imbedded submanifold of $\mathbb{R}^n$. By Theorem 3.5, φ is a regular imbedding. $\qquad\square$

§1–4 Frobenius' Theorem

The concept of a tangent vector $X_p \in T_p$ at any point p in a manifold M has been defined in §1–2. Theorem 2.3 interprets a tangent vector X_p as a real-valued function $X_p : C^\infty_p \longrightarrow \mathbb{R}$ where C^∞_p is the set of smooth functions at $p \in M$. If for an arbitrary point p in M a tangent vector X_p on M is assigned at p, then X is called a **tangent vector field** on M. For $f \in C^\infty(M)$, let

$$(Xf)(p) = X_p f. \tag{4.1}$$

Then Xf is a real-valued function on the manifold M.

Definition 4.1. Suppose X is a tangent vector field on a smooth manifold M, and for any $f \in C^\infty(M)$, we have $Xf \in C^\infty(M)$. Then X is called a **smooth** tangent vector field on M.

We see that a smooth tangent vector field X is an operator from $C^\infty(M)$ to itself. Applying Theorem 2.3, we obtain the following properties of X. Suppose $f, g \in C^\infty(M)$, $\alpha, \beta \in \mathbb{R}$, then

1) $X(\alpha f + \beta g) = \alpha(Xf) + \beta(Xg)$;
2) $X(f \cdot g) = f \cdot Xg + g \cdot Xf$.

In order to study a tangent vector field, we need to elucidate its local behavior. Suppose X is a smooth tangent vector field on a manifold M. Then for any nonempty open subset U, the restriction of X on U, $X\big|_U$, is a smooth tangent vector field on the open submanifold U. To demonstrate the smoothness of $X\big|_U$ it is sufficient to show that for any $f \in C^\infty(U)$, $X\big|_U f$ is also a smooth function on U. Choose any point $p \in U$. Then there is a coordinate neighborhood V such that $\overline{V}$ is compact and $\overline{V} \subset U$. By Lemma 3 in §1–3, there exists a smooth function $g \in C^\infty(M)$ such that

$$g|_V = 1, \qquad g|_{M-U} = 0.$$

Let

$$\tilde{f}(x) = \begin{cases} f(x) \cdot g(x), & x \in U, \\ 0, & x \notin U. \end{cases} \tag{4.2}$$

Then $\tilde{f} \in C^\infty(M)$ and $\tilde{f}\Big|_V = f|_V$. Hence

$$(X|_U f)(x) = X_x f = (X\tilde{f})(x), \quad \forall x \in V. \tag{4.3}$$

Since X is a smooth tangent vector field on M, $X\tilde{f} \in C^\infty(M)$. Therefore $X|_U f$ is a smooth function at the point $p \in U$, and hence is a smooth function on U.

Theorem 4.1. *A necessary and sufficient condition for a tangent vector field X on a smooth manifold M to be a smooth tangent vector field is that, for any point $p \in M$, there exists a local coordinate system $(U; u^i)$ such that the restriction of X on U can be expressed as*

$$X|_U = \sum_{i=1}^m \xi^i \frac{\partial}{\partial u^i}, \tag{4.4}$$

where $\xi^i, 1 \leq i \leq m$, are smooth functions on U.

Proof. Sufficiency is obvious, so we only prove necessity. Since X is a smooth tangent vector field on M, $X|_U$ is a smooth tangent vector field on the submanifold U. The tangent vector field $X|_U$ can be expressed as

$$X|_U = \sum_{i=1}^m \xi^i \frac{\partial}{\partial u^i},$$

under the natural basis $\left\{ \dfrac{\partial}{\partial u^i} \right\}$. Since the coordinates u^i are smooth functions on U,

$$X|_U u^i = \xi^i$$

are also smooth functions on U. $\qquad\square$

Suppose X and Y are two smooth tangent vector fields on M. Their **Poisson bracket product** is defined by

$$[X, Y] = XY - YX. \tag{4.5}$$

Thus $[X, Y]$ is an operator on $C^\infty(M)$, and for any $f \in C^\infty(M)$ we have

$$[X, Y]f = X(Yf) - Y(Xf). \tag{4.6}$$

It is easy to verify that, for any $f, g \in C^\infty(M)$, the following hold:

1) $[X, Y](f + g) = [X, Y]f + [X, Y]g$;
2) $[X, Y](fg) = f \cdot [X, Y]g + g \cdot [X, Y]f$.

This implies that $[X, Y]$ is a smooth tangent vector field on the manifold M. Later we will use local coordinates to describe this product and demonstrate smoothness. Before doing this, we list some additional properties of the Poisson bracket in the following theorem:

Theorem 4.2. *Suppose X, Y, Z are smooth tangent vector fields on M, and $f, g \in C^\infty(M)$. Then*

1) $[X, Y] = -[Y, X]$;
2) $[X + Y, Z] = [X, Z] + [Y, Z]$;
3) $[fX, gY] = f \cdot (Xg)Y - g \cdot (Yf)X + f \cdot g[X, Y]$;
4) $[X, [Y, Z]] + [Y, [Z, X]] + [Z, [X, Y]] = 0.$.

Proof. Each of these properties can be verified directly by using the definition of the Poisson bracket product. For example, to prove 3), suppose $h \in C^\infty(M)$. Then

$$\begin{aligned}
[fX, gY]h &= (fX)((gY)h) - (gY)((fX)h) \\
&= f \cdot X(g \cdot Yh) - g \cdot Y(f \cdot Xh) \\
&= f \cdot (Xg)(Yh) + f \cdot g \cdot X(Yh) - g \cdot (Yf)(Xh) - g \cdot f \cdot Y(Xh) \\
&= (f(Xg) \cdot Y - g(Yf) \cdot X + f \cdot g[X, Y])h.
\end{aligned}$$

Hence 3) is true. $\qquad\square$

The formula in 4) is called **Jacobi's identity**.

Now we will use local coordinates to express $[X, Y]$. Suppose $(U; u^i)$ is a local coordinate system on a manifold M. Then we may assume

$$X|_U = \sum_{i=1}^{m} \xi^i \frac{\partial}{\partial u^i}, \quad Y|_U = \sum_{i=1}^{m} \eta^i \frac{\partial}{\partial u^i}, \tag{4.7}$$

where ξ^i, η^i are smooth functions on U. Since $\left[\frac{\partial}{\partial u^i}, \frac{\partial}{\partial u^j}\right] = 0, 1 \le i, j \le m$, it follows from property 3) of Theorem 4.2 that

$$
\begin{aligned}
[X, Y]|_U \;&=\; [X|_U, Y|_U] \\
&=\; \sum_{j=1}^{m}\sum_{i=1}^{m}\left(\xi^i \frac{\partial \eta^j}{\partial u^i} - \eta^i \frac{\partial \xi^j}{\partial u^i}\right)\frac{\partial}{\partial u^j}.
\end{aligned}
\tag{4.8}
$$

Therefore $[X, Y]$ is a smooth tangent vector field on M.

Definition 4.2. Suppose X is a smooth tangent vector field on M and $p \in M$. If $X_p = 0$, then p is called a **singular point** of X.

The properties of a vector field X near a singular point p may be very complicated [see, for example, Chern 1967(a)]. The singularities of a smooth tangent vector field are closely related to the topological properties of the manifold (see the Corollary to Theorem 4.1 in Chapter 5). For example, there are no singularity-free smooth tangent vector fields on an even-dimensional sphere, but such a tangent vector field exists on a torus.

The nature of a smooth tangent vector field near nonsingular points is quite simple. We have the following theorem:

Theorem 4.3. *Suppose X is a smooth tangent vector field on a manifold M. If $X_p \ne 0$ at a point $p \in M$, then there exists a local coordinate system $(W; w^i)$ such that*

$$
X|_W = \frac{\partial}{\partial w^1}.
\tag{4.9}
$$

Proof. By Theorem 4.1, there exists a local coordinate system $(U; u^i)$, $u^i(p) = 0$, such that the restriction of X on U can be expressed as

$$
X|_U = \sum_{i=1}^{m} \xi^i \frac{\partial}{\partial u^i},
\tag{4.10}
$$

where the ξ^i are smooth functions on U. Since $X_p \ne 0$, we may assume that $\xi^1(p) \ne 0$. By the continuity of ξ^1, we may assume that U is a sufficiently small neighborhood of p such that ξ^1 is nonzero in U. Consider the system of ordinary differential equations

$$
\frac{du^\alpha}{du^1} = \frac{\xi^\alpha(u^1, \ldots, u^m)}{\xi^1(u^1, \ldots, u^m)}, \quad 2 \le \alpha \le m,
\tag{4.11}
$$

where u^1 is the independent variable and $u^\alpha, 2 \le \alpha \le m$, are unknown functions. By the theory of ordinary differential equations (see Hurewicz 1966), there exists a positive number δ such that $\{(u^1, \ldots, u^m) \mid |u^i| < \delta\} \subset U$, and for any given initial value $(v^2, \ldots, v^m)$, $|v^\alpha| < \delta, 2 \leq \alpha \leq m$, the system (4.11) has a unique solution

$$u^\alpha = \varphi^\alpha(u^1; v^2, \ldots, v^m), \quad -\delta < u^1 < \delta, \tag{4.12}$$

which satisfies the initial conditions

$$\varphi^\alpha(0; v^2, \ldots, v^m) = v^\alpha, \tag{4.13}$$

where the functions φ^α depend on u^1 and the initial values v^α smoothly.

Consider the change of coordinates

$$\begin{cases} u^1 = v^1, \\ u^\alpha = \varphi^\alpha(v^1; v^2, \ldots, v^m), & 2 \leq \alpha \leq m. \end{cases} \tag{4.14}$$

Then its Jacobian is

$$\frac{\partial(u^1, \ldots, u^m)}{\partial(v^1, \ldots, v^m)}\bigg|_{v^1=0} = 1. \tag{4.15}$$

Hence there exists a neighborhood $W \subset U$ of p that has $v^i, 1 \leq i \leq m$, as its local coordinate system. Under this system,

$$\begin{aligned} X|_W &= \sum_{i=1}^m \xi^i \frac{\partial}{\partial u^1} \\ &= \xi^1 \cdot \sum_{i=1}^m \frac{\partial u^i}{\partial v^1} \frac{\partial}{\partial u^i} \\ &= \xi^1 \frac{\partial}{\partial v^1}. \end{aligned} \tag{4.16}$$

Let

$$\begin{cases} w^1 = \displaystyle\int_0^{v^1} \frac{dv^1}{\xi^1}, \\ w^\alpha = v^\alpha, & 2 \leq \alpha \leq m. \end{cases} \tag{4.17}$$

Then $w^i, 1 \leq i \leq m$, is a local coordinate system of W and

$$X|_W = \frac{\partial}{\partial w^1}.$$

This completes the proof. $\qquad\qquad\qquad\qquad\qquad\qquad\qquad\qquad\qquad\quad \square$

A generalization of the problem dealt with in Theorem 4.3 is the following. Suppose at every point $p \in M$, an h-dimensional subspace $L^h(p)$ of the tangent space T_p is assigned. That is, L^h is an h-dimensional tangent subspace field of M. More precisely, if for any point $p \in M$, there exists a neighborhood U of p and smooth tangent vector fields $X_1, \ldots, X_h$ which are linearly independent at every point in U (so that at any point $q \in U$, $L^h(q)$ is spanned by vectors $X_1(q), \ldots, X_h(q)$), then L^h is called an h-dimensional **smooth tangent subspace field**, or an h-dimensional **smooth distribution** on M, denoted by

$$L^h\big|_U = \{X_1, \ldots, X_h\}. \tag{4.18}$$

The tangent vector fields $X_1, \ldots, X_h$ are determined by L^h up to a nondegenerate linear transformation with functional coefficients. In fact, if we let

$$Y_\alpha = \sum_{\beta=1}^{h} a_\alpha^\beta X_\beta, \qquad 1 \le \alpha \le h, \tag{4.19}$$

where the a_α^β are smooth functions on U, and $\det(a_\alpha^\beta)$ is nonzero everywhere in U, then $L^h\big|_U$ is also spanned by $Y_1, \ldots, Y_h$. That is,

$$L^h\big|_U = \{Y_1, \ldots, Y_h\}. \tag{4.20}$$

The problem is: given an h-dimensional distribution L^h on M, is there a local coordinate system $(W; w^i)$ such that

$$L^h\big|_W = \left\{ \frac{\partial}{\partial w^1}, \cdots, \frac{\partial}{\partial w^h} \right\}? \tag{4.21}$$

When (4.21) is true, the tangent vector field X_α can be expressed as

$$X_\alpha = \sum_{\beta=a}^{h} a_\alpha^\beta \frac{\partial}{\partial w^\beta}. \tag{4.22}$$

Since $\left[\frac{\partial}{\partial w^\alpha}, \frac{\partial}{\partial w^\beta} \right] = 0$,

$$[X_\alpha, X_\beta] = \sum_{\gamma=1}^{h} C_{\alpha\beta}^\gamma X_\gamma, \tag{4.23}$$

where

$$C_{\alpha\beta}^\gamma = \sum_{\delta,\eta=1}^{h} \left(a_\alpha^\delta \frac{\partial a_\beta^\eta}{\partial w^\delta} - a_\beta^\delta \frac{\partial a_\alpha^\eta}{\partial w^\delta} \right) (a^{-1})_\eta^\gamma,$$

and a^{-1} denotes the inverse of the matrix $a = (a^\alpha_\beta)$. Obviously if the tangent vector fields $Y_1, \ldots, Y_h$ span the same h-dimensional distribution L^h, then $[Y_\alpha, Y_\beta]$ can be written as a linear combination of the Y_γ, when $X_\alpha, 1 \leq \alpha \leq h$, satisfy condition (4.23).

Definition 4.3. Suppose $L^h = \{X_1, \ldots X_h\}$ is an h-dimensional distribution spanned by the X's. If $[X_\alpha, X_\beta]$ is a linear combination of $X_\gamma, 1 \leq \alpha, \beta, \gamma \leq h$, then the distribution is said to satisfy the **Frobenius condition**.

Theorem 4.4. *Suppose $L^h = \{X_1, \ldots, X_h\}$ is an h-dimensional distribution in an open set U. A necessary and sufficient condition for the existence of a local coordinate system $(W; w^i), W \subset U$ such that*

$$L^h\big|_W = \left\{ \frac{\partial}{\partial w^1}, \cdots, \frac{\partial}{\partial w^h} \right\}$$

is that L^h satisfies the Frobenius condition.

This theorem is usually called the **Frobenius Theorem**.

Proof. Necessity has already been proved. For sufficiency, we will use induction on h, the dimension of the distribution.

For $h = 1$, the theorem follows from Theorem 4.3. Assume that sufficiency is true for $(h - 1)$-dimensional distributions. Suppose the distribution L^h is spanned by linearly independent tangent vector fields $X_1, \ldots, X_h$ on U, and

$$[X_\alpha, X_\beta] \equiv 0 \bmod X_\gamma, \quad 1 \leq \alpha, \beta, \gamma \leq h.$$

By Theorem 4.3, there exists a coordinate system $(y^1, \ldots, y^m)$ such that

$$X_h = \frac{\partial}{\partial y^h}. \tag{4.24}$$

Let

$$X'_\lambda = X_\lambda - (X_\lambda y^h) X_h, \tag{4.25}$$

with $1 \leq \lambda \leq h - 1$ (also for the rest of this proof). Obviously,

$$X'_\lambda y^h = 0, \quad X_h y^h = 1. \tag{4.26}$$

Since $L^h = \{X'_1, \ldots, X'_{h-1}, X_h\}$, these h tangent vector fields still satisfy the Frobenius condition. Therefore we may assume that

$$[X'_\lambda, X'_\mu] \equiv a_{\lambda\mu} X_h \bmod X'_\nu, \quad 1 \leq \lambda, \mu, \nu \leq h - 1.$$

If we apply the operators on both sides of this equivalence on the functions y^h, we get $a_{\lambda\mu} = 0$. Hence the $(h-1)$-dimensional distribution $L'^{h-1} = \{X'_1, \ldots, X'_{h-1}\}$ satisfies the Frobenius condition. By the induction hypothesis, there exists a local coordinate system $(z^1, \ldots, z^m)$ at p such that

$$L'^{h-1} = \left\{ \frac{\partial}{\partial z^1}, \cdots, \frac{\partial}{\partial z^{h-1}} \right\}. \tag{4.27}$$

Because the $\frac{\partial}{\partial z^\lambda}$ and the X'_μ differ only by a nondegenerate linear transformation, (4.26) gives

$$\frac{\partial}{\partial z^\lambda} y^h = 0. \tag{4.28}$$

Since $L^h = \left\{ \frac{\partial}{\partial z^1}, \cdots, \frac{\partial}{\partial z^{h-1}}, X_h \right\}$, the Frobenius condition allows us to assume

$$\left[\frac{\partial}{\partial z^\lambda}, X_h \right] \equiv b_\lambda X_h \bmod \frac{\partial}{\partial z^\mu}.$$

Applying the operators on both sides on y^h, we get $b_\lambda = 0$. Hence

$$\left[\frac{\partial}{\partial z^\lambda}, X_h \right] = \sum_{\mu=1}^{h-1} C_\lambda^\mu \frac{\partial}{\partial z^\mu}. \tag{4.29}$$

Suppose in the local coordinate system $(z^1, \ldots, z^m)$, X_h can be expressed as

$$X_h = \sum_{i=1}^m \xi^i \frac{\partial}{\partial z^i}. \tag{4.30}$$

Then

$$\left[\frac{\partial}{\partial z^\lambda}, X_h \right] = \sum_{i=1}^m \frac{\partial \xi^i}{\partial z^\lambda} \frac{\partial}{\partial z^i}. \tag{4.31}$$

Comparing the above equation with (4.29), we get

$$\frac{\partial \xi^\rho}{\partial z^\lambda} = 0, \quad 1 \le \lambda \le h-1, \quad h \le \rho \le m. \tag{4.32}$$

This implies that ξ^ρ is a function of $z^h, \ldots, z^m$ only. Let

$$X'_h = \sum_{\rho=h}^m \xi^\rho \frac{\partial}{\partial z^\rho}. \tag{4.33}$$

Then we still have

$$L^h = \left\{ \frac{\partial}{\partial z^1}, \cdots, \frac{\partial}{\partial z^{h-1}}, X'_h \right\}.$$

By Theorem 4.3, there is a change of local coordinates from $(z^h, \ldots, z^m)$ to $(w^h, \ldots, w^m)$ such that

$$X'_h = \frac{\partial}{\partial w^h}. \tag{4.34}$$

The above change of coordinates does not involve $z^1, \ldots, z^{h-1}$. Now let $w^\lambda = z^\lambda$, $1 \le \lambda \le h - 1$. $(w^1, \ldots, w^m)$ then becomes a local coordinate system at p and under this system,

$$L^h = \left\{ \frac{\partial}{\partial w^1}, \cdots, \frac{\partial}{\partial w^h} \right\}.$$

$\square$

Remark. In various applications, it is convenient to express the Frobenius Theorem in its dual form using exterior derivatives. (See Theorem 2.4 in Chapter 3.)

Chapter 2

Multilinear Algebra

§2–1   Tensor Products

This chapter provides an algebraic preparation for an advanced study in differentiable manifolds. First we review some vector space concepts.

Let $\mathbb{F}$ be a field. In this book, $\mathbb{F}$ usually refers to the real number field $\mathbb{R}$ or the complex number field $\mathbb{C}$. A **vector space** V over $\mathbb{F}$ is a set with the following two operations:

1) addition;
2) scalar multiplication with respect to $\mathbb{F}$ (with the element of $\mathbb{F}$ written on the left),

which are required to satisfy the following two conditions:

1) The set V is a commutative group with respect to the operation of addition, and the identity element is denoted by 0 (zero vector);
2) For $\alpha, \beta \in \mathbb{F}, x, y, \in V$,

$$a) \quad \alpha(x + y) = \alpha x + \alpha y,$$

$$b) \quad (\alpha + \beta)x = \alpha x + \beta x,$$

$$c) \quad (\alpha\beta)x = \alpha(\beta x),$$

$$d) \quad 0 \cdot x = 0,\ 1 \cdot x = x.$$

The elements of V are called **vectors** and those of $\mathbb{F}$ are called **scalars**.

If there exist n elements $a_1, \ldots, a_n$ of V such that any element of V can be uniquely expressed by a linear combination of these n elements, then we call

V an n-**dimensional** vector space. The set $\{a_1, \dots, a_n\}$ is called a **basis** of V. Obviously, after a basis is chosen, V can be viewed as the vector space of ordered n-tuples $\{\alpha^1, \dots, \alpha^n\}$, $\alpha^i \in \mathbb{F}$.

Remark. If we replace the field $\mathbb{F}$ above by a ring R, then we can define linear spaces over the ring R analogously. A linear space over a ring R is called a (left) R-**module**. For example, the sections of a vector bundle defined in §3–1 form a $C^\infty(M)$-module.

Suppose $f : V \longrightarrow \mathbb{F}$ is an $\mathbb{F}$-valued function on V. If for any $v_1, v_2 \in V$ and $\alpha^1, \alpha^2 \in \mathbb{F}$,

$$f\left(\alpha^1 v_1 + \alpha^2 v_2\right) = \alpha^1 f(v_1) + \alpha^2 f(v_2), \tag{1.1}$$

then f is called an $\mathbb{F}$-valued **linear function** on V. Obviously, if f, g are $\mathbb{F}$-valued linear functions on V and $\alpha \in \mathbb{F}$, then $f + g$ and αf are also $\mathbb{F}$-valued linear functions on V. Thus the set of all $\mathbb{F}$-valued linear functions on V forms a vector space over $\mathbb{F}$, called the **dual space** of V, denoted by V^*.

If V is an n-dimensional vector space over $\mathbb{F}$, then V^* is also an n-dimensional vector space over $\mathbb{F}$. To see this, suppose $\{a_1, \dots, a_n\}$ is a basis of V, and

$$v = \sum_{i=1}^{n} v^i a_i \in V, \qquad f \in V^*.$$

Then

$$f(v) = \sum_{i=1}^{n} v^i f(a_i). \tag{1.2}$$

Therefore the linear function f is determined by its values $f(a_i)$, $1 \le i \le n$, on the basis. We may define linear functions $a^{*i} \in V^*$, $1 \le i \le n$, such that

$$a^{*i}(a_j) = \delta_j^i, \quad 1 \le j \le n. \tag{1.3}$$

Then $a^{*i}(v) = v^i$. Hence

$$f(v) = \sum_{i=1}^{n} f_i v^i = \sum_{i=1}^{n} f_i a^{*i}(v),$$

$$f = \sum_{i=1}^{n} f_i a^{*i}, \tag{1.4}$$

where $f_i = f(a_i)$. Equation (1.4) says that any element in V^* can be expressed as a linear combination of $\{a^{*i}, 1 \leq i \leq n\}$. It is easy to see that the expression is unique, and therefore $\{a^{*i}, 1 \leq i \leq n\}$ is a basis of V^*, which is called the **dual basis** of $\{a_i, 1 \leq i \leq n\}$. Thus V^* is also an n-dimensional vector space over $\mathbb{F}$.

Remark. For a fixed basis $\{a_1, \ldots, a_n\}$ of a vector space V, a^{*i} are the coordinate functions on V. In fact, $a^{*i}(v) = v^i$, i.e., $a^{*i}(v)$ is the i-th component of the vector v with respect to the fixed basis.

V^* and V are dual spaces of each other. Define

$$\langle v, v^* \rangle = v^*(v), \qquad v \in V, \quad v^* \in V^*. \tag{1.5}$$

Then $\langle \ , \ \rangle$ is an $\mathbb{F}$-valued function defined on $V \times V^*$. It is linear in each variable, i.e., for any $v, v_1, v_2 \in V$, $v^*, v^{*1}, v^{*2} \in V^*$ and $\alpha_1, \alpha_2 \in \mathbb{F}$ we have

$$\begin{cases} \langle \alpha_1 v_1 + \alpha_2 v_2 , v^* \rangle = \alpha_1 \langle v_1, v^* \rangle + \alpha_2 \langle v_2, v^* \rangle, \\ \langle v , \alpha_1 v^{*1} + \alpha_2 v^{*2} \rangle = \alpha_1 \langle v, v^{*1} \rangle + \alpha_2 \langle v, v^{*2} \rangle. \end{cases} \tag{1.6}$$

If we fix a vector $v \in V$ in (1.5), then $\langle v, \cdot \rangle$ is an $\mathbb{F}$-valued linear function on V^*. Conversely, any $\mathbb{F}$-valued linear function can be expressed in this way. Suppose φ is an $\mathbb{F}$-valued linear function on V^*. Simply let $v = \sum_{i=1}^n \varphi(a^{*i}) a_i$. Then we have, for any $v^* \in V^*$,

$$\begin{aligned} \langle v, v^* \rangle &= \sum_{i=1}^n \varphi(a^{*i}) \langle a_i, v^* \rangle \\ &= \sum_{i=1}^n v^*(a_i) \cdot \varphi(a^{*i}) \\ &= \varphi\left(\sum_{i=1}^n v^*(a_i) \cdot a^{*i} \right) \\ &= \varphi(v^*). \end{aligned} \tag{1.7}$$

Therefore V can be viewed as a vector space formed by all $\mathbb{F}$-valued linear functions on V^*. In other words, V is the dual space of V^*.

Now we generalize the discussion above. Assume that V, W, Z are all finite-dimensional vector spaces over the field $\mathbb{F}$.

Definition 1.1. A map $f : V \longrightarrow Z$ is called **linear** if for any $v_1, v_2 \in V$, $\alpha^1, \alpha^2 \in \mathbb{F}$,

$$f(\alpha^1 v_1 + \alpha^2 v_2) = \alpha^1 f(v_1) + \alpha^2 f(v_2). \tag{1.8}$$

A map $f : V \times W \longrightarrow Z$ is called **bilinear** if for any $v, v_1, v_2 \in V$, $w, w_1, w_2 \in W$ and $\alpha^1, \alpha^2 \in \mathbb{F}$,

$$\begin{cases} f(\alpha^1 v_1 + \alpha^2 v_2 \, , \, w) = \alpha^1 f(v_1, w) + \alpha^2 f(v_2, w), \\ f(v \, , \, \alpha^1 w_1 + \alpha^2 w_2) = \alpha^1 f(v, w_1) + \alpha^2 f(v, w_2). \end{cases} \tag{1.9}$$

Similarly we can define an r-**linear** map $f : V_1 \times \cdots \times V_r \longrightarrow Z$ where $V_1, \ldots V_r$ are vector spaces over $\mathbb{F}$.

When $Z = \mathbb{F}$ (viewed as a one-dimensional vector space over $\mathbb{F}$), the objects defined in Definition 1.1 become $\mathbb{F}$-valued linear functions, $\mathbb{F}$-valued bilinear functions, and $\mathbb{F}$-valued r-linear functions, respectively.

Denote the set of all r-linear maps from $V_1 \times \cdots \times V_r$ to Z by $\mathcal{L}(V_1, \ldots, V_r; Z)$. Suppose $f, g \in \mathcal{L}(V_1, \ldots, V_r; Z), \alpha \in \mathbb{F}$. For any $v_i \in V_i$, $1 \leq i \leq r$, define

$$\begin{cases} (f + g)(v_1, \ldots, v_r) = f(v_1, \ldots, v_r) + g(v_1, \ldots, v_r), \\ (\alpha f)(v_1, \ldots, v_r) = \alpha \cdot f(v_1, \ldots, v_r). \end{cases} \tag{1.10}$$

Then $f + g$ and αf are also in $\mathcal{L}(V_1, \ldots, V_r; Z)$. Obviously, the set $\mathcal{L}(V_1, \ldots, V_r)$ is a vector space over $\mathbb{F}$ with respect to the two operations.

The structure of the space $\mathcal{L}(V; Z)$ is simpler. Choose a basis $\{a_1, \ldots, a_n\}$ of V and a basis $\{b_1, \ldots, b_m\}$ of Z. Then $f \in \mathcal{L}(V; Z)$ is determined by its action on the basis $\{a_i\}$. Suppose

$$f(a_i) = \sum_{j=1}^{m} f_{ij} b_j, \quad 1 \leq i \leq n. \tag{1.11}$$

Then f corresponds to the $n \times m$ matrix (f_{ij}). It is not difficult to see that the space $\mathcal{L}(V; Z)$ and the vector space of all $n \times m$ matrices with elements in $\mathbb{F}$ are isomorphic. Usually we use the notation

$$\mathcal{L}(V; Z) = \text{Hom}(V; Z). \tag{1.12}$$

Now the problem is to transform a bilinear map on $V \times W$ into a linear map. More precisely, given two vector spaces V and W, we will construct a vector space Y and a bilinear map $h : V \times W \longrightarrow Y$ such that they depend on V and W only and satisfy the following: for any bilinear map $f : V \times W \longrightarrow Z$, there exists a unique linear map $g : Y \longrightarrow Z$ such that

$$f = g \circ h : V \times W \longrightarrow Z, \tag{1.13}$$

i.e., the following diagram commutes:

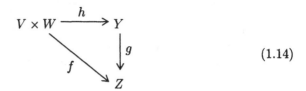

$$(1.14)$$

The space Y to be constructed is called the **tensor product** of V and W.

To describe the construction more precisely, let us consider the tensor product of the dual spaces V^* and W^* first. Suppose $v^* \in V^*, w^* \in W^*$. The tensor product $v^* \otimes w^*$, of linear functions v^* and w^*, is defined by

$$v^* \otimes w^*(v, w) = v^*(v) \cdot w^*(w)$$
$$= \langle v, v^* \rangle \cdot \langle w, w^* \rangle, \qquad (1.15)$$

where $v \in V$, $w \in W$. Then $v^* \otimes w^*$ is a bilinear map on $V \times W$, i.e., $v^* \otimes w^* \in \mathcal{L}(V, W; \mathbb{F})$. The operation $\otimes$ is a bilinear map from $V^* \times W^*$ to $\mathcal{L}(V, W; \mathbb{F})$. In fact, the following is true for any $v_1^*, v_2^* \in V^*$, $w^* \in W^*$ and $\alpha_1, \alpha_2 \in \mathbb{F}$.

$$(\alpha_1 v_1^* + \alpha_2 v_2^*) \otimes w^*(v, w) = \langle v, \alpha_1 v_1^* + \alpha_2 v_2^* \rangle \cdot \langle w, w^* \rangle$$
$$= \alpha_1 \langle v, v_1^* \rangle \langle w, w^* \rangle + \alpha_2 \langle v, v_2^* \rangle \langle w, w^* \rangle$$
$$= (\alpha_1 (v_1^* \otimes w^*) + \alpha_2 (v_2^* \otimes w^*))(v, w),$$

i.e.,

$$(\alpha_1 v_1^* + \alpha_2 v_2^*) \otimes w^* = \alpha_1 (v_1^* \otimes w^*) + \alpha_2 (v_2^* \otimes w^*). \qquad (1.16)$$

Similarly, the operation $\otimes$ is also linear in the second variable.

The tensor product $V^* \otimes W^*$ of the vector spaces V^* and W^* refers to the vector space generated by all elements of the form $v^* \otimes w^*$, $v^* \in V^*, w^* \in W^*$. It is a subspace of $\mathcal{L}(V, W; \mathbb{F})$. We need to point out that any element in $V^* \otimes W^*$ is a finite linear combination of elements of the form $v^* \otimes w^*$, but generally cannot be written as a single term $v^* \otimes w^*$ (the reader should construct examples). An element in $V^* \otimes W^*$ which can be written in the form $v^* \otimes w^*$ is called **reducible**.

Choose bases $\{a^{*i}\}_{1 \leq i \leq n}$ and $\{b^{*j}\}_{1 \leq j \leq m}$ for V^* and W^*, respectively. Because $\otimes$ is bilinear, we have

$$v^* \otimes w^* = \sum_{i,j} v^*(a_i) w^*(b_j) a^{*i} \otimes b^{*j}, \qquad (1.17)$$

where $\{a_i\}$ and $\{b_j\}$ are dual bases for V and W, respectively. Therefore any element in $V^* \otimes W^*$ can be expressed as a linear combination of the $a^{*i} \otimes b^{*j}$. It is easy to show that $a^{*i} \otimes b^{*j}$, $1 \le i \le n$, $1 \le j \le m$, are linearly independent, and hence form a basis of $V^* \otimes W^*$. Therefore $V^* \otimes W^*$ is an $(n \times m)$-dimensional vector space. We can also show that any $\mathbb{F}$-valued bilinear function $f : V \times W \longrightarrow \mathbb{F}$ can be expressed as a linear combination of $a^{*i} \otimes b^{*j}$. Therefore

$$V^* \otimes W^* = \mathcal{L}(V, W; \mathbb{F}).$$

Since V and W are the dual spaces of V^* and W^*, respectively, we can analogously define the tensor product $V \otimes W$. The following is also true:

$$V \otimes W = \mathcal{L}(V^*, W^*; \mathbb{F}).$$

$V \otimes W$ and $V^* \otimes W^*$ are dual spaces of each other if we require

$$\langle v \otimes w, v^* \otimes w^* \rangle = \langle v, v^* \rangle \cdot \langle w, w^* \rangle. \tag{1.18}$$

Particularly,

$$\begin{aligned}
\langle a_i \otimes b_j, a^{*p} \otimes b^{*q} \rangle &= \delta_i^p \delta_j^q \\
&= \begin{cases} 1 & \text{if } (i, j) = (p, q), \\ 0 & \text{if } (i, j) \ne (p, q). \end{cases}
\end{aligned} \tag{1.19}$$

Therefore $\{a_i \otimes b_j, 1 \le i \le n, 1 \le j \le m\}$ and $\{a^{*i} \otimes b^{*j}, 1 \le i \le n, 1 \le j \le m\}$ are dual bases. Thus

$$V^* \otimes W^* = (V \otimes W)^*.$$

Theorem 1.1. *Suppose $h : V \times W \longrightarrow V \otimes W$ is the bilinear map obtained from the tensor product $\otimes$, i.e., for $v \in V, w \in W$,*

$$h(v, w) = v \otimes w. \tag{1.20}$$

Then for any bilinear map $f : V \times W \longrightarrow Z$, there exists a unique linear map $g : V \otimes W \longrightarrow Z$ such that

$$f = g \circ h : V \times W \longrightarrow Z. \tag{1.21}$$

Proof. Define a linear map $g : V \otimes W \longrightarrow Z$ such that its action on a basis is

$$g(a_i \otimes b_j) = f(a_i, b_j), \quad 1 \le i \le n, \quad 1 \le j \le m. \tag{1.22}$$

Suppose

$$v = \sum_{i=1}^{n} v^i a_i \in V, \qquad w = \sum_{j=1}^{m} w^j b_j \in W.$$

Then

$$g(v \otimes w) = \sum_{i,j} v^i w^j g(a_i \otimes b_j)$$
$$= \sum_{i,j} v^i w^j f(a_i, b_j)$$
$$= f(v, w).$$

Therefore $f = g \circ h : V \times W \longrightarrow Z$, and it is clear that g is unique. $\square$

Corollary 1. *The vector spaces $\mathcal{L}(V, W; Z)$ and $\mathcal{L}(V \otimes W; Z)$ are isomorphic.*

Proof. Define a map

$$\varphi : \mathcal{L}(V \otimes W; Z) \longrightarrow \mathcal{L}(V, W; Z),$$

such that

$$\varphi(g) = g \circ h, \quad g \in \mathcal{L}(V \otimes W; Z), \tag{1.23}$$

where h is defined as in (1.20). Theorem 1.1 implies that φ is a one to one surjection. It is obvious that φ is linear. Therefore, φ is an isomorphism. $\square$

The operation of the tensor product of linear functions can be generalized to arbitrary multilinear functions. Suppose $f \in \mathcal{L}(V_1, \ldots V_s; \mathbb{F}), g \in \mathcal{L}(W_1, \ldots, W_r; \mathbb{F})$. Their tensor product $f \otimes g$ is defined as follows. Let $v_i \in V_i, 1 \leq i \leq s; w_j \in W_j, 1 \leq j \leq r$. Then

$$f \otimes g(v_1, \ldots v_s, w_1, \ldots, w_r) = f(v_1, \ldots v_s) \cdot g(w_1, \ldots w_r). \tag{1.24}$$

Obviously, $f \otimes g$ is an $\mathbb{F}$-valued $(r + s)$-linear function on $V_1 \times \cdots \times V_s \times W_1 \times \cdots \times W_r$, the tensor product operation $\otimes$ being a bilinear map from $\mathcal{L}(V_1, \ldots, V_s; \mathbb{F}) \times \mathcal{L}(W_1, \ldots, W_r; \mathbb{F})$ to $\mathcal{L}(V_1, \ldots, V_s, W_1, \ldots, W_r; \mathbb{F})$.

Theorem 1.2. *The tensor product $\otimes$ is associative, that is, for any $\varphi \in \mathcal{L}(V_1, \ldots, V_s; \mathbb{F}), \psi \in \mathcal{L}(W_1, \ldots, W_r; \mathbb{F}), \xi \in \mathcal{L}(Z_1, \ldots, Z_t; \mathbb{F})$, the following holds:*

$$(\varphi \otimes \psi) \otimes \xi = \varphi \otimes (\psi \otimes \xi). \tag{1.25}$$

Proof. To simplify the proof, we only consider the case $s = r = t = 1$. The general cases are similar. Let $v \in V_1, w \in W_1, z \in Z_1$. We have

$$(\varphi \otimes \psi) \otimes \xi(v, w, z) = \varphi \otimes \psi(v, w) \cdot \xi(z)$$
$$= \varphi(v) \cdot \psi(w) \cdot \xi(z).$$

Similarly,

$$\varphi \otimes (\psi \otimes \xi)(v, w, z) = \varphi(v) \cdot \psi(w) \cdot \xi(z).$$

Therefore

$$(\varphi \otimes \psi) \otimes \xi = \varphi \otimes (\psi \otimes \xi).$$

Thus the notation $\varphi \otimes \psi \otimes \xi$ is meaningful. It is called the tensor product of the three elements.   □

We denote the vector space generated by elements of the form $v \otimes w \otimes z, v \in V, w \in W, z \in Z$, by $V \otimes W \otimes Z$ and call it the **tensor product** of V, W, and Z. Here v, w, and z are viewed as $\mathbb{F}$-valued linear functions on V^*, W^*, and Z^*, respectively.

Similarly, if $V_1, \ldots, V_s$ are vector spaces over $\mathbb{F}$, then we can define their tensor product $V_1 \otimes \cdots \otimes V_s$. Suppose

$$\left\{ a_1^{(i)}, \ldots, a_{n_i}^{(i)} \right\}$$

is a basis of V_i. Then

$$a_{j_1}^{(1)} \otimes a_{j_2}^{(2)} \otimes \cdots \otimes a_{j_s}^{(s)}, \quad 1 \leq j_i \leq n_i, \quad i = 1, \ldots, s \qquad (1.26)$$

is a basis of $V_1 \otimes \cdots \otimes V_s$. Therefore

$$\dim(V_1 \otimes \cdots \otimes V_s) = \dim V_1 \cdot \dim V_2 \cdot \cdots \cdot \dim V_s. \qquad (1.27)$$

It is easy to show that

$$V_1 \otimes \cdots \otimes V_s = \mathcal{L}(V_1^*, \ldots, V_s^*; \mathbb{F}).$$

Theorem 1.3. *Suppose $h : V_1 \times \cdots \times V_s \longrightarrow V_1 \otimes \cdots \otimes V_s$ is the s-linear map defined by the tensor product $\otimes$, i.e., for any $v_i \in V_i$, $1 \leq i \leq s$,*

$$h(v_1, \ldots, v_s) = v_1 \otimes \cdots \otimes v_s. \qquad (1.28)$$

Then for any $f \in \mathcal{L}(V_1, \ldots, V_s; Z)$, there exists a unique linear map $g \in \mathcal{L}(V_1 \otimes \cdots \otimes V_s; Z)$ such that

$$f = g \circ h : V_1 \times \cdots \times V_s \longrightarrow Z. \qquad (1.29)$$

The proof is similar to that of Theorem 1.1, and should be carried out by the reader.

§2–2 Tensors

We have discussed some general tensor product concepts in the last section. In differential geometry the tensor products of copies of a vector space and its dual space are often used.

Definition 2.1. Suppose V is an n-dimensional vector space over $\mathbb{F}$ with dual space V^*. The elements in the tensor product

$$V_s^r = \underbrace{V \otimes \cdots \otimes V}_{r \text{ terms}} \otimes \underbrace{V^* \otimes \cdots \otimes V^*}_{s \text{ terms}} \tag{2.1}$$

are called (r, s)-**type tensors**, where r is the **contravariant order** and s is the **covariant order**.

In particular, the elements in V_0^r are called **contravariant tensors** of **order** r, and those in V_s^0 are called **covariant tensors** of **order** s. We also have the following conventions: $V_0^0 = \mathbb{F}, V_0^1 = V, V_1^0 = V^*$. The elements of V are called **contravectors**, and the elements of V^* are called **covectors**.

Remark. In practice, the elements in V and in V^* may appear alternately in the tensor product V_s^r. Writing them down in the order as shown in (2.1) is done here just for convenience of notation.

By the discussion in §2–1, $\dim V_s^r = n^{r+s}$, and

$$V_s^r = \mathcal{L}(\underbrace{V^*, \ldots, V^*}_{r \text{ terms}}, \underbrace{V, \ldots, V}_{s \text{ terms}}; \mathbb{F}).$$

This shows that (r, s)-type tensors are $\mathbb{F}$-valued $(r+s)$-linear functions defined on

$$\underbrace{V^* \times \cdots \times V^*}_{r \text{ terms}} \times \underbrace{V \times \cdots \times V}_{s \text{ terms}}.$$

Suppose $\{e_i\}_{1 \le i \le n}$ and $\{e^{*i}\}_{1 \le i \le n}$ are dual bases in V and V^*, respectively. Then

$$e_{i_1} \otimes \cdots \otimes e_{i_r} \otimes e^{*k_1} \otimes \cdots \otimes e^{*k_s}, \quad 1 \le i_1, \ldots, i_r, k_1, \ldots, k_s \le n \tag{2.2}$$

form a basis of V_s^r. Therefore an (r, s)-type tensor x can be uniquely expressed as

$$x = \sum_{\substack{i_1 \cdots i_r \\ k_1 \cdots k_s}} x_{k_1 \cdots k_s}^{i_1 \cdots i_r} e_{i_1} \otimes \cdots \otimes e_{i_r} \otimes e^{*k_1} \otimes \cdots \otimes e^{*k_s}, \tag{2.3}$$

where the $x^{i_1\cdots i_r}_{k_1\cdots k_s}$ are called the **components** of the tensor x under the basis (2.2). It is obvious that

$$x^{i_1\cdots i_r}_{k_1\cdots k_s} = x(e^{*i_1},\ldots,e^{*i_r},e_{k_1},\ldots,e_{k_s})$$
$$= \langle e^{*i_1}\otimes\cdots\otimes e^{*i_r}\otimes e_{k_1}\otimes\cdots\otimes e_{k_s} , x\rangle. \qquad (2.4)$$

In working with tensors, we often use the summation convention of Einstein: if an index occurs as both a subscript and superscript in the same term, then the term is summed over the range of the repeated index, and the summation sign is omitted. For example, there should be $(r+s)$ summation signs in (2.3), but by this convention it can be written as

$$x = x^{i_1\cdots i_r}_{k_1\cdots k_s}e_{i_1}\otimes\cdots\otimes e_{i_r}\otimes e^{*k_1}\otimes\cdots\otimes e^{*k_s}, \qquad (2.5)$$

When the basis of the vector space V is changed, the components of a tensor are changed according to specific rules. Suppose $\{\bar{e}_i\}_{1\le i\le n}$ is another basis of V with dual basis $\{\bar{e}^{*i}\}_{1\le i\le n}$. We may assume that under the original basis,

$$\bar{e}_i = \alpha^j_i e_j , \qquad (2.6)$$

where $\alpha = (\alpha^j_i)$ is a nonsingular $n\times n$ matrix. Therefore

$$\bar{e}^{*i} = \beta^i_j e^{*j} , \qquad (2.7)$$

where $\beta = (\beta^j_i)$ is the inverse matrix of α, that is,

$$\alpha^j_i\beta^k_j = \beta^j_i\alpha^k_j = \delta^k_i. \qquad (2.8)$$

If the components of a tensor x under the new basis are denoted by $\bar{x}^{i_1\cdots i_r}_{k_1\cdots k_s}$, then

$$x = \bar{x}^{i_1\cdots i_r}_{k_1\cdots k_s}\bar{e}_{i_1}\otimes\cdots\otimes\bar{e}_{i_r}\otimes\bar{e}^{*k_1}\otimes\cdots\otimes\bar{e}^{*k_s}$$
$$= \bar{x}^{i_1\cdots i_r}_{k_1\cdots k_s}\alpha^{j_1}_{i_1}\cdots\alpha^{j_r}_{i_r}\beta^{k_1}_{l_1}\cdots\beta^{k_s}_{l_s}e_{j_1}\otimes\cdots\otimes e_{j_r}\otimes e^{*l_1}\otimes\cdots\otimes e^{*l_s}.$$

Therefore

$$x^{j_1\cdots j_r}_{l_1\cdots l_s} = \bar{x}^{i_1\cdots i_r}_{k_1\cdots k_s}\alpha^{j_1}_{i_1}\cdots\alpha^{j_r}_{i_r}\beta^{k_1}_{l_1}\cdots\beta^{k_s}_{l_s}. \qquad (2.9)$$

In classical tensor analysis, (2.9) is used to define tensors.

The space V^r_s of all (r,s)-type tensors is a vector space. Thus two tensors of the same type can be added, and any tensor can be multiplied by scalars. Also tensors admit the operations of multiplication and contraction. The product of two tensors is their tensor product when they are viewed as multilinear functions.

Definition 2.2. Suppose x is an (r_1, s_1)-type tensor and y is an (r_2, s_2)-type tensor. Then their **tensor product** $x \otimes y$ is an $(r_1 + r_2, s_1 + s_2)$-type tensor given by

$$x \otimes y(v^{*1}, \ldots, v^{*r_1+r_2}, v_1, \ldots, v_{s_1+s_2})$$

$$= x(v^{*1}, \ldots, v^{*r_1}, v_1, \ldots, v_{s_1}) \tag{2.10}$$

$$\cdot y(v^{*r_1+1}, \ldots, v^{*r_1+r_2}, v_{s_1+1}, \ldots, v_{s_1+s_2}).$$

When a basis is chosen, the components of $x \otimes y$ are the products of the components of x and y, i.e.,

$$(x \otimes y)^{i_1 \cdots i_{r_1+r_2}}_{k_1 \cdots k_{s_1+s_2}} = x^{i_1 \cdots i_{r_1}}_{k_1 \cdots k_{s_1}} \cdot y^{i_{r_1+1} \cdots i_{r_1+r_2}}_{k_{s_1+1} \cdots k_{s_1+s_2}}. \tag{2.11}$$

As discussed in §2–1, the multiplication of tensors satisfies the distributive and associative laws (see Theorem 1.2).

Definition 2.3. Choose two indices λ, μ, $1 \leq \lambda \leq r$, $1 \leq \mu \leq s$. For any reducible (r, s)-type tensor

$$x = v_1 \otimes \cdots \otimes v_r \otimes v^{*1} \otimes \cdots \otimes v^{*s} \in V^r_s, \tag{2.12}$$

let

$$C_{\lambda\mu}(x) = \langle v_\lambda, v^{*\mu} \rangle v_1 \otimes \cdots \otimes \hat{v}_\lambda \otimes \cdots \otimes v_r \otimes v^{*1} \otimes \cdots \otimes \hat{v}^{*\mu} \otimes \cdots \otimes v^{*s}, \tag{2.13}$$

where the notation "$\hat{v}_\lambda$" means that we omit the term v_λ. Then $C_{\lambda\mu}(x) \in V^{r-1}_{s-1}$. We extend the map $x \mapsto C_{\lambda\mu}(x)$ linearly to get a linear map $C_{\lambda\mu} : V^r_s \longrightarrow V^{r-1}_{s-1}$, called a **tensor contraction** map.

Suppose x can be expressed in components as

$$x = x^{i_1 \cdots i_r}_{k_1 \cdots k_s} e_{i_1} \otimes \cdots \otimes e_{i_r} \otimes e^{*k_1} \otimes \cdots \otimes e^{*k_s}. \tag{2.14}$$

By the definition of tensor contraction, we get

$$C_{\lambda\mu}(x) = x^{i_1 \cdots i_r}_{k_1 \cdots k_s} C_{\lambda\mu}(e_{i_1} \otimes \cdots \otimes e_{i_r} \otimes e^{*k_1} \otimes \cdots \otimes e^{*k_s})$$

$$= x^{i_1 \cdots i_{\lambda-1} \, j \, i_\lambda \cdots i_{r-1}}_{k_1 \cdots k_{\mu-1} \, j \, k_\mu \cdots k_{s-1}} e_{i_1} \otimes \cdots \otimes e_{i_{r-1}} \otimes e^{*k_1} \otimes \cdots \otimes e^{*k_{s-1}}. \tag{2.15}$$

Thus from the viewpoint of components, the tensor contraction $C_{\mu\lambda}$ is the sum with respect to equated values of the λ-th upper index and the μ-th lower index. A contraction lowers the order of a tensor, and is a very basic operation.

For instance, suppose $x = \xi^i_j e_i \otimes e^{*j}$ is a $(1,1)$-type tensor. Then the tensor contraction of x is the trace $\sum_{i=1}^n \xi^i_i$ of the matrix (ξ^i_j), and hence a scalar independent of the choice of coordinate systems.

Suppose

$$T^r(V) = V^r_0 = \underbrace{V \otimes \cdots \otimes V}_{r \text{ terms}}.$$

Consider the direct sum $T(V) = \sum_{r \geq 0} T^r(V)$. Any element x in it can be expressed as the formal sum

$$x = \sum_{r \geq 0} x^r, \quad x^r \in T^r(V), \tag{2.16}$$

where all but finitely many terms are zero. Thus $T(V)$ is an infinite-dimensional vector space. With the distributive law, the multiplication between tensors can be extended to multiplication in $T(V)$. Therefore the vector space $T(V)$ becomes an algebra with respect to this multiplication, and is called the **tensor algebra** of V.

Similarly, the tensor algebra of V^* is $T(V^*) = \sum_{r \geq 0} V^0_r$.

As discussed in §2–1, the vector spaces $T^r(V^*)$ and $T^r(V)$ are dual to each other, and their pairing is given by

$$\langle v_1 \otimes \cdots \otimes v_r \,,\, v^{*1} \otimes \cdots \otimes v^{*r} \rangle = \langle v_1 \,,\, v^{*1} \rangle \cdots \langle v_r \,,\, v^{*r} \rangle, \tag{2.17}$$

where $v_i \in V, v^{*i} \in V^*$.

Denote the permutation group of the set of natural numbers $\{1, \ldots, r\}$ by $S(r)$. Any element σ in $S(r)$ determines an automorphism of the vector space $T^r(V)$. Suppose $x \in T^r(V)$. We define

$$\sigma x(v^{*1}, \ldots, v^{*r}) = x(v^{*\sigma(1)}, \ldots, v^{*\sigma(r)}), \tag{2.18}$$

where $v^{*i} \in V^*$. It is easy to see that if $x = v_1 \otimes \cdots \otimes v_r$, then

$$\sigma x = v_{\sigma^{-1}(1)} \otimes \cdots \otimes v_{\sigma^{-1}(r)}, \tag{2.19}$$

where $\sigma^{-1} \in S(r)$ is the inverse of σ.

Definition 2.4. Suppose $x \in T^r(V)$. If for any $\sigma \in S(r)$ we have

$$\sigma x = x, \tag{2.20}$$

then we call x a **symmetric** contravariant tensor of order r. If for any $\sigma \in \mathcal{S}(r)$, we have

$$\sigma x = \text{sgn } \sigma \cdot x, \tag{2.21}$$

where sgn σ denotes the sign of the permutation σ, i.e.,

$$\text{sgn } \sigma = \begin{cases} 1 & \text{if } \sigma \text{ is an even permutation,} \\ -1 & \text{if } \sigma \text{ is an odd permutation,} \end{cases} \tag{2.22}$$

then we call x an **alternating** contravariant tensor of order r.

Theorem 2.1. *Suppose $x \in T^r(V)$. A necessary and sufficient condition for x to be a symmetric tensor is that all its components are symmetric with respect to all indices. A necessary and sufficient condition for x to be an alternating tensor is that all its components are alternating with respect to all indices.*

Proof. Suppose $\{e_1, \ldots, e_n\}$ is a basis of V, and x is a symmetric tensor. Then for any $\sigma \in \mathcal{S}(r)$,

$$\begin{aligned} x^{i_1 \cdots i_r} &= x(e^{*i_1}, \ldots, e^{*i_r}) \\ &= \sigma x(e^{*i_1}, \ldots, e^{*i_r}) \\ &= x(e^{*i_{\sigma(1)}}, \ldots, e^{*i_{\sigma(r)}}) \\ &= x^{i_{\sigma(1)} \cdots i_{\sigma(r)}}. \end{aligned} \tag{2.23}$$

On the other hand, if x is alternating, then for any $\sigma \in \mathcal{S}(r)$,

$$\begin{aligned} x^{i_1 \cdots i_r} &= x(e^{*i_1}, \ldots, e^{*i_r}) \\ &= \text{sgn } \sigma \cdot \sigma x(e^{*i_1}, \ldots, e^{*i_r}) \\ &= \text{sgn } \sigma \cdot x^{i_{\sigma(1)} \cdots i_{\sigma(r)}}. \end{aligned} \tag{2.24}$$

The converses are also true. □

Denote the set of all symmetric contravariant tensors of order r by $P^r(V)$, and the set of all alternating contravariant tensors of order r by $\Lambda^r(V)$. Because the permutation σ is an automorphism on $T^r(V)$, the sum of symmetric tensors is still symmetric and the sum of alternating tensors is still alternating. Therefore $P^r(V)$ and $\Lambda^r(V)$ are linear subspaces of $T^r(V)$.

Definition 2.5. For any $x \in T^r(V)$, let

$$S_r(x) = \frac{1}{r!} \sum_{\sigma \in \mathcal{S}(r)} \sigma x, \tag{2.25}$$

$$A_r(x) = \frac{1}{r!} \sum_{\sigma \in \mathcal{S}(r)} \text{sgn } \sigma \cdot \sigma x. \tag{2.26}$$

Then $S_r(x)$, $A_r(x) \in T^r(V)$. Obviously, both maps $S_r, A_r : T^r(V) \longrightarrow T^r(V)$ are automorphisms of $T^r(V)$, called the **symmetrizing map** and the **alternating map** , respectively, on contravariant tensors of order r.

Theorem 2.2.

$$P^r(V) = S_r(T^r(V)),$$
$$\Lambda^r(V) = A_r(T^r(V)).$$

Proof. First we show that the image of a tensor x under the symmetrizing mapping is a symmetric tensor, and the image of a tensor x under the alternating mapping is an alternating tensor. Suppose $x \in T^r(V)$. Then for any $\tau \in S(r)$,

$$\tau(S_r(x)) = \frac{1}{r!} \sum_{\sigma \in S(r)} \tau(\sigma(x))$$

$$= S_r(x), \tag{2.27}$$

$$\tau(A_r(x)) = \frac{1}{r!} \sum_{\sigma \in S(r)} \text{sgn}\ \sigma \cdot \tau(\sigma(x))$$

$$= \text{sgn}\ \tau \cdot \frac{1}{r!} \sum_{\sigma \in S(r)} \text{sgn}\ (\tau \circ \sigma) \cdot (\tau \circ \sigma(x))$$

$$= \text{sgn}\ \tau \cdot A_r(x). \tag{2.28}$$

Therefore $S_r(T^r(V)) \subset P^r(V)$, $A_r(T^r(V)) \subset \Lambda^r(V)$.

Furthermore, it is easy to show that a symmetric tensor is invariant under the symmetrizing mapping and an alternating tensor is invariant under the alternating mapping. Therefore $P^r(V) = S_r(P^r(V))$, $\Lambda^r(V) = A_r(\Lambda^r(V))$. Thus

$$P^r(V) = S_r(T^r(V)),$$
$$\Lambda^r(V) = A_r(T^r(V)).$$

$\square$

The above discussion about symmetric and alternating contravariant tensors can be applied analogously to covariant tensors. The set of all symmetric covariant tensors of order r is denoted by $P^r(V^*)$, and the set of all alternating covariant tensors of order r by $\Lambda^r(V^*)$.

§2–3   Exterior Algebra

Due to E. Cartan's systematic development of the method of exterior differentiation, the alternating tensors have played an important role in the study of manifolds. An alternating contravariant tensor of order r is also called an **exterior vector** of **degree** r or an exterior r-vector. The space $\Lambda^r(V)$ is called the **exterior space** of V of degree r. For convenience, we have the following conventions: $\Lambda^1(V) = V$, $\Lambda^0(V) = \mathbb{F}$.

More importantly, there exists an operation, the exterior (wedge) product, for exterior vectors such that the product of two exterior vectors is another exterior vector.

Definition 3.1. Suppose ξ is an exterior k-vector, and η an exterior l-vector. Let

$$\xi \wedge \eta = A_{k+l}(\xi \otimes \eta), \tag{3.1}$$

where A_{k+l} is the alternating mapping defined in Definition 2.5. Then $\xi \wedge \eta$ is an exterior $(k + l)$-vector, called the **exterior (wedge) product** of ξ and η.

Theorem 3.1. *The exterior product satisfies the following rules. Suppose* $\xi, \xi_1, \xi_2, \in \Lambda^k(V)$, $\eta, \eta_1, \eta_2, \in \Lambda^l(V)$, $\zeta \in \Lambda^h(V)$. *Then we have*

1) *Distributive Law* $(\xi_1 + \xi_2) \wedge \eta = \xi_1 \wedge \eta + \xi_2 \wedge \eta,$

                                          $\xi \wedge (\eta_1 + \eta_2) = \xi \wedge \eta_1 + \xi \wedge \eta_2;$

2) *Anticommutative Law* $\xi \wedge \eta = (-1)^{kl} \eta \wedge \xi;$

3) *Associative Law* $(\xi \wedge \eta) \wedge \zeta = \xi \wedge (\eta \wedge \zeta).$

Proof.   1) The distributive law is an obvious consequence of the linearity of the tensor product and of the alternating mapping.

2) Since $\xi \wedge \eta$ is an alternating tensor, for any $\tau \in \mathcal{S}(k + l)$,

$$\tau(\xi \wedge \eta) = \operatorname{sgn} \tau \cdot \xi \wedge \eta.$$

Choose

$$\tau = \begin{pmatrix} 1 & \cdots & k & k+1 & \cdots & k+l \\ l+1 & \cdots & k+l & 1 & \cdots & l \end{pmatrix}.$$

Then $\operatorname{sgn} \tau = (-1)^{kl}$. Thus for any $v^{*1}, \dots, v^{*k+l} \in V^*$,

$$
\begin{aligned}
&\xi \wedge \eta(v^{*1}, \dots, v^{*k+l}) \\
&= (-1)^{kl} \xi \wedge \eta(v^{*\tau(1)}, \dots, v^{*\tau(k+l)}) \\
&= \frac{(-1)^{kl}}{(k+l)!} \sum_{\sigma \in \mathcal{S}(k+l)} \operatorname{sgn} \sigma \cdot \xi(v^{*\sigma\circ\tau(1)}, \dots, v^{*\sigma\circ\tau(k)}) \times \\
&\quad \eta(v^{*\sigma\circ\tau(k+1)}, \dots, v^{*\sigma\circ\tau(k+l)}) \\
&= \frac{(-1)^{kl}}{(k+l)!} \sum_{\sigma \in \mathcal{S}(k+l)} \operatorname{sgn} \sigma \cdot \eta(v^{*\sigma(1)}, \dots, v^{*\sigma(l)}) \times \\
&\quad \xi(v^{*\sigma(l+1)}, \dots, v^{*\sigma(k+l)}) \\
&= (-1)^{kl} \eta \wedge \xi(v^{*1}, \dots, v^{*k+l}).
\end{aligned}
$$

3) Suppose $v^{*1}, \dots, v^{*k+l+h} \in V^*$. Then, by definition,

$$
\begin{aligned}
&(\xi \wedge \eta) \wedge \zeta(v^{*1}, \dots, v^{*k+l+h}) \\
&= \frac{1}{(k+l+h)!} \cdot \frac{1}{(k+l)!} \sum_{\sigma \in \mathcal{S}(k+l+h)} \sum_{\tau \in \mathcal{S}(k+l)} (\operatorname{sgn} \sigma \cdot \operatorname{sgn} \tau) \times \\
&\quad \xi(v^{*\sigma\circ\tau(1)}, \dots, v^{*\sigma\circ\tau(k)}) \times \\
&\quad \eta(v^{*\sigma\circ\tau(k+1)}, \dots, v^{*\sigma\circ\tau(k+l)}) \times \\
&\quad \zeta(v^{*\sigma\circ\tau(k+l+1)}, \dots, v^{*\sigma\circ\tau(k+l+h)}) \\
&= \frac{1}{(k+l+h)!} \sum_{\sigma \in \mathcal{S}(k+l+h)} \Big(\operatorname{sgn} \sigma \cdot \xi(v^{*\sigma(1)}, \dots, v^{*\sigma(k)}) \times \\
&\quad \eta(v^{*\sigma(k+1)}, \dots, v^{*\sigma(k+l)}) \times \\
&\quad \zeta(v^{*\sigma(k+l+1)}, \dots, v^{*\sigma(k+l+h)}) \Big).
\end{aligned}
$$

Therefore

$$
(\xi \wedge \eta) \wedge \zeta = A_{k+l+h}(\xi \otimes \eta \otimes \zeta). \tag{3.2}
$$

Similarly, we obtain

$$
\xi \wedge (\eta \wedge \zeta) = A_{k+l+h}(\xi \otimes \eta \otimes \zeta) = (\xi \wedge \eta) \wedge \zeta.
$$

$\square$

Remark. Suppose $\xi, \eta \in V = \Lambda^1(V)$. Then the anticommutative law implies

$$
\begin{aligned}
\xi \wedge \eta &= -\eta \wedge \xi, \\
\xi \wedge \xi &= \eta \wedge \eta = 0.
\end{aligned} \tag{3.3}
$$

Generally, if there are repeated exterior 1-vectors in a polynomial wedge product, then the product is zero.

Suppose $\{e_1, \ldots, e_n\}$ is a basis of V. Then according to the proof of the associative law

$$e_{i_1} \wedge \cdots \wedge e_{i_r} = A_r(e_{i_1} \otimes \cdots \otimes e_{i_r}), \quad 1 \le i_1, \ldots, i_r \le n. \tag{3.4}$$

By the above remark, an exterior vector $e_{i_1} \wedge \cdots \wedge e_{i_r}$ is nonzero only if $i_1, \ldots, i_r$ are distinct. Particularly, if $r > n$, then there must be repeated indices among $i_1, \ldots, i_r$, and therefore the corresponding exterior vector is zero.

Suppose ξ is an exterior r-vector with the following expression in components:

$$\xi = \xi^{i_1 \cdots i_r} e_{i_1} \otimes \cdots \otimes e_{i_r}. \tag{3.5}$$

Since any alternating mapping is linear,

$$
\begin{aligned}
\xi &= A_r \xi \\
&= \xi^{i_1 \cdots i_r} A_r(e_{i_1} \otimes \cdots \otimes e_{i_r}) \\
&= \xi^{i_1 \cdots i_r} e_{i_1} \wedge \cdots \wedge e_{i_r}.
\end{aligned}
$$

Therefore any exterior vector of degree greater than n is zero, i.e.,

$$\Lambda^r(V) = 0 \qquad \forall r > n. \tag{3.6}$$

Suppose $r \le n$. Then by Theorem 2.1 $\xi^{i_1 \cdots i_r}$, the components of ξ, are antisymmetric with respect to all upper indices. Therefore ξ can be expressed as follows:

$$\xi = r! \sum_{i_1 < \cdots < i_r} \xi^{i_1 \cdots i_r} e_{i_1} \wedge \cdots \wedge e_{i_r}. \tag{3.7}$$

Now we prove that for $r \le n$, $\{e_{i_1} \wedge \cdots \wedge e_{i_r}, \ 1 \le i_1 < \cdots < i_r \le n\}$ forms a basis of $\Lambda^r(V)$. To see this, we need only show that these

$$\binom{n}{r} = \frac{n!}{r!(n-r)!}$$

exterior vectors are linearly independent. First we derive a formula for evaluating $e_{i_1} \wedge \cdots \wedge e_{i_r}$.

Suppose $v^{*1}, \ldots, v^{*r}$ are r arbitrary elements of V^*. Then

$$e_{i_1} \wedge \cdots \wedge e_{i_r}(v^{*1}, \ldots, v^{*r}) = \frac{1}{r!} \sum_{\sigma \in \mathcal{S}(r)} \text{sgn } \sigma \cdot \left\langle e_{i_1}, v^{*\sigma(1)} \right\rangle \cdots \left\langle e_{i_r}, v^{*\sigma(r)} \right\rangle$$

$$= \frac{1}{r!} \begin{vmatrix} \langle e_{i_1}, v^{*1} \rangle & \cdots & \langle e_{i_1}, v^{*r} \rangle \\ \langle e_{i_2}, v^{*1} \rangle & \cdots & \langle e_{i_2}, v^{*r} \rangle \\ \vdots & \ddots & \vdots \\ \langle e_{i_r}, v^{*1} \rangle & \cdots & \langle e_{i_r}, v^{*r} \rangle \end{vmatrix}.$$

$$\tag{3.8}$$

The above formula is called the **evaluation formula** for $e_{i_1} \wedge \cdots \wedge e_{i_r}$. In particular,

$$e_{i_1} \wedge \cdots \wedge e_{i_r}(e^{*j_1}, \ldots, e^{*j_r}) = \frac{1}{r!} \det \left(\langle e_{i_\alpha}, e^{*j_\beta} \rangle \right)$$

$$= \frac{1}{r!} \delta^{j_1 \cdots j_r}_{i_1 \cdots i_r}, \tag{3.9}$$

where

$$\delta^{j_1 \cdots j_r}_{i_1 \cdots i_r} = \begin{cases} 1 & \text{if } i_1, \ldots, i_r \text{ are distinct, and } \{j_1, \ldots, j_r\} \\ & \text{is an even permutation of } \{i_1, \ldots, i_r\}; \\ -1 & \text{if } i_1, \ldots, i_r \text{ are distinct, and } \{j_1, \ldots, j_r\} \\ & \text{is an odd permutation of } \{i_1, \ldots, i_r\}; \\ 0 & \text{otherwise,} \end{cases} \tag{3.10}$$

is called the **generalized Kronecker δ-symbol.**
From (3.9) we obtain

$$e_1 \wedge \cdots \wedge e_n(e^{*1}, \ldots, e^{*n}) = \frac{1}{n!}. \tag{3.11}$$

Thus $e_1 \wedge \cdots \wedge e_n \neq 0$. For $r < n$, if $\{e_{i_1} \wedge \cdots \wedge e_{i_r}, \ 1 \le i_1 < \cdots < i_r \le n\}$ is linearly dependent, then there exist scalars $a^{i_1 \cdots i_r} \in \mathbb{F}$, not all zero, such that

$$\sum_{1 \le i_1 < \cdots < i_r \le n} a^{i_1 \cdots i_r} e_{i_1} \wedge \cdots \wedge e_{i_r} = 0. \tag{3.12}$$

We may assume one of the nonzero scalars to be $a^{j_1 \cdots j_r}$, $1 \leq j_1 < \cdots < j_r \leq n$, with the remaining index set $k_1 < \cdots < k_{n-r}$. That is, $\{j_1, \ldots, j_r, k_1, \ldots, k_{n-r}\}$ is a permutation of $\{1, \ldots, n\}$. Wedge multiplying both sides of (3.12) by $e_{k_1} \wedge \cdots \wedge e_{k_{n-r}}$, we get

$$a^{j_1 \cdots j_r} e_{j_1} \wedge \cdots \wedge e_{j_r} \wedge e_{k_1} \wedge \cdots \wedge e_{k_{n-r}} = \pm a^{j_1 \cdots j_r} e_1 \wedge \cdots \wedge e_n$$
$$= 0.$$

Therefore

$$a^{j_1 \cdots j_r} = 0,$$

which is a contradiction. We conclude that $\{e_{i_1} \wedge \cdots \wedge e_{i_r}, 1 \leq i_1 < \cdots < i_r \leq n\}$ is linearly independent, and forms a basis of $\Lambda^r(V)$. This implies that the dimension of $\Lambda^r(V)$ is

$$\binom{n}{r} = \frac{n!}{r!(n-r)!}.$$

Definition 3.2. Denote the formal sum $\sum\limits_{r=0}^{n} \Lambda^r(V)$ by $\Lambda(V)$. Then $\Lambda(V)$ is a 2^n-dimensional vector space. Let

$$\xi = \sum_{r=0}^{n} \xi^r, \qquad \eta = \sum_{s=0}^{n} \eta^s, \tag{3.13}$$

where $\xi^r \in \Lambda^r(V), \eta^s \in \Lambda^s(V)$. Define the **exterior (wedge) product** of ξ and η by

$$\xi \wedge \eta = \sum_{r,s=0}^{n} \xi^r \wedge \eta^s. \tag{3.14}$$

Then $\Lambda(V)$ becomes an algebra with respect to the exterior product, and is called the **exterior algebra** or **Grassman algebra** of V.

The set $\{1, e_i (1 \leq i \leq n), e_{i_1} \wedge e_{i_2} (1 \leq i_1 < i_2 \leq n), \ldots, e_1 \wedge \cdots \wedge e_n\}$ is a basis of the vector space $\Lambda(V)$.

Similarly, we have an exterior algebra for the dual space V^*

$$\Lambda(V^*) = \sum_{0 \leq r \leq n} \Lambda^r(V^*).$$

An element of $\Lambda^r(V^*)$ is called an **exterior form** of degree r, or **exterior r-form** on V; it is an alternating $\mathbb{F}$-valued r-linear function on V.

The vector spaces $\Lambda^r(V)$ and $\Lambda^r(V^*)$ are dual to each other by a certain pairing. Suppose

$$v_1 \wedge \cdots \wedge v_r \in \Lambda^r(V),$$
$$v^{*1} \wedge \cdots \wedge v^{*r} \in \Lambda^r(V^*).$$

Then

$$\langle v_1 \wedge \cdots \wedge v_r, v^{*1} \wedge \cdots \wedge v^{*r} \rangle = \det\left(\langle v_\alpha, v^{*\beta} \rangle \right). \tag{3.15}$$

Thus $\{e_{i_1} \wedge \cdots \wedge e_{i_r}, 1 \le i_1 < \cdots < i_r \le n\}$ and $\{e^{*j_1} \wedge \cdots \wedge e^{*j_r}, 1 \le j_1 < \cdots < j_r \le n\}$, the bases of $\Lambda^r(V)$ and $\Lambda^r(V^*)$, respectively, satisfy the following relationship:

$$\langle e_{i_1} \wedge \cdots \wedge e_{i_r}, e^{*j_1} \wedge \cdots \wedge e^{*j_r} \rangle = \det\left(\langle e_{i_\alpha}, e^{*j_\beta} \rangle \right)$$
$$= \delta^{j_1 \cdots j_r}_{i_1 \cdots i_r}$$
$$= \begin{cases} 1 & \text{if } \{j_1, \ldots, j_r\} = \{i_1, \ldots, i_r\}, \\ 0 & \text{if } \{j_1, \ldots, j_r\} \ne \{i_1, \ldots, i_r\}. \end{cases}$$
$$\tag{3.16}$$

Thus these two bases are dual to each other.

Remark. In §2–2, we have already defined the pairing between the tensor spaces $T^r(V)$ and $T^r(V^*)$ by (2.17). Since $\Lambda^r(V)$ and $\Lambda^r(V^*)$ are subspaces of $T^r(V)$ and $T^r(V^*)$, respectively, equation (2.17) induces a pairing between $\Lambda^r(V)$ and $\Lambda^r(V^*)$ which is the same as that defined by (3.15) up to a factor $r!$. Therefore we have defined pairings on the tensor space and exterior vector space separately so that the dual bases have simple forms without redundant scalar factors. Although we use the same notation $\langle\,,\,\rangle$ to represent two different pairings, no confusion will arise if we keep in mind which space we are dealing with.

Suppose $f : V \longrightarrow W$ is a linear map from a vector space V to another vector space W. Then f induces a linear map from the exterior space $\Lambda^r(W^*)$ to the exterior space $\Lambda^r(V^*)$. Suppose $\varphi \in \Lambda^r(W^*)$. For arbitrary $v_1, \ldots, v_r \in V$, let

$$f^*\varphi(v_1, \ldots, v_r) = \varphi(f(v_1), \ldots, f(v_r)). \tag{3.17}$$

It is easy to show (see below) that f^* is linear and commutes with the exterior product. Thus f^* is a homomorphism from the exterior algebra $\Lambda(W^*)$ to the exterior algebra $\Lambda(V^*)$.

Theorem 3.2. *Suppose* $f : V \longrightarrow W$ *is a linear map. Then* f^* *commutes with the exterior product, that is, for any* $\varphi \in \Lambda^r(W^*)$ *and* $\psi \in \Lambda^s(W^*)$,

$$f^*(\varphi \wedge \psi) = f^*\varphi \wedge f^*\psi. \tag{3.18}$$

Proof. Choose any $v_1, \ldots, v_{r+s} \in V$. Then

$$
\begin{aligned}
f^*(\varphi \wedge \psi)(v_1, \ldots, v_{r+s}) &= \varphi \wedge \psi(f(v_1), \ldots, f(v_{r+s})) \\
&= \frac{1}{(r+s)!} \sum_{\sigma \in \mathcal{S}(r+s)} \operatorname{sgn} \sigma \cdot \varphi\left(f(v_{\sigma(1)}), \ldots, f(v_{\sigma(r)})\right) \\
&\qquad \cdot \psi\left(f(v_{\sigma(r+1)}), \ldots, f(v_{\sigma(r+s)})\right) \\
&= \frac{1}{(r+s)!} \sum_{\sigma \in \mathcal{S}(r+s)} \operatorname{sgn} \sigma \cdot f^*\varphi\left(v_{\sigma(1)}, \ldots, v_{\sigma(r)}\right) \\
&\qquad \cdot f^*\psi\left(v_{\sigma(r+1)}, \ldots v_{\sigma(r+s)}\right) \\
&= f^*\varphi \wedge f^*\psi(v_1, \ldots, v_{r+s}).
\end{aligned}
$$

Therefore

$$f^*(\varphi \wedge \psi) = f^*\varphi \wedge f^*\psi.$$

$\square$

The concept of an exterior algebra was originally introduced by H. Grassman for the purpose of studying linear subspaces. Subsequently Elie Cartan developed the theory of exterior differentiation, and successfully applied it to the study of differential geometry and differential equations. More recently, exterior algebras have become powerful and irreplaceable tools in the study of differentiable manifolds. We shall discuss a few useful propositions below.

Theorem 3.3. *A necessary and sufficient condition for the vectors* $v_1, \ldots, v_r \in V$ *to be linearly dependent is*

$$v_1 \wedge \cdots \wedge v_r = 0. \tag{3.19}$$

Proof. If $v_1, \ldots, v_r$ are linearly dependent, then we may assume without loss of generality that v_r can be expressed as a linearly combination of $v_1, \ldots, v_{r-1}$:

$$v_r = a_1 v_1 + \cdots + a_{r-1} v_{r-1}.$$

Then

$$
\begin{aligned}
v_1 \wedge \cdots \wedge v_{r-1} \wedge v_r &= v_1 \wedge \cdots \wedge v_{r-1} \wedge (a_1 v_1 + \cdots + a_{r-1} v_{r-1}) \\
&= 0.
\end{aligned}
$$

Conversely, if $v_1, \ldots, v_r$ are linearly independent, then they can be extended to a basis $\{v_1, \ldots v_r, v_{r+1}, \ldots, v_n\}$ of V. Then

$$v_1 \wedge \cdots \wedge v_r \wedge v_{r+1} \wedge \cdots \wedge v_n \neq 0.$$

Therefore

$$v_1 \wedge \cdots \wedge v_r \neq 0.$$

$\square$

Theorem 3.4 (Cartan's Lemma). *Suppose* $\{v_1, \ldots v_r\}$ *and* $\{w_1, \ldots, w_r\}$ *are two sets of vectors in V such that*

$$\sum_{\alpha=1}^{r} v_\alpha \wedge w_\alpha = 0. \tag{3.20}$$

If $v_1, \ldots, v_r$ are linearly independent, then the w_α can be expressed as linear combinations of the v_β:

$$w_\alpha = \sum_{\beta=1}^{r} a_{\alpha\beta} v_\beta, \qquad 1 \leq \alpha \leq r, \tag{3.21}$$

with

$$a_{\alpha\beta} = a_{\beta\alpha}. \tag{3.22}$$

Proof. Since $v_1, \ldots, v_r$ are linearly independent, they can be extended to a basis $\{v_1, \ldots, v_r, v_{r+1}, \ldots, v_n\}$ of V. Therefore we may assume that

$$w_\alpha = \sum_{\beta=1}^{r} a_{\alpha\beta} v_\beta + \sum_{i=r+1}^{n} a_{\alpha i} v_i. \tag{3.23}$$

Plugging this into (3.20), we get

$$0 = \sum_{\alpha\beta=1}^{r} a_{\alpha\beta} v_\alpha \wedge v_\beta + \sum_{\alpha=1}^{r}\sum_{i=r+1}^{n} a_{\alpha i} v_\alpha \wedge v_i$$

$$= \sum_{1 \leq \alpha < \beta \leq r} (a_{\alpha\beta} - a_{\beta\alpha}) v_\alpha \wedge v_\beta + \sum_{\alpha=1}^{r}\sum_{i=r+1}^{n} a_{\alpha i} v_\alpha \wedge v_i. \tag{3.24}$$

Since $\{v_i \wedge v_j, 1 \leq i < j \leq n\}$ is a basis of $\Lambda^2(V)$, we can obtain from (3.24) that

$$\begin{cases} a_{\alpha\beta} - a_{\beta\alpha} &= 0, \\ a_{\alpha i} &= 0, \end{cases}$$

that is,

$$w_\alpha = \sum_{\beta=1}^{r} a_{\alpha\beta} v_\beta,$$

and

$$a_{\alpha\beta} = a_{\beta\alpha}.$$

□

Theorem 3.5. *Suppose $v_1, \ldots, v_r$ are r linearly independent vectors in V, and $w \in \Lambda^p(V)$. A necessary and sufficient condition for w to be expressible in the form*

$$w = v_1 \wedge \psi_1 + \cdots + v_r \wedge \psi_r, \qquad (3.25)$$

where $\psi_1, \ldots, \psi_r \in \Lambda^{p-1}(V)$, is that

$$v_1 \wedge \cdots \wedge v_r \wedge w = 0. \qquad (3.26)$$

Proof. When $p + r > n$, (3.25) and (3.26) are trivially true. In the following we assume that $p + r \leq n$.

Necessity is obvious, so we need only show sufficiency. Extend $v_1, \ldots, v_r$ to a basis $\{v_1, \ldots, v_r, v_{r+1}, \ldots, v_n\}$ of V. Then w can be expressed as

$$w = v_1 \wedge \psi_1 + \cdots + v_r \wedge \psi_r + \sum_{r+1 \leq \alpha_1 < \cdots < \alpha_p \leq n} \xi^{\alpha_1 \cdots \alpha_p} v_{\alpha_1} \wedge \cdots \wedge v_{\alpha_p}, \quad (3.27)$$

where $\psi_1, \ldots, \psi_r \in \Lambda^{p-1}(V)$. Plugging into (3.26) we get

$$\sum_{r+1 \leq \alpha_1 < \cdots < \alpha_p \leq n} \xi^{\alpha_1 \cdots \alpha_p} v_1 \wedge \cdots \wedge v_r \wedge v_{\alpha_1} \wedge \cdots \wedge v_{\alpha_p} = 0. \qquad (3.28)$$

Inside the summation, the terms $v_1 \wedge \cdots \wedge v_r \wedge v_{\alpha_1} \wedge \cdots \wedge v_{\alpha_p}$ ($r + 1 \leq \alpha_1 < \cdots < \alpha_p \leq n$) are all basis vectors of $\Lambda^{p+r}(V)$. Therefore (3.28) gives

$$\xi^{\alpha_1 \cdots \alpha_p} = 0, \qquad r + 1 \leq \alpha_1 < \cdots < \alpha_p \leq n,$$

i.e.,

$$w = v_1 \wedge \psi_1 + \cdots + v_r \wedge \psi_r.$$

□

Usually we represent (3.25) by the statement $w \equiv 0 \bmod (v_1, \ldots, v_r)$.

Theorem 3.6. *Suppose* v_α, w_α ; $v'_\alpha, w'_\alpha (1 \le \alpha \le k)$ *are two sets of vectors in* V. *If* $\{v_\alpha, w_\alpha, 1 \le \alpha \le k\}$ *is linearly independent, and*

$$\sum_{\alpha=1}^{k} v_\alpha \wedge w_\alpha = \sum_{\alpha=1}^{k} v'_\alpha \wedge w'_\alpha, \qquad (3.29)$$

then v'_α, w'_α *are linear combinations of* $v_1, \ldots, v_k, w_1, \ldots, w_k$, *and are also linearly independent.*

Proof. Wedge-multiply (3.29) by itself k times to get

$$k!(v_1 \wedge w_1 \wedge \cdots \wedge v_k \wedge w_k) = k!(v'_1 \wedge w'_1 \wedge \cdots \wedge v'_k \wedge w'_k). \qquad (3.30)$$

Since $\{v_\alpha, w_\alpha,\ 1 \le \alpha \le k\}$ is linearly independent, the left hand side of (3.30) is not equal to zero, that is, $\{v'_\alpha, w'_\alpha,\ 1 \le \alpha \le k\}$ is also linearly independent (Theorem 3.3). We can also obtain from (3.30) that

$$v_1 \wedge w_1 \wedge \cdots \wedge v_k \wedge w_k \wedge v'_\alpha = 0,$$

which means $\{v_1, w_1, \ldots, v_k, w_k, v'_\alpha\}$ is linearly dependent. Therefore v'_α can be expressed as a linear combination of $v_1, \ldots, v_k, w_1, \ldots, w_k$. The above conclusion is also true for w'_α. $\qquad \square$

Remark. For a geometrical application of Theorem 3.6, refer to Chern 1944(a).

Exterior algebra is closely related to the determinant function, as seen in the evaluation formula for exterior vectors in (3.8) expressed by a determinant. Suppose $v_1, \ldots, v_k \in V$, and $w_1, \ldots, w_k$ are linear combinations of them, i.e.,

$$w_\alpha = \sum_{\beta=1}^{k} t_\alpha^\beta v_\beta. \qquad (3.31)$$

Then

$$w_1 \wedge \cdots \wedge w_k = \det(t_\alpha^\beta) v_1 \wedge \cdots \wedge v_k. \qquad (3.32)$$

Thus the only difference between the exterior vectors $w_1 \wedge \cdots \wedge w_k$ and $v_1 \wedge \cdots \wedge v_k$ is a scalar factor which is a determinant.

Denote the set of all the k-dimensional subspaces L^k of an n-dimensional vector space V by $G(k, n)$. There exists a natural differentiable structure[a] on $G(k, n)$ which makes $G(k, n)$ a $k(n - k)$-dimensional differentiable manifold,

---

[a]The concept of a Grassman manifold is very important. The reader may refer to Boothby 1975 and Kobayashi and Nomizu 1963 and 1969 for descriptions of its differentiable structure.

called the **Grassman manifold**. When $k = 1$, $G(k, n)$ is precisely the $(n-1)$-dimensional **projective space** P^{n-1}. Exterior vectors define the so-called Plücker–Grassman coordinates for $G(k, n)$ (see below).

Choose any $L^k \in G(k, n)$ and suppose $v_1, \ldots, v_k$ are k linearly independent vectors which span the subspace L^k. By (3.32), $\xi = v_1 \wedge \cdots \wedge v_k$ is determined up to a nonzero scalar factor. The exterior vector ξ defines the **Plücker–Grassman** coordinates of the subspace L^k in $G(k, n)$, which is a generalization of the homogeneous coordinates in projective spaces.

For example, let us consider $G(2, 4)$. Choose a basis $\{a_1, a_2, a_3, a_4\}$ for V. For arbitrary $L^2 \in G(2, 4)$, we may choose linearly independent vectors $v, w \in L^2$. Let

$$v = \sum_{i=1}^{4} v^i a_i, \qquad w = \sum_{i=1}^{4} w^i a_i. \tag{3.33}$$

Then

$$\xi = v \wedge w = \sum_{i<k} p^{ik} a_i \wedge a_k, \tag{3.34}$$

where $p^{ik} = v^i w^k - v^k w^i$. The scalars $\{p^{ik}, \ i < k\}$ are determined up to a nonzero scalar factor. They are precisely the Plücker–Grassman coordinates in $G(2, 4)$.

Since $v \wedge w \wedge v \wedge w = 0$,

$$(p^{12}p^{34} + p^{13}p^{42} + p^{14}p^{23})a_1 \wedge a_2 \wedge a_3 \wedge a_4 = 0,$$

That is, p^{ik} should satisfy the **Plücker equation**

$$p^{12}p^{34} + p^{13}p^{42} + p^{14}p^{23} = 0. \tag{3.35}$$

Thus, there are only 4 independent variables among the 6 numbers p^{ik} ($1 \leq i < k \leq 4$), which implies that the dimension of $G(2, 4)$ is 4. Conversely, the Plücker equation (3.35) is also a sufficient condition [b] for the exterior vector

$$\sum_{i<k} p^{ik} a_i \wedge a_k$$

to be reducible. Thus any set of scalars p^{ik} with elements not all zero which satisfies (3.35) will determine an element in $G(2, 4)$.

---

[b] We may assume that $p^{12} \neq 0$. Then

$$p^{34} = \frac{p^{13}p^{24}}{p^{12}} - \frac{p^{14}p^{23}}{p^{12}}.$$

Suppose

$$p^{12} = x^1 + y^1, \quad p^{13} = x^2 + y^2, \quad p^{14} = x^3 + y^3,$$
$$p^{34} = x^1 - y^1, \quad p^{42} = x^2 - y^2, \quad p^{23} = x^3 - y^3. \tag{3.36}$$

Then (3.35) becomes

$$(x^1)^2 + (x^2)^2 + (x^3)^2 = (y^1)^2 + (y^2)^2 + (y^3)^2. \tag{3.37}$$

Multiply p^{ik} by an appropriate scalar so that each side in (3.37) has the value 1. Then (x^1, x^2, x^3) and (y^1, y^2, y^3) represent points on the unit sphere $S^2 \subset \mathbb{R}^3$. By the discussion above, (3.36) provides a surjective map

$$\pi : S^2 \times S^2 \longrightarrow G(2,4), \tag{3.38}$$

with

$$\pi(x, y) = \pi(-x, -y), \qquad (x, y) \in S^2 \times S^2, \tag{3.39}$$

where $-x$ means the antipodal point of x on S^2. Therefore $S^2 \times S^2$ is a double covering space of $G(2,4)$.[c] Since $S^2 \times S^2$ is simply connected, it is a universal covering space of $G(2,4)$. Thus the **fundamental group** $\pi_1(G(2,4))$ of $G(2,4)$ is $\mathbb{Z}^2$.

Denote the manifold of all oriented 2-dimensional subspaces in a 4-dimensional vector space V by $\tilde{G}(2,4)$. Then $\tilde{G}(2,4)$ is also a double covering space of $G(2,4)$, and hence homeomorphic to $S^2 \times S^2$.

By direct calculation, we get

$$
\begin{aligned}
\sum_{i<j} p^{ij} a_i \wedge a_j &= a_1 \wedge (p^{12} a_2 + p^{13} a_3 + p^{14} a_4) \\
&\quad - \tfrac{p^{23}}{p^{12}} a_3 \wedge (p^{12} a_2 + p^{13} a_3 + p^{14} a_4) \\
&\quad - \tfrac{p^{24}}{p^{12}} a_4 \wedge (p^{12} a_2 + p^{13} a_3 + p^{14} a_4) \\
&= \left(a_1 - \tfrac{p^{23}}{p^{12}} a_3 - \tfrac{p^{24}}{p^{12}} a_4 \right) \wedge (p^{12} a_2 + p^{13} a_3 + p^{14} a_4).
\end{aligned}
$$

Therefore when (3.35) is satisfied, the exterior vector $\sum_{i<j} p^{ij} a_i \wedge a_j$ is reducible. For a general discussion of the Plücker equation, refer to Cartan 1945.

[c]For the concepts of covering spaces and fundamental groups, refer to Munkres 1975.

Chapter 3

Exterior Differential Calculus

§3–1   Tensor Bundles and Vector Bundles

The product of two manifolds (see §1–1) is a very basic concept. For example, the graph $(x, f(x))$ of a real-valued function $f(x)$ on a manifold M is a map from M to the product manifold $M \times \mathbb{R}$. In terms of fiber bundles, it is a section of the fiber bundle $M \times \mathbb{R}$. Fiber bundles are generalizations of product manifolds. In differential geometry, we study mainly a special class of fiber bundles—vector bundles—defined such that any vector field on a manifold is a section of some vector bundle. In this section, we first discuss concrete tensor bundles, then more general vector bundles.

Suppose M is an m-dimensional smooth manifold, T_p and T_p^* are the tangent and cotangent spaces of M at p. Then there is an (r, s)-type tensor space

$$T_s^r(p) = \underbrace{T_p \otimes \cdots \otimes T_p}_{r \text{ terms}} \otimes \underbrace{T_p^* \otimes \cdots \otimes T_p^*}_{s \text{ terms}}, \tag{1.1}$$

of M at p which is an m^{r+s}-dimensional vector space. Let

$$T_s^r = \bigcup_{p \in M} T_s^r(p). \tag{1.2}$$

We will introduce a topology on T_s^r so that it becomes a Hausdorff space with a countable basis, and then define a C^∞ differentiable structure to make it a smooth manifold. The smooth manifold obtained in this way is locally diffeomorphic to a product manifold. We call T_s^r an (r, s)-type tensor bundle on M.

Let us first introduce some notation. Suppose V is an m-dimensional vector space over $\mathbb{R}$. Denote the group of linear automorphisms of V by $GL(V)$. Choose a basis $\{e_1, \ldots, e_m\}$ in V. Then V is isomorphic to $\mathbb{R}^m$. Write an element y in V as a coordinate row

$$y = (y^1, \ldots, y^m). \tag{1.3}$$

Thus $GL(V)$ is a multiplicative group of nonsingular $m \times m$ matrices, i.e., $GL(V)$ is the general linear group $GL(m; \mathbb{R})$. Define an action by the group $GL(V)$ on V as a multiplication on the right, with the matrix representation given by

$$y \cdot a = (y^1, \ldots, y^m) \cdot \begin{pmatrix} a_1^1 & \cdots & a_1^m \\ \vdots & \ddots & \vdots \\ a_m^1 & \cdots & a_m^m \end{pmatrix}, \tag{1.4}$$

where $\det a = \det(a_i^j) \neq 0$.

V_s^r represents the space of all (r,s)-type tensors on the vector space V (see Definition 2.1 in Chapter 2) and has a basis

$$e_{i_1} \otimes \cdots \otimes e_{i_r} \otimes e^{*j_1} \otimes \cdots \otimes e^{*j_s}, \quad 1 \leq i_\alpha, j_\beta \leq m. \tag{1.5}$$

Thus the elements of V_s^r can be expressed by components. If we denote the elements of V_s^r by coordinate rows as in (1.3), then the basis elements in (1.5) must be ordered in some fashion. For example, the index system $(i_1, \ldots, i_r, j_1, \ldots, j_s)$ may be arranged according to the lexicographic order.

Consider a coordinate neighborhood U on M with local coordinates $u^1 \ldots, u^m$. Then for any point $p \in U$, there exist natural bases

$$\left\{ \left(\frac{\partial}{\partial u^1}\right)_p, \ldots, \left(\frac{\partial}{\partial u^m}\right)_p \right\} \quad \text{and} \quad \{(du^1)_p, \ldots, (du^m)_p\},$$

dual to each other, for T_p and T_p^*, respectively. Therefore

$$\left(\frac{\partial}{\partial u^{i_1}}\right)_p \otimes \cdots \otimes \left(\frac{\partial}{\partial u^{i_r}}\right)_p \otimes (du^{j_1})_p \otimes \cdots \otimes (du^{j_s})_p,$$

$$1 \leq i_\alpha, j_\beta \leq m \tag{1.6}$$

is a basis for $T_s^r(p)$. Now we can define a map

$$\varphi_U : U \times V_s^r \longrightarrow \bigcup_{p \in U} T_s^r(p) \tag{1.7}$$

such that for any $p \in U, y \in V_s^r$, $\varphi_U(p, y)$ is an element of $T_s^r(p)$, and the components of $\varphi_U(p, y)$ with respect to (1.6) are the same as those of y with respect to (1.5). Obviously, such a φ_U is a one to one map.

Choose a coordinate covering $\{U, W, \ldots\}$ of M, and suppose the maps corresponding to (1.7) are $\{\varphi_U, \varphi_W, \ldots\}$. Choose the set of images of all open subsets of $U \times V_s^r$ etc. under the map φ_U to be a topological basis [a] for T_s^r. Such a topology makes T_s^r a Hausdorff space with a countable basis and every map defined by (1.7) a homeomorphism.

Fix a point $p \in U$. Then we can define a map $\varphi_{U,p} : V_s^r \longrightarrow T_s^r(p)$ such that

$$\varphi_{U,p}(y) = \varphi_U(p, y), \qquad y \in V_s^r. \tag{1.8}$$

By the definition of φ_U, $\varphi_{U,p}$ is a linear isomorphism from the vector space V_s^r to $T_s^r(p)$.

If W is another coordinate neighborhood of M containing p with local coordinates $w^1, \ldots, w^m$, let

$$g_{UW}(p) = \varphi_{W,p}^{-1} \circ \varphi_{U,p} : V_s^r \longrightarrow V_s^r. \tag{1.9}$$

Then $g_{UW}(p)$ is an automorphism of the vector space V_s^r, i.e., $g_{UW}(p) \subset GL(V_s^r)$. As defined in (1.4), the operation of any element in $GL(V_s^r)$ on V_s^r is a multiplication on the right. Thus a necessary and sufficient condition for any $y, y' \in V_s^r$ to satisfy

$$\varphi_U(p, y) = \varphi_W(p, y') \tag{1.10}$$

is

$$y' = y \cdot g_{UW}(p). \tag{1.11}$$

For any two coordinate neighborhoods U, W of M, if $U \cap W \neq \varnothing$, then the map

$$g_{UW} : U \cap W \longrightarrow GL(V_s^r) \tag{1.12}$$

defined by (1.9) will be shown to be smooth. Without loss of generality we discuss only the case when $r = s = 1$.

---

[a] We need to show that such a set satisfies the conditions of a topological basis. In fact, if $U \cap W \neq \varnothing$ then

$$\left(\bigcup_{p \in U} T_s^r(p) \right) \cap \left(\bigcup_{q \in W} T_s^r(q) \right) = \bigcup_{p \in U \cap W} T_s^r(p),$$

which is the image of $U \cap W$ under φ_W, and also the image under φ_U. The reader should fill in the details of the proof.

The natural bases for the tangent space T_p and the cotangent space T_p^* are $\left\{ \left(\frac{\partial}{\partial w^1} \right)_p, \cdots, \left(\frac{\partial}{\partial w^m} \right)_p \right\}$ and $\{ (dw^1)_p, \ldots, (dw^m)_p \}$, respectively, with respect to the local coordinates $w^1, \ldots, w^m$. Suppose $y, y' \in V_1^1$ are expressed in components by

$$y = y_j^i e_i \otimes e^{*j}, \quad y' = y_j'^i e_i \otimes e^{*j}. \tag{1.13}$$

Then

$$\begin{cases} \varphi_U(p, y) = y_j^i \left(\frac{\partial}{\partial u^i} \right)_p \otimes (du^j)_p, \\[2mm] \varphi_W(p, y') = y_j'^i \left(\frac{\partial}{\partial w^i} \right)_p \otimes (dw^j)_p. \end{cases} \tag{1.14}$$

On $U \cap W$, the following relations between the natural bases are satisfied:

$$\begin{cases} du^i = \dfrac{\partial u^i}{\partial w^j} dw^j, \\[3mm] \dfrac{\partial}{\partial u^i} = \dfrac{\partial w^j}{\partial u^i} \dfrac{\partial}{\partial w^j}, \end{cases} \tag{1.15}$$

where $J_{UW} = \left(\frac{\partial w^i}{\partial u^k} \right)$ is the Jacobian matrix of the change of local coordinates. Plugging (1.15) into (1.14) and (1.10), we get

$$y_j'^i = y_l^k \left(\frac{\partial w^i}{\partial u^k} \right)_p \left(\frac{\partial u^l}{\partial w^j} \right)_p,$$

i.e.,

$$(y \cdot g_{UW}(p))_j^i = y_l^k \left(\frac{\partial w^i}{\partial u^k} \right)_p \left(\frac{\partial u^l}{\partial w^j} \right)_p. \tag{1.16}$$

If we denote y by the coordinate row $(y_1^1, y_2^1, \ldots, y_1^m, y_2^m, \ldots, y_m^m)$, then the above equation shows that the $(m^2 \times m^2)$ nondegenerate matrix representing $g_{UW}(p)$ is precisely the tensor product [b] of the Jacobian matrix, J_{UW}, and its

---

[b] The tensor product of the matrices $A = \left(a_j^i \right)$ and $B = \left(b_\beta^\alpha \right)$ is the block matrix

$$A \otimes B = \begin{pmatrix} a_1^1 B & \cdots & a_1^m B \\ \vdots & \ddots & \vdots \\ a_m^1 B & \cdots & a_m^m B \end{pmatrix},$$

with elements $a_j^i \cdot b_\beta^\alpha$.

inverse:

$$g_{UW} = J_{UW} \otimes J_{UW}^{-1}. \tag{1.17}$$

Since both Jacobian matrices J_{UW} and $J_{UW}^{-1} = J_{WU}$ are composed of smooth functions on $U \cap W$, g_{UW} is also smooth on $U \cap W$.

By the topological structure of T_s^r, $\{\varphi_U(U \times V_s^r), \varphi_W(W \times V_s^r), \dots\}$ forms an open coordinate covering of T_s^r. The coordinates of a point $\varphi_U(p, y)$ in the coordinate neighborhood $\varphi_U(U \times V_s^r)$ are

$$(u^i(p), y_{j_1 \cdots j_s}^{i_1 \cdots i_r}), \tag{1.18}$$

where u^i is a local coordinate in the coordinate neighborhood U of the manifold M, and $y_{j_1 \cdots j_s}^{i_1 \cdots i_r}$ is the component of $y \in V_s^r$ with respect to the basis in (1.5). When $U \cap W \neq \varnothing$, since $g_{UW} : U \cap W \longrightarrow GL(V_s^r)$ is smooth, equation (1.11) implies that the coordinate covering of T_s^r given above is C^∞-compatible. T_s^r thus becomes a smooth manifold. Obviously, the natural projection

$$\pi : T_s^r \longrightarrow M \tag{1.19}$$

maps elements in $T_s^r(p)$ to the point $p \in M$. It is a smooth surjection. We call the smooth manifold T_s^r a **type (r, s)-tensor bundle** on M, π the **bundle projection**, and $T_s^r(p)$ the **fiber** of the bundle T_s^r at the point p.

Letting $r = 1$, $s = 0$, we get the **tangent bundle** of M, denoted by $T(M)$. Letting $r = 0$, $s = 1$, we get the **cotangent bundle** of M, denoted by $T^*(M)$. Following the above procedure to construct tensor bundles, we can analogously construct **exterior vector bundles** and **exterior form bundles** on M, denoted respectively by

$$\begin{cases} \Lambda^r(M) &= \bigcup_{p \in M} \Lambda^r(T_p), \\ \Lambda^r(M^*) &= \bigcup_{p \in M} \Lambda^r(T_p^*). \end{cases} \tag{1.20}$$

Suppose $f : M \longrightarrow T_s^r$ is a smooth map. If

$$\pi \circ f = \mathrm{id} : M \longrightarrow M,$$

that is, for any $p \in M$, $f(p) \in T_s^r(p)$, then f is called a **smooth section** of the tensor bundle T_s^r, or a **type (r, s)-smooth tensor field** on M. A section of a tangent bundle is a tangent vector field on M, and a section of a cotangent bundle is a differential 1-form. A smooth section of the exterior form bundle $\Lambda^r(M^*)$ is called an **exterior differential form** of degree r on M.

Generalizing the structure of tensor bundles, we obtain the concept of general vector bundles. Vector bundles and connections discussed in the next chapter form the mathematical basis for the theory of gauge fields in physics.

Definition 1.1. Suppose E, M are two smooth manifolds, and $\pi : E \longrightarrow M$ is a smooth surjective map. Let $V = \mathbb{R}^q$ be a q-dimensional vector space. If an open covering $\{U, W, Z, \dots\}$ of M and a set of maps $\{\varphi_U, \varphi_W, \varphi_Z, \dots\}$ satisfy all of the following conditions, then (E, M, π) is called a (real) q-dimensional **vector bundle** on M, where E is called the **bundle space**, M is called the **base space**, π is called the **bundle projection**, and $V = \mathbb{R}^q$ is called the **typical fiber**:

1) Every map φ_U is a diffeomorphism from $U \times \mathbb{R}^q$ to $\pi^{-1}(U)$, and for any $p \in U$, $y \in \mathbb{R}^q$,

$$\pi \circ \varphi_U(p, y) = p. \tag{1.21}$$

2) For any fixed $p \in U$, let

$$\varphi_{U,p}(y) = \varphi_U(p, y), \qquad y \in \mathbb{R}^q. \tag{1.22}$$

Then $\varphi_{U,p} : \mathbb{R}^q \longrightarrow \pi^{-1}(p)$ is a homeomorphism. When $U \cap W \neq \varnothing$, for any $p \in U \cap W$,

$$g_{UW}(p) = \varphi_{W,p}^{-1} \circ \varphi_{U,p} : \mathbb{R}^q \longrightarrow \mathbb{R}^q \tag{1.23}$$

is a linear automorphism of $V = \mathbb{R}^q$, i.e., $g_{UW}(p) \in GL(V)$.

3) When $U \cap W \neq \varnothing$, the map $g_{UW} : U \cap W \longrightarrow GL(V)$ is smooth.

From condition 2), we know that a necessary and sufficient condition for elements y_U, y_W in V to satisfy

$$\varphi_U(p, y_U) = \varphi_W(p, y_W) \tag{1.24}$$

is

$$y_U \cdot g_{UW}(p) = y_W, \tag{1.25}$$

where $g_{UW}(p)$ is viewed as a nondegenerate $(q \times q)$ matrix.

For any $p \in M$, define $E_p = \pi^{-1}(p)$ and call it the **fiber** of the vector bundle E at the point p. Suppose U is a coordinate neighborhood of M containing p. Then the linear structure of the typical fiber V can be transported to the fiber E_p through the map $\varphi_{U,p}$, making E_p a q-dimensional vector space. By condition 2), the linear structure of E_p is independent of the choices of U and φ_U. (The reader should verify this.) A vector bundle E can therefore be viewed intuitively as the result of pasting together product manifolds of the form $U \times \mathbb{R}^q$ along corresponding fibers at the same point $p \in M$ (U being a coordinate neighborhood of M), in such a way that the linear relationship between the fibers is preserved.

The product manifold $M \times \mathbb{R}^q = E$ is the simplest example of a vector bundle, called the **trivial bundle** over M, or the **product bundle**. Obviously, all the tensor bundles T_s^r mentioned previously are vector bundles.

Remark. If V is a q-dimensional complex vector space, then Definition 1.1 defines a q-dimensional complex vector bundle on M. In this case, $GL(V)$ is isomorphic to $GL(q; \mathbb{C})$, and the fiber $\pi^{-1}(p)$, $p \in M$, is a q-dimensional complex vector space. Even though the contents in this section are developed for real vector bundles, they can be applied to complex vector bundles after appropriate adjustments.

The map $g_{UW} : U \cap W \longrightarrow GL(V)$ defined in condition 2) satisfies the following compatibility conditions:

1) for $p \in U$, $g_{UU}(p) = \text{id} : V \longrightarrow V$;
2) if $p \in U \cap W \cap Z \neq \varnothing$, then

$$g_{UW}(p) \cdot g_{WZ}(p) \cdot g_{ZU}(p) = \text{id} : V \longrightarrow V.$$

The set $\{g_{UW}\}$ is called the family of **transition functions** of the vector bundle (E, M, π), and the above compatibility conditions are necessary and sufficient conditions for $\{g_{UW}\}$ to be such a family. More precisely, we have the following theorem:

Theorem 1.1. *Suppose M is an m-dimensional smooth manifold, $\{U_\alpha\}_{\alpha \in A}$ is an open covering of M, and V is a q-dimensional vector space. If for any pair of indices, $\alpha, \beta \in A$ where $U_\alpha \cap U_\beta \neq \varnothing$, there exists a smooth map $g_{\alpha\beta} : U_\alpha \cap U_\beta \longrightarrow GL(V)$ that satisfies both compatibility conditions 1) and 2), then there exists a q-dimensional vector bundle (E, M, π) which has $\{g_{\alpha\beta}\}$ as its transition functions.*

For a detailed proof of Theorem 1.1, see p.14 of Steenrod 1951. The idea of the proof is to paste the local products $U_\alpha \times V$ along the corresponding fibers. To describe it briefly, let

$$\tilde{E} = \bigcup_{\alpha \in A} \{\alpha\} \times U_\alpha \times V, \tag{1.26}$$

which is naturally a differentiable manifold. Define an equivalence relation $\sim$ in $\tilde{E}$ as follows. For any $(\alpha, p, y), (\beta, p', y') \in \tilde{E}$, a necessary and sufficient condition for $(\alpha, p, y) \sim (\beta, p', y')$ is that

$$\begin{aligned} p &= p' \in U_\alpha \cap U_\beta, \quad \text{and} \\ y' &= y \cdot g_{\alpha\beta}(p). \end{aligned} \tag{1.27}$$

Let $e = \tilde{E}/\sim$ denote the quotient space of $\tilde{E}$ with respect to the equivalence relation $\sim$. Then it is also a smooth manifold. Denote the equivalence class of (α, p, y) by $[\alpha, p, y]$, and define the projection $\pi : E \longrightarrow M$ by

$$\pi([\alpha, p, y]) = p, \tag{1.28}$$

which is a smooth map. We can show that (E, M, π) is a q-dimensional vector bundle on M, and its transition functions are precisely $\{g_{\alpha\beta}\}$.

By the theorem, we know that the family of transition functions describes the essence of a vector bundle. To construct a vector bundle, we need only specify its transition functions.

Example 1 (The dual bundle E^* of a vector bundle E). Suppose V^* is the dual space of V, E^* is the vector bundle on M with V^* as its typical fiber, and the bundle projection is denoted by $\tilde{\pi}$. The structure of the local products of the bundle E^* is given by $\{(U, \psi_U), (W, \psi_W), (Z, \psi_Z), \dots\}$. If for any $p \in U \cap W \neq \varnothing$, and $y_U, y_W \in V$, $\lambda_U, \lambda_W \in V^*$ satisfying

$$
\begin{aligned}
\varphi_U(p, y_U) &= \varphi_W(p, y_W), \\
\psi_U(p, \lambda_U) &= \psi_W(p, \lambda_W),
\end{aligned}
\tag{1.29}
$$

it is always true that

$$
< y_U, \lambda_U > = < y_W, \lambda_W >,
\tag{1.30}
$$

then we can define a pairing between the fibers $\pi^{-1}(p)$ and $\tilde{\pi}^{-1}(p)$ such that they become vector spaces dual to each other. The pairing between the fibers is defined by

$$
\langle \varphi_U(p, y_U), \psi_U(p, \lambda_U) \rangle = < y_U, \lambda_U >,
\tag{1.31}
$$

which is independent of the choice of U. We call the vector bundle E^* the **dual bundle** of E.

If we choose dual bases of V and V^*, and then denote any element y in V by a coordinate row and any element λ in V^* by a coordinate column, then the pairing between V and V^* can be expressed as multiplication of matrices:

$$
< y, \lambda > = y \cdot \lambda.
\tag{1.32}
$$

By the first equation in (1.29),

$$
y_W = y_U \cdot g_{UW}(p).
$$

Substituting into (1.30), we get

$$
y_U \cdot \lambda_U = y_U \cdot g_{UW}(p) \cdot \lambda_W.
$$

Therefore

$$
\lambda_U = g_{UW}(p) \cdot \lambda_W.
\tag{1.33}
$$

If we also denote the elements in V^* by coordinate rows, then elements of $GL(V^*)$ operate on V^* on the right, and the transition functions of E^* are

$$h_{UW} = {}^t(g_{UW}^{-1}) = {}^t g_{WU}. \qquad (1.34)$$

When E is the tangent bundle of M, the family $\{J_{UW}\}$ of its transition functions is composed of the Jacobian matrices of coordinate transformations. Any transition function of the cotangent bundle is the transpose of the inverse matrix of some J_{UW}. Hence the cotangent bundle is the dual bundle to the tangent bundle.

Example 2 (The direct sum $E \oplus E'$ of E and E'). Suppose E and E' are vector bundles on a manifold M with typical fibers V and V', transition functions $\{g_{UW}\}$ and $\{g'_{UW}\}$, respectively. Let

$$h_{UW} = \begin{pmatrix} g_{UW} & 0 \\ 0 & g'_{UW} \end{pmatrix}. \qquad (1.35)$$

Then h_{UW} is a linear automorphism on $V \oplus V'$ operating on the right, and $\{h_{UW}\}$ satisfies the compatibility conditions 1) and 2) for families of transition functions. The vector bundle on M with typical fiber $V \oplus V'$ and transition function family $\{h_{UW}\}$ is called the **direct sum** of E and E', denoted by $E \oplus E'$.

Example 3 (The tensor product $E \otimes E'$ of vector bundles E and E'). Suppose E and E' are the same as in Example 2. Let h_{UW} be the tensor product of g_{UW} and g'_{UW}, that is, $h_{UW} = g_{UW} \otimes g'_{UW}$, with its operation on $V \otimes V'$ being defined by

$$(v \otimes v') \cdot h_{UW} = \left(v \cdot g_{UW} \right) \otimes \left(v' \cdot g'_{UW} \right), \qquad (1.36)$$

where $v \in V$, $v' \in V'$. Obviously, $\{h_{UW}\}$ also satisfies the compatibility conditions of transition functions. The vector bundle on M with transition function family $\{h_{UW}\}$ and typical fiber $V \otimes V'$ is called the **tensor product** of E and E', denoted by $E \otimes E'$. It is easy to see that an (r, s)-type tensor bundle on M is the tensor product of r tangent bundles and s cotangent bundles.

Definition 1.2. Suppose $s : M \longrightarrow E$ is a smooth map. If

$$\pi \circ s = \mathrm{id} : M \longrightarrow M, \qquad (1.37)$$

then s is called a **smooth section** of the vector bundle (E, M, π). We denote the set of all smooth sections of the vector bundle (E, M, π) by $\Gamma(E)$.

Since every fiber of a vector bundle is a vector space isomorphic to V, we can define in a pointwise fashion addition and scalar multiplication of sections. Suppose $s, s_1, s_2 \in \Gamma(E)$, and α is a smooth real-valued function on M. For any $p \in M$, let

$$(s_1 + s_2)(p) = s_1(p) + s_2(p),$$

$$(\alpha s)(p) = \alpha(p) \cdot s(p).$$

Then $s_1 + s_2$ and αs are also smooth sections of the vector bundle E. This shows that $\Gamma(E)$ is a $C^\infty(M)$-module, and of course a real vector space.

Remark. A smooth section of a vector bundle E which is nonzero everywhere does not always exist. The existence of such a section reflects certain specific topological properties of the manifold M.

§3–2 Exterior Differentiation

Suppose M is an m-dimensional smooth manifold. The bundle of exterior r-forms on M

$$\Lambda^r(M^*) = \bigcup_{p \in M} \Lambda^r(T_p^*)$$

is a vector bundle on M. Use $A^r(M)$ to denote the space of the smooth sections of the exterior form bundle $\Lambda^r(M^*)$:

$$A^r(M) = \Gamma(\Lambda^r(M^*)). \tag{2.1}$$

$A^r(M)$ is a $C^\infty(M)$-module. The elements of $A^r(M)$ are called **exterior differential r-forms** on M. Therefore, an exterior differential r-form on M is a smooth skew-symmetric covariant tensor field of order r on M.

Similarly, the exterior form bundle $\Lambda(M^*) = \bigcup_{p \in M} \Lambda(T_p^*)$ is also a vector bundle on M. The elements of the space of its sections $A(M)$ are called exterior differential forms on M. Obviously $A(M)$ can be expressed as the direct sum

$$A(M) = \sum_{r=0}^{m} A^r(M), \tag{2.2}$$

i.e., every differential form ω can be written as

$$\omega = \omega^0 + \omega^1 + \cdots + \omega^m, \tag{2.3}$$

where ω^i is an exterior differential i-form. The wedge product of exterior forms can be extended to the space of exterior differential forms $A(M)$. Suppose $\omega_1, \omega_2 \in A(M)$. For any $p \in M$, let

$$\omega_1 \wedge \omega_2(p) = \omega_1(p) \wedge \omega_2(p), \tag{2.4}$$

where the right hand side is a wedge product of two exterior forms. It is obvious that $\omega_1 \wedge \omega_2 \in A(M)$. The space $A(M)$ then becomes an algebra with respect to addition, scalar multiplication and the wedge product. Moreover, it is a **graded algebra**. This means that $A(M)$ is a direct sum (2.2) of a sequence of vector spaces, and the wedge product $\wedge$ defines a map

$$\wedge : A^r(M) \times A^s(M) \longrightarrow A^{r+s}(M), \tag{2.5}$$

where $A^{r+s}(M)$ is zero when $r + s > m$.

Remark. The tensor algebras $T(V)$ and $T(V^*)$, with respect to the tensor product $\otimes$, and the exterior algebra $\Lambda(V)$, with respect to the exterior product $\wedge$, are all graded algebras.

Under the local coordinates $u^1, \ldots, u^m$, the restriction of the exterior differential r-form ω in the coordinate neighborhood U can be written as

$$\omega = a_{i_1, \ldots, i_r} du^{i_1} \wedge \cdots \wedge du^{i_r}, \tag{2.6}$$

where $a_{i_1, \ldots, i_r}$ is a smooth function on U which is skew-symmetric with respect to the lower indices.

The exterior r-vector bundle $\Lambda^r(M)$ and the exterior r-form bundle $\Lambda^r(M^*)$ are dual to each other. As described in Example 1 in §3–1, the pairing of the fibers of $\Lambda^r(M)$ and $\Lambda^r(M^*)$ at $p \in M$ is induced from the pairing of $\Lambda^r(V)$ and $\Lambda^r(V^*)$. Therefore it follows from (3.16) of Chapter 2 that

$$\left\langle \frac{\partial}{\partial u^{i_1}} \wedge \cdots \wedge \frac{\partial}{\partial u^{i_r}}, du^{j_1} \wedge \cdots \wedge du^{j_r} \right\rangle = \delta_{i_1 \cdots i_r}^{j_1 \cdots j_r}. \tag{2.7}$$

Thus the components $a_{i_1 \cdots i_r}$ of ω in the local coordinate system u^i can be expressed as

$$a_{i_1 \cdots i_r} = \frac{1}{r!} \left\langle \frac{\partial}{\partial u^{i_1}} \wedge \cdots \wedge \frac{\partial}{\partial u^{i_r}}, \omega \right\rangle. \tag{2.8}$$

The space $A(M)$ of exterior differential forms plays a crucial role in manifold theory, due to the existence of an exterior derivative operator d on $A(M)$ which gives zero on operating twice.

Theorem 2.1. *Suppose M is an m-dimensional smooth manifold. Then there exists a unique map $d : A(M) \longrightarrow A(M)$ such that $d(A^r(M)) \subset A^{r+1}(M)$ and such that d satisfies the following:*

1) *For any $\omega_1, \omega_2 \in A(M)$, $d(\omega_1 + \omega_2) = d\omega_1 + d\omega_2$.*
2) *Suppose ω_1 is an exterior differential r-form. Then*

$$d(\omega_1 \wedge \omega_2) = d\omega_1 \wedge \omega_2 + (-1)^r \omega_1 \wedge d\omega_2.$$

3) *If f is a smooth function on M, i.e., $f \in A^0(M)$, then df is precisely the differential of f.*
4) *If $f \in A^0(M)$, then $d(df) = 0$.*

The map d defined above is called the **exterior derivative**.

Proof. First we show that if the exterior derivative operator d exists, then d is a local operator. This means that, assuming $\omega_1, \omega_2 \in A(M)$, if ω_1 is the same as ω_2 in an open set U on M, then the restriction of $d\omega_1$ and $d\omega_2$ in U are the same. To see this using condition 1), we need only show that $\omega|_U = 0$ implies $(d\omega)|_U = 0$. Choose any point $p \in U$. By local compactness of manifolds, there is an open neighborhood W containing p such that $p \in W \subset \overline{W} \subset U$. By Lemma 3 of §1–3, there exists a smooth function h on M such that

$$h(p') = \begin{cases} 1, & p' \in W, \\ \\ 0, & p' \notin U. \end{cases}$$

Thus $h\omega \in A(M)$ and $h\omega \equiv 0$. Therefore

$$dh \wedge \omega + h d\omega = 0,$$

$$d\omega|_W = 0.$$

Due to the arbitrariness of p in U, the restriction of $d\omega$ in U must be zero.

Suppose ω is an exterior differential form defined on the open set U. Using Lemma 3 in §1–3, for any point $p \in U$, there is a coordinate neighborhood $U_1 \subset U$ of p and an exterior differential form $\tilde{\omega}$ defined on M such that

$$\tilde{\omega}|_{U_1} = \omega|_{U_1}. \tag{2.9}$$

Thus we can define

$$d\omega|_{U_1} = d\tilde{\omega}|_{U_1}. \tag{2.10}$$

Since d is a local differential operator, the above definition is independent of the choice of $\tilde{\omega}$. $d\omega$ is therefore well-defined.

Now we show the uniqueness of the exterior derivative d within a local coordinate neighborhood. By condition 1) we only need to show this for a monomial. Suppose in a coordinate neighborhood U, ω is expressed by

$$\omega = a\,du^1 \wedge \cdots \wedge du^r, \tag{2.11}$$

where a is a smooth function on U. The action of d on an exterior differential form defined on U still satisfies conditions 1)–4). Therefore[c]

$$d\omega = da \wedge du^1 \wedge \cdots \wedge du^r, \tag{2.12}$$

where da is the differential of the function a. Thus $d\omega$ restricted to the coordinate neighborhood U has a completely determined form.

Suppose

$$\omega|_U = a_{i_1 \cdots i_r}\,du^{i_1} \wedge \cdots \wedge du^{i_r}. \tag{2.13}$$

Then we can define

$$d(\omega|_U) = da_{i_1 \cdots i_r} \wedge du^{i_1} \wedge \cdots \wedge du^{i_r}. \tag{2.14}$$

Obviously, $d(\omega|_U)$ is an exterior differential $(r + 1)$-form on U satisfying conditions 1) and 3). To show that 2) holds, we need only consider any two monomials

$$\begin{aligned}
\alpha_1 &= a\,du^{i_1} \wedge \cdots \wedge du^{i_r} \\
\alpha_2 &= b\,du^{j_1} \wedge \cdots \wedge du^{j_s}.
\end{aligned}$$

By the definition (2.14), we have

$$\begin{aligned}
d(\alpha_1 \wedge \alpha_2) &= (b\,da + a\,db) \wedge du^{i_1} \wedge \cdots \wedge du^{i_r} \wedge du^{j_1} \wedge \cdots \wedge du^{j_s} \\
&= \left(da \wedge du^{i_1} \wedge \cdots \wedge du^{i_r}\right) \wedge \left(b\,du^{j_1} \wedge \cdots \wedge du^{j_s}\right) + \\
&\quad (-1)^r \left(a\,du^{i_1} \wedge \cdots \wedge du^{i_r}\right) \wedge \left(db \wedge du^{j_1} \wedge \cdots \wedge du^{j_s}\right) \\
&= d\alpha_1 \wedge \alpha_2 + (-1)^r \alpha_1 \wedge d\alpha_2.
\end{aligned}$$

Property 2) is therefore established.

We now prove condition 4). Suppose f is a smooth function on M. Then on U it satisfies

$$df = \frac{\partial f}{\partial u^i}\,du^i. \tag{2.15}$$

---

[c]This equality is obtained by using the following fact: the coordinate functions u^i are smooth functions on U. Hence by condition 4), $d(du^i) = 0$.

Since f is C^∞, its higher than first order partial derivatives are independent of the order taken, i.e.,

$$\frac{\partial^2 f}{\partial u^i \partial u^j} = \frac{\partial^2 f}{\partial u^j \partial u^i}. \qquad (2.16)$$

Therefore

$$\begin{aligned}
d(df) &= d\left(\frac{\partial f}{\partial u^i}\right) \wedge du^i \\
&= \frac{\partial^2 f}{\partial u^i \partial u^j} du^j \wedge du^i \\
&= \frac{1}{2}\left(\frac{\partial^2 f}{\partial u^i \partial u^j} - \frac{\partial^2 f}{\partial u^j \partial u^i}\right) du^j \wedge du^i \\
&= 0.
\end{aligned}$$

If W is another coordinate neighborhood, we obtain by the local property of the exterior derivative operator and its uniqueness in a local coordinate neighborhood that

$$\begin{aligned}
d(\omega|_U)|_{U \cap W} &= d(\omega|_{U \cap W}) \\
&= d(\omega|_W)|_{U \cap W}. \qquad (2.17)
\end{aligned}$$

Hence the exterior derivative operator d is uniformly defined by (2.14) on $U \cap W$, i.e., d is an operator defined on M globally. This proves the existence of the operator d satisfying the conditions of the theorem. □

Theorem 2.2 (Poincaré's Lemma). $d^2 = 0$, *i.e., for any exterior differential form* ω, $d(d\omega) = 0$.

Proof. Since d is a linear operator, we need only prove the lemma when ω is a monomial. By the local properties of d, it is sufficient to assume that

$$\omega = a\, du^1 \wedge \cdots \wedge du^r.$$

Hence

$$d\omega = da \wedge du^1 \wedge \cdots \wedge du^r.$$

Differentiating one more time and applying conditions 2) and 4), we have

$$\begin{aligned}
d(d\omega) &= d(da) \wedge du^1 \wedge \cdots \wedge du^r \\
&\quad -da \wedge d(du^1) \wedge \cdots \wedge du^r \\
&\quad + \cdots \\
&= 0.
\end{aligned}$$

□

Example 1. Suppose the Cartesian coordinates in $\mathbb{R}^3$ are given by (x, y, z).
1) If f is a smooth function on $\mathbb{R}^3$, then

$$df = \frac{\partial f}{\partial x} dx + \frac{\partial f}{\partial y} dy + \frac{\partial f}{\partial z} dz.$$

The vector formed by its coefficients $\left(\dfrac{\partial f}{\partial x}, \dfrac{\partial f}{\partial y}, \dfrac{\partial f}{\partial z} \right)$ is the gradient of f, denoted by grad f.
2) Suppose $a = A dx + B dy + C dz$, where A, B, C are smooth functions on $\mathbb{R}^3$. Then

$$
\begin{aligned}
da &= dA \wedge dx + dB \wedge dy + dC \wedge dz \\
&= \left(\frac{\partial C}{\partial y} - \frac{\partial B}{\partial z} \right) dy \wedge dz + \left(\frac{\partial A}{\partial z} - \frac{\partial C}{\partial x} \right) dz \wedge dx \\
&\quad + \left(\frac{\partial B}{\partial x} - \frac{\partial A}{\partial y} \right) dx \wedge dy.
\end{aligned}
$$

Let X be the vector (A, B, C), then the vector

$$\left(\frac{\partial C}{\partial y} - \frac{\partial B}{\partial z}, \frac{\partial A}{\partial z} - \frac{\partial C}{\partial x}, \frac{\partial B}{\partial x} - \frac{\partial A}{\partial y} \right)$$

formed by the coefficients of da is just the curl of the vector field X, denoted by curl X.
3) Suppose $a = A dy \wedge dz + B dz \wedge dx + C dx \wedge dy$. Then

$$
\begin{aligned}
da &= \left(\frac{\partial A}{\partial x} + \frac{\partial B}{\partial y} + \frac{\partial C}{\partial z} \right) dx \wedge dy \wedge dz \\
&= \operatorname{div} X dx \wedge dy \wedge dz,
\end{aligned}
$$

where div X means the divergence of the vector field $X = (A, B, C)$.

From Poincaré's Lemma, two fundamental formulas in vector calculus follow immediately. Suppose f is a smooth function on $\mathbb{R}^3$ and X is a smooth tangent vector field on $\mathbb{R}^3$. Then

$$
\begin{cases}
\operatorname{curl} \ (\operatorname{grad} f) &= \ 0, \\
\operatorname{div} \ (\operatorname{curl} X) &= \ 0.
\end{cases}
\tag{2.18}
$$

Theorem 2.3. *Suppose ω is a differential 1-form on a smooth manifold M; X and Y are smooth tangent vector fields on M. Then*

$$\langle X \wedge Y, d\omega \rangle = X \langle Y, \omega \rangle - Y \langle X, \omega \rangle - \langle [X, Y], \omega \rangle. \tag{2.19}$$

Proof. Since both sides in (2.19) are linear with respect to ω, we may assume that ω is a monomial:

$$\omega = g\,df, \tag{2.20}$$

where f, g are smooth functions on M. Therefore

$$d\omega = dg \wedge df. \tag{2.21}$$

By (3.15) in Chapter 2, the left hand side of (2.19) is

$$\langle X \wedge Y, dg \wedge df \rangle = \begin{vmatrix} \langle X, dg \rangle & \langle X, df \rangle \\ \langle Y, dg \rangle & \langle Y, df \rangle \end{vmatrix} \tag{2.22}$$

$$= Xg \cdot Yf - Xf \cdot Yg.$$

Since

$$\langle X, \omega \rangle = \langle X, g\,df \rangle = g \cdot Xf,$$

we have

$$Y \langle X, \omega \rangle = Yg \cdot Xf + g \cdot Y(Xf). \tag{2.23}$$

Similarly,

$$X \langle Y, \omega \rangle = Xg \cdot Yf + g \cdot X(Yf). \tag{2.24}$$

Hence the right hand side of (2.19) is

$$X \langle Y, \omega \rangle \quad - Y \langle X, \omega \rangle - \langle [X, Y], \omega \rangle$$

$$= Xg \cdot Yf - Yg \cdot Xf + g(X(Yf) - Y(Xf)) - g \langle [X, Y], df \rangle$$

$$= Xg \cdot Yf - Yg \cdot Xf. \tag{2.25}$$

Therefore (2.19) holds. □

Remark. For an exterior differential form ω with arbitrary degree, the following formula holds. Suppose $\omega \in A^r(M)$ and $X_1, \ldots, X_{r+1}$ are smooth tangent vector fields on M. Then

$$\langle X_1 \wedge \cdots \wedge X_{r+1}, d\omega \rangle = \sum_{i=1}^{r+1} (-1)^{i+1} X_i \left\langle X_1 \wedge \cdots \wedge \hat{X}_i \wedge \cdots \wedge X_{r+1}, \omega \right\rangle$$

$$+ \sum_{1 \le i < j \le r+1} (-1)^{i+j} \left\langle [X_i, X_j] \wedge \cdots \wedge \hat{X}_i \wedge \cdots \wedge \hat{X}_j \wedge \cdots \wedge X_{r+1}, \omega \right\rangle. \tag{2.26}$$

The reader should prove (2.26).

Using Theorem 2.3, the Frobenius condition for r-dimensional distributions mentioned in Chapter 1 can be rephrased in its dual form. Suppose $L^r = \{X_1, \ldots, X_r\}$ is a smooth r-dimensional distribution on M. Then for any point $p \in M$, $L^r(p)$ is an r-dimensional linear subspace of T_p. Let

$$(L^r(p))^\perp = \{\omega \in T_p^* \mid \langle X, \omega \rangle = 0 \text{ for any } X \in L^r(p)\}. \qquad (2.27)$$

$(L^r(p))^\perp$ is certainly an $(m - r)$-dimensional subspace of T_p^*, called the **annihilator subspace** of $L^r(p)$. In a neighborhood of an arbitrary point, there exist $m - r$ linearly independent differential 1-forms $\omega_{r+1}, \ldots, \omega_m$ that span the annihilator subspace $(L^r(p))^\perp$ at any point p in the neighborhood. In fact, L^r is spanned by r linearly independent smooth tangent vector fields $X_1, \ldots, X_r$ in a neighborhood. Therefore there exist $m - r$ smooth tangent vector fields $X_{r+1}, \ldots, X_m$ such that $\{X_1, \ldots, X_r, X_{r+1}, \ldots, X_m\}$ is linearly independent everywhere in that neighborhood. Suppose $\{\omega_1, \ldots, \omega_r, \omega_{r+1}, \ldots, \omega_m\}$ are the dual differential 1-forms in that neighborhood. Then at every point p, $(L^r(p))^\perp$ is spanned by $\omega_{r+1}, \ldots, \omega_m$. Locally the distribution L^r is therefore equivalent to the system of equations

$$\omega_s = 0, \qquad r + 1 \le s \le m, \qquad (2.28)$$

often called a **Pfaffian** system of equations.

By (2.19), we have

$$
\begin{aligned}
\langle X_\alpha \wedge X_\beta, d\omega_s \rangle &= X_\alpha \langle X_\beta, \omega_s \rangle - X_\beta \langle X_\alpha, \omega_s \rangle - \langle [X_\alpha, X_\beta], \omega_s \rangle \\
&= -\langle [X_\alpha, X_\beta], \omega_s \rangle.
\end{aligned}
$$

Hence the distribution $L^r = \{X_1, \ldots, X_r\}$ satisfies the Frobenius condition

$$[X_\alpha, X_\beta] \in L^r, \qquad 1 \le \alpha, \beta \le r \qquad (2.29)$$

if and only if

$$\langle X_\alpha \wedge X_\beta, d\omega_s \rangle = 0, \quad 1 \le \alpha, \beta \le r, \quad r + 1 \le s \le m. \qquad (2.30)$$

We will show that (2.30) is equivalent to

$$d\omega_s \equiv 0 \bmod (\omega_{r+1}, \ldots, \omega_m), \quad r + 1 \le s \le m. \qquad (2.31)$$

In fact, $d\omega_s$ can be expressed in terms of $\omega_i (1 \le i \le m)$:

$$d\omega_s = \sum_{t=r+1}^{m} \psi_{st} \wedge \omega_t + \sum_{\alpha,\beta=1}^{r} a_{\alpha\beta}^s \omega_\alpha \wedge \omega_\beta, \qquad (2.32)$$

where ψ_{st} is a differential 1-form, and $a^s_{\alpha\beta}$ is a smooth function which is skew-symmetric with respect to the lower indices. Since

$$\langle X_i, \omega_j \rangle = \delta_{ij},$$

plugging (2.32) into (2.30) gives

$$a^s_{\alpha\beta} = 0, \quad 1 \le \alpha, \beta \le r, \ r+1 \le s \le m,$$

that is,

$$dw_s = \sum_{t=r+1}^{m} \psi_{st} \wedge \omega_t, \tag{2.33}$$

which is (2.31). We call (2.31) the Frobenius condition satisfied by the Pfaffian system (2.28).

If there exists a local coordinate system u^i such that the submanifolds

$$u^s = \text{const}, \ r+1 \le s \le m \tag{2.34}$$

satisfy the Pfaffian system (2.28), then the Pfaffian system is said to be **completely integrable**. Thus for a completely integrable Pfaffian system (2.28), there exists a local coordinate system u^i such that the system is equivalent to

$$du^s = 0, \quad r+1 \le s \le m. \tag{2.35}$$

In this situation the distribution L^r is spanned precisely by the tangent vector fields $\partial/\partial u^1, \dots, \partial/\partial u^r$. The converse is also true. Thus the Frobenius Theorem in Chapter 1 can be rephrased as

Theorem 2.4. *A necessary and sufficient condition for the Pfaffian system of equations*

$$\omega_\alpha = 0, \ 1 \le \alpha \le r, \tag{2.36}$$

to be completely integrable is

$$d\omega_\alpha \equiv 0 \bmod (\omega_1, \dots, \omega_r), \quad 1 \le \alpha \le r. \tag{2.37}$$

Example 2. Suppose

$$\Omega \equiv P dx + Q dy + R dz = 0 \tag{2.38}$$

is a total differential equation in $\mathbb{R}^3$, where P, Q and R are smooth functions on $\mathbb{R}^3$. Complete integrability of (2.38) means that in every sufficiently small neighborhood there exists a smooth function F such that

$$F = \text{const}$$

is a first integral of (2.38). By Theorem 2.4 a necessary and sufficient condition for (2.38) to be completely integrable is

$$d\Omega \wedge \Omega = 0,$$

that is

$$P\left(\frac{\partial R}{\partial y} - \frac{\partial Q}{\partial z}\right) + Q\left(\frac{\partial P}{\partial z} - \frac{\partial R}{\partial x}\right) + R\left(\frac{\partial Q}{\partial x} - \frac{\partial P}{\partial y}\right) = 0. \qquad (2.39)$$

In other words, (2.39) is a necessary and sufficient condition for (2.38) to have an integrating factor.

We should point out that the Frobenius Theorem is a local theorem, and describes the solution of the Pfaffian system of equations in a neighborhood of some point. Yet starting from this "local" theorem, we can study the global integral manifolds. To do this, we need to introduce a new topology on M. Suppose L^r is an r-dimensional distribution on M satisfying the Frobenius condition. If we define the "open sets" of M to be the unions of integral manifolds of L^r, then the collection of all these "open sets" forms a topology O on M.

Theorem 2.5. *Suppose L^r is an r-dimensional distribution satisfying the Frobenius condition on a manifold M. Then through any point $p \in M$, there exists a maximal integral manifold $\mathcal{L}(p)$ of L^r such that any integral manifold of L^r through p is an open submanifold of $\mathcal{L}(p)$ with respect to the topology O.*

The term **maximal integral manifold** in this theorem means that it is not a proper subset of another integral manifold. The reader may refer to p. 92 of Chevalley 1946 for details.

Suppose $f : M \longrightarrow N$ is a smooth map from a smooth manifold M to a smooth manifold N. Then it induces a linear map between the spaces of exterior differential forms:

$$f^* : A(N) \longrightarrow A(M).$$

In fact, f induces a tangent mapping $f_* : T_p(M) \longrightarrow T_{f(p)}(N)$ at every point $p \in M$, and the definition of the map $f^* : A(N) \longrightarrow A(M)$ for each homogeneous part of $A(N)$ and $A(M)$ is as follows:

If $\beta \in A^r(N)$, $r \geq 1$, then $f^*\beta \in A^r(M)$ such that for any r smooth tangent vector fields $X_1, \ldots, X_r$ on M,

$$\langle X_1 \wedge \cdots \wedge X_r, f^*\beta \rangle_p = \langle f_* X_1 \wedge \cdots \wedge f_* X_r, \beta \rangle_{f(p)}, p \in M, \qquad (2.40)$$

where $\langle \, , \, \rangle$ is the pairing defined in (2.7).

If $\beta \in A^0(N)$, we define

$$f^*\beta = \beta \circ f \in A^0(M). \tag{2.41}$$

By Theorem 3.2 in Chapter 2, the map f^* distributes over the exterior product, that is, for any $\omega, \eta \in A(N)$,

$$f^*(\omega \wedge \eta) = f^*\omega \wedge f^*\eta. \tag{2.42}$$

The importance of the induced map f^* also rests on the fact that it commutes with the exterior derivative d.

Theorem 2.6. *Suppose $f : M \longrightarrow N$ is a smooth map from a smooth manifold M to a smooth manifold N. Then the induced map $f^* : A(N) \longrightarrow A(M)$ commutes with the exterior derivative d, that is,*

$$f^* \circ d = d \circ f^* : A(N) \longrightarrow A(M). \tag{2.43}$$

In other words, the following diagram commutes.

$$
\begin{array}{ccc}
A(N) & \xrightarrow{\ d\ } & A(N) \\
f^* \downarrow & & \downarrow f^* \\
A(M) & \xrightarrow{\ d\ } & A(M)
\end{array}
$$

Proof. Since both f^* and d are linear, we need only consider the operation of both sides of (2.43) on a monomial β.

First suppose β is a smooth function on N, i.e., $\beta \in A^0(N)$. Choose any smooth tangent vector field X on M. Then it follows from (2.40) that

$$
\begin{aligned}
\langle X, f^*(d\beta) \rangle &= \langle f_*X, d\beta \rangle \\
&= f_*X(\beta) \\
&= X(\beta \circ f) \\
&= \langle X, d(f^*\beta) \rangle.
\end{aligned}
$$

Therefore

$$f^*(d\beta) = d(f^*\beta).$$

Next suppose $\beta = u\,dv$, where u, v are smooth functions on N. Then

$$
\begin{aligned}
f^*(d\beta) &= f^*(du \wedge dv) \\
&= f^*du \wedge f^*dv \\
&= d(f^*u) \wedge d(f^*v) \\
&= d(f^*\beta).
\end{aligned}
$$

Now assume that (2.43) holds for exterior differential forms of degree $< r$. We need to show that it also holds for exterior differential r-forms. Suppose β is a monomial of degree r, and $\beta \in A^r(N)$ is written

$$\beta = \beta_1 \wedge \beta_2,$$

where β_1 is a differential 1-form on N, and β_2 is an exterior differential $(r-1)$-form on N. Then by the induction hypothesis we have

$$
\begin{aligned}
d \circ f^*(\beta_1 \wedge \beta_2) &= d(f^*\beta_1 \wedge f^*\beta_2) \\
&= d(f^*\beta_1) \wedge f^*\beta_2 - f^*\beta_1 \wedge d(f^*\beta_2) \\
&= f^*(d\beta_1 \wedge \beta_2) - f^*(\beta_1 \wedge d\beta_2) \\
&= f^* \circ d(\beta_1 \wedge \beta_2).
\end{aligned}
$$

$\square$

§3-3 Integrals of Differential Forms

The simplest device connecting the local and global properties of a manifold is the integration of exterior differential forms on the manifold. To define these integrals we need some prerequisites.

Definition 3.1. An m-dimensional smooth manifold M is called **orientable** if there exists a continuous and nonvanishing exterior differential m-form ω on M. If M is given such an ω, then M is said to be **oriented**. If two such forms are given on M such that they differ by a function factor which is always positive, then we say that they assign the same **orientation** to M.

There exist exactly two orientations on a connected orientable manifold. The reason is that if ω, ω' are two exterior differential m-forms giving orientations to M, then there is a nonvanishing continuous function f such that

$$\omega' = f\omega. \tag{3.1}$$

Since M is connected, f retains the same sign on all of M. Therefore the orientation given by ω' is either identical to the one given by ω or the one given by $-\omega$.

Suppose M is oriented by the exterior differential form ω, and $(U; u^i)$ is any local coordinate system on M. Then $du^1 \wedge \cdots \wedge du^m$ and $\omega|_U$ are the same up to a non-zero factor. If the factor is positive, then we say that $(U; u^i)$ is a coordinate system **consistent** with the orientation of M. It is obvious that for any oriented manifold there exists a coordinate covering which is consistent with the orientation of the manifold, and the Jacobian of the change of coordinates between any two coordinate neighborhoods with nonempty intersection will always be positive. Conversely, using the partition of unity theorem below, we can prove that if there exists a compatible coordinate covering such that the Jacobian of the change of coordinates in the intersection of two neighborhoods is always positive, then M is orientable. (We leave the proof to the reader.)

Definition 3.2. Suppose $f : M \longrightarrow \mathbb{R}$ is a real function on M. The **support** of f is the closure of the set of points at which f is nonzero:

$$\text{supp } f = \overline{\{p \in M | f(p) \neq 0\}}. \tag{3.2}$$

If ϕ is an exterior differential form, then the support of ϕ is

$$\text{supp } \phi = \overline{\{p \in M | \phi(p) \neq 0\}}. \tag{3.3}$$

Obviously, the compliment of $\text{supp } \phi$ is exactly the largest open subset of M on which $\phi = 0$.

Definition 3.3. Suppose Σ_0 is an open covering of M. If every compact subset of M intersects only finitely many elements of Σ_0, then Σ_0 is called a **locally finite** open covering of M.

Theorem 3.1. *Suppose Σ is a topological basis of the manifold M. Then there is a subset Σ_0 of Σ such that Σ_0 is a locally finite open covering of M.*

Proof. By the definition of manifolds, M is locally compact. Since we have assumed that M satisfies the second countability axiom, there exists a countable open covering $\{U_i\}$ of M such that the closure $\overline{U}_i$ of every U_i is compact. Let

$$P_i = \bigcup_{1 \leq r \leq i} \overline{U}_r. \tag{3.4}$$

Then P_i is compact, $P_i \subset P_{i+1}$, and $\bigcup_{i=1}^{\infty} P_i = M$. Now we construct another sequence of compact sets Q_i satisfying

$$P_i \subset Q_i \subset \overset{\circ}{Q}_{i+1}, \tag{3.5}$$

where $\overset{\circ}{Q}_{i+1}$ means the interior of Q_{i+1}.

By induction, assume that $Q_1, \ldots, Q_i$ have been constructed. Since $Q_i \cup P_{i+1}$ is compact, there exist finitely many elements $U_\alpha, 1 \leq \alpha \leq s$ of $\{U_j\}$ which together form a covering of $Q_i \cup P_{i+1}$. Let

$$Q_{i+1} = \bigcup_{1 \leq \alpha \leq s} \overline{U}_\alpha. \tag{3.6}$$

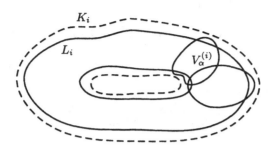

FIGURE 9.

Then Q_{i+1} satisfies condition (3.5). First, Q_{i+1} is compact, and $P_{i+1} \subset Q_{i+1}$. Furthermore

$$Q_i \subset \bigcup_{1 \le \alpha \le s} U_\alpha \subset \overset{\circ}{Q}_{i+1} . \tag{3.7}$$

Obviously, $\bigcup_{i=1}^{\infty} Q_i = M$.

Now let

$$L_i = Q_i - \overset{\circ}{Q}_{i-1}, \quad K_i = \overset{\circ}{Q}_{i+1} - Q_{i-2}, \tag{3.8}$$

where $1 \le i < +\infty$, and $Q_{-1} = Q_0 = \varnothing$. (see Figure 9). Then L_i is compact, K_i is open, and $L_i \subset K_i$.

By assumption, Σ is a topological basis of M. Thus K_i can be expressed as a union of elements of Σ. Since L_i is compact, and $L_i \subset K_i$, there exist for each i finitely many elements $V_\alpha^{(i)}$, $1 \le \alpha \le r_i$, in Σ such that

$$L_i \subset \bigcup_{1 \le \alpha \le r_i} V_\alpha^{(i)} \subset K_i. \tag{3.9}$$

Because $\bigcup_{i=1}^{\infty} L_i = M$,

$$\Sigma_0 = \{V_\alpha^{(i)}, \ 1 \le \alpha \le r_i, \ 1 \le i < +\infty\} \tag{3.10}$$

is a subcovering of Σ.

We now show that Σ_0 is locally finite. Suppose A is an arbitrary compact set. By (3.4) we know that there exists a sufficiently large integer i such that $A \subset P_i \subset Q_i$. For $k \ge i+2$,

$$K_k = \overset{\circ}{Q}_{k+1} - Q_{k-2} \subset \overset{\circ}{Q}_{k+1} - Q_i, \tag{3.11}$$

hence $K_k \cap Q_i = \varnothing$. Therefore

$$V_\alpha^{(k)} \cap A \subset K_k \cap Q_i = \varnothing, \quad 1 \le \alpha \le r_k, \quad k \ge i + 2, \tag{3.12}$$

that is, only finitely many elements of Σ_0 intersect A. $\qquad\qquad\qquad\square$

Theorem 3.2 (Partition of Unity Theorem). *Suppose Σ is an open covering of a smooth manifold M. Then there exists a family of smooth functions $\{g_\alpha\}$ on M satisfying the following conditions:*

   1) *$0 \le g_\alpha \le 1$, and supp g_α is compact for each α. Moreover, there exists an open set $W_i \in \Sigma$ such that supp $g_\alpha \subset W_i$;*
   2) *For each point $p \in M$, there is a neighborhood U that intersects supp g_α for only finitely many α;*
   3) *$\sum_\alpha g_\alpha = 1$.*

Because of condition 2), for any point $p \in M$, there are only finitely many nonzero terms on the left hand side of condition 3). Thus the summation is meaningful. The family $\{g_\alpha\}$ is called a **partition of unity** subordinate to the open covering Σ.

Proof. Because M is a manifold, there is a topological basis $\Sigma_0 = \{U_\alpha\}$ such that each element U_α is a coordinate neighborhood, $\overline{U}_\alpha$ is compact, and there also exists $W_i \in \Sigma$ such that $\overline{U}_\alpha \subset W_i$. By Theorem 3.1, Σ_0 has a locally finite subcovering, so we may assume that Σ_0 itself is a locally finite open covering of M and has countably many elements. It is not difficult to show by induction that we can obtain V_α by a contraction of U_α [d] such that $\overline{V}_\alpha \subset U_\alpha$ and $\{V_\alpha\}$ is also an open covering for M.

By Lemma 3 of §1–3, there exist smooth functions h_α, with $0 \le h_\alpha \le 1$ on M such that

$$h_\alpha(p) = \begin{cases} 1, & p \in V_\alpha \\[2mm] 0, & p \notin U_\alpha. \end{cases} \tag{3.13}$$

Clearly supp $h_\alpha \subset \overline{U}_\alpha$. For any point $p \in M$, there exists a neighborhood U such that $\overline{U}$ is compact. By the local finiteness of Σ_0, $\overline{U}$ intersects only finitely many elements of Σ_0, and there are only finitely many nonzero terms in the sum $\sum_\alpha h_\alpha(p)$. Thus $h = \sum_\alpha h_\alpha$ is a smooth function on M. Since

---

[d]The way to contract is as follows: Let $W = \bigcup_{i \ne \alpha} U_i$. Then $M - W$ is a closed set contained in U_α. Since $\overline{U}_\alpha$ is compact, $M - W$ is also compact. Thus there are finitely many coordinate neighborhoods W_s, $1 \le s \le r$, such that $\overline{W}_s \subset U_\alpha$ and $\bigcup_{s=1}^r W_s \subset M - W$. Now let $V_\alpha = \bigcup_{s=1}^r W_s$.

$\{V_\alpha\}$ forms a covering for M, the point p must lie in some V_α, i.e., $h(p) \geq 1$. Let

$$g_\alpha = \frac{h_\alpha}{h}. \tag{3.14}$$

Then g_α is a smooth function on M. It is easy to verify that the family $\{g_\alpha\}$ satisfies all the conditions of the theorem. □

With the above background, we can proceed to define the integration of exterior differential forms on a manifold M. Suppose M is an m-dimensional smooth manifold, and φ is an exterior differential m-form on M with a compact support. Choose any coordinate covering $\Sigma = \{W_i\}$ which is consistent with the orientation of M, and suppose that $\{g_\alpha\}$ is a partition of unity subordinate to Σ. Then

$$\varphi = \left(\sum_\alpha g_\alpha \right) \cdot \varphi = \sum_\alpha (g_\alpha \cdot \varphi). \tag{3.15}$$

Clearly, supp $(g_\alpha \cdot \varphi) \subset$ supp g_α is contained in some coordinate neighborhood $W_i \in \Sigma$. Therefore we can define

$$\int_M g_\alpha \cdot \varphi = \int_{W_i} g_\alpha \cdot \varphi, \tag{3.16}$$

where the right hand side is the usual Riemann integral, that is, if $g_\alpha \cdot \varphi$ with respect to the coordinate system $u^1, \ldots, u^m$ in W_i is expressed as

$$f(u^1, \ldots, u^m) du^1 \wedge \cdots \wedge du^m,$$

then the integral on the right hand side in (3.16) is

$$\int_{W_i} f(u^i, \ldots, u^m) du^1 \cdots du^m. \tag{3.17}$$

To show that (3.16) is well-defined, we need only show that the right hand side is independent of the choice of W_i. Suppose supp $(g_\alpha \cdot \varphi)$ is contained in two coordinate neighborhoods W_i and W_j, and suppose the local coordinates consistent with the orientation of M are u^k and v^k, respectively. Then the Jacobian of the change of coordinates satisfies

$$J = \frac{\partial(v^1, \ldots, v^m)}{\partial(u^1, \ldots, u^m)} > 0. \tag{3.18}$$

Suppose $g_\alpha \cdot \varphi$ is expressed in W_i and W_j by

$$
\begin{aligned}
g_\alpha \cdot \varphi &= f du^1 \wedge \cdots \wedge du^m \\
&= f' dv^1 \wedge \cdots \wedge dv^m,
\end{aligned}
\tag{3.19}
$$

respectively. Then

$$
f = f' \cdot J = f' \cdot |J|,
\tag{3.20}
$$

and supp $f = $ supp $f' = $ supp $(g_\alpha \cdot \varphi) \subset W_i \cap W_j$. By the formula for the change of variables in the Riemann integral, we have

$$
\begin{aligned}
\int_{W_i \cap W_j} f' dv^1 \cdots dv^m &= \int_{W_i \cap W_j} f' |J| du^1 \cdots du^m \\
&= \int_{W_i \cap W_j} f\, du^1 \cdots du^m,
\end{aligned}
$$

i.e.,

$$
\int_{W_i} g_\alpha \cdot \varphi = \int_{W_j} g_\alpha \cdot \varphi.
\tag{3.21}
$$

Since supp φ is compact, it only intersects finitely many supp g_α by condition 2) of the Partition of Unity Theorem. Therefore the right hand side of (3.15) is a sum of only finitely many terms. Let

$$
\int_M \varphi = \sum_\alpha \int_M g_\alpha \cdot \varphi.
\tag{3.22}
$$

For any given partition of unity $\{g_\alpha\}$ subordinate to Σ, the right hand side of (3.22) is completely determined. Now we show that (3.22) is independent of the choice of the partition of unity $\{g_\alpha\}$.

Suppose $\{g'_\beta\}$ is another partition of unity subordinate to Σ. Then

$$
\begin{aligned}
\sum_\beta \int_M g'_\beta \cdot \varphi &= \sum_\beta \sum_\alpha \int_M g_\alpha \cdot g'_\beta \cdot \varphi \\
&= \sum_\alpha \int_M \sum_\beta g'_\beta \cdot g_\alpha \cdot \varphi \\
&= \sum_\alpha \int_M g_\alpha \cdot \varphi.
\end{aligned}
$$

Definition 3.4. Suppose M is an m-dimensional oriented smooth manifold and φ is an exterior differential m-form on M with compact support. The numerical value $\int_M \varphi$ defined in (3.22) is called the **integral** of the exterior differential form φ on M.

If $\varphi, \varphi_1, \varphi_2$ are exterior differential m-forms on M with compact support, then $\varphi_1 + \varphi_2$ has a compact support, as has $c\varphi$, for any real number c. By the definition of the integral, it is obvious that

$$\int_M (\varphi_1 + \varphi_2) = \int_M \varphi_1 + \int_M \varphi_2,$$

$$\int_M c\varphi = c \int_M \varphi. \tag{3.23}$$

Therefore the integral $\int_M$ is a linear functional on the set of all exterior differential m-forms on M with compact support.

If supp φ happens to be inside a coordinate neighborhood U with local coordinates u^i consistent with the orientation of M, then φ can be expressed as

$$\varphi = f(u^1, \ldots, u^m) du^1 \wedge \cdots \wedge du^m, \tag{3.24}$$

and $\int_M \varphi$ is precisely the Riemann integral

$$\int_M \varphi = \int_U f du^1 \cdots du^m. \tag{3.25}$$

We see that Definition 3.4 is a generalization of the Riemann integral.

If φ is an exterior differential r-form, $r < m$, with compact support, then we can define the integral of φ on any r-dimensional submanifold N of M. Suppose

$$h : N \longrightarrow M \tag{3.26}$$

is an r-dimensional imbedding of N into M. Then $h^*\varphi$ is an exterior differential r-form on the r-dimensional smooth manifold N, and it has compact support. Therefore the integral $\int_N h^*\varphi$ is well-defined. We define the integral of φ on the submanifold $h(N)$ as

$$\int_{h(N)} \varphi = \int_N h^*\varphi. \tag{3.27}$$

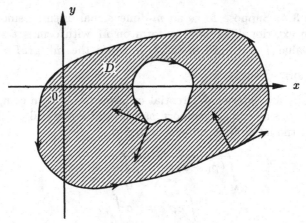

FIGURE 10.

§3–4   Stokes' Formula

The relationship between integrals over a domain and integrals over its boundary is at the heart of the calculus. Let us look at a few examples first.

Example 1. Suppose $D = [a,b]$ is a closed interval in $\mathbb{R}^1$ and f is a continuously differentiable function on D. Then the Fundamental Theorem of Calculus holds:

$$\int_D df = f(b) - f(a). \tag{4.1}$$

We denote the oriented boundary $\{b\} - \{a\}$ of D by ∂D. Then the right hand side above can be denoted by $\int_{\partial D} f$.

Example 2. Suppose D is a bounded domain in $\mathbb{R}^2$ whose orientation is consistent with that of $\mathbb{R}^2$. Use ∂D to denote the oriented boundary of D with the orientation induced by D, that is, the positive orientation of ∂D together with the normal vector pointing towards the interior of D form a coordinate system which is consistent with the orientation of $\mathbb{R}^2$ (see Figure 10). Suppose P and Q are continuously differentiable functions on D. Then the Green's Formula holds:

$$\int_{\partial D} P\,dx + Q\,dy = \int_D \left(\frac{\partial Q}{\partial x} - \frac{\partial P}{\partial y} \right) dx\,dy. \tag{4.2}$$

If we let $\omega = P\,dx + Q\,dy$, then

$$d\omega = \left(\frac{\partial Q}{\partial x} - \frac{\partial P}{\partial y}\right) dx \wedge dy.$$

Thus (4.2) can be written as

$$\int_{\partial D} \omega = \int_D d\omega. \tag{4.3}$$

Example 3. Suppose D is a bounded domain in $\mathbb{R}^3$ whose orientation is consistent with that of $\mathbb{R}^3$. The outward normal as the positive direction then induces an orientation on the boundary ∂D. Suppose P, Q, and R are continuously differentiable functions on D. Then Gauss' formula gives

$$\int_{\partial D} P\,dy\,dz + Q\,dz\,dx + R\,dx\,dy = \int_D \left(\frac{\partial P}{\partial x} + \frac{\partial Q}{\partial y} + \frac{\partial R}{\partial z}\right) dx\,dy\,dz, \tag{4.4}$$

or,

$$\int_{\partial D} \varphi - \int_D d\varphi, \tag{4.5}$$

where $\varphi = P\,dy \wedge dz + Q\,dz \wedge dx + R\,dx \wedge dy$.

Example 4. Suppose Σ is an oriented surface in $\mathbb{R}^3$ whose boundary $\partial\Sigma$ is an oriented closed curve. The positive orientation of $\partial\Sigma$ along with any positive normal vector to Σ satisfy the right hand rule (assuming $\mathbb{R}^3$ to be oriented by a right-handed system). Suppose P, Q, and R are continuously differentiable functions on a domain containing Σ. Then the Stokes' formula holds:

$$\int_{\partial\Sigma} P\,dx + Q\,dy + R\,dz = \iint_\Sigma \left\{ \left(\frac{\partial R}{\partial y} - \frac{\partial Q}{\partial z}\right) dy\,dz \right. \tag{4.6}$$

$$\left. + \left(\frac{\partial P}{\partial z} - \frac{\partial R}{\partial x}\right) dz\,dx + \left(\frac{\partial Q}{\partial x} - \frac{\partial P}{\partial y}\right) dx\,dy \right\}.$$

If we denote

$$\omega = P\,dx + Q\,dy + R\,dz,$$

then the above formula can be rewritten as

$$\int_{\partial\Sigma} \omega = \int_\Sigma d\omega. \tag{4.7}$$

We see that the above four formulas assume a unified form in the notation of exterior differential forms. The Stokes formula considered in this section is the generalization of the above formulas for manifolds. First we need to explain some essential concepts.

Definition 4.1. Suppose M is an m-dimensional smooth manifold. A **region** D **with boundary** is a subset of the manifold M with two kinds of points:

    1) *Interior points*, each of which has a neighborhood in M contained in D.

    2) *Boundary points* p, for each of which there exists a coordinate chart $(U; u^i)$ such that $u^i(p) = 0$ and

$$U \cap D = \{q \in U | u^m(q) \geq 0\}. \tag{4.8}$$

A coordinate system u^i with the above property is called an **adapted coordinate system** for the boundary point p.

The set of all the boundary points of D is called the **boundary** of D, denoted by B.

Theorem 4.1. *The boundary B of a region D with boundary is a regular imbedded closed submanifold. If M is orientable, then B is also orientable.*

Proof. The boundary B of the region D is obviously a closed subset of M. Suppose $(U; u^i)$ is an adapted coordinate neighborhood. Then

$$U \cap B = \{q \in U | u^m(q) = 0\}. \tag{4.9}$$

By Definition 3.2 in Chapter 1, B is a regular imbedded closed submanifold of M.

Suppose M is an oriented manifold. Choose an adapted coordinate neighborhood $(U; u^i)$ which is consistent with the orientation of M at an arbitrary point $p \in B$. Then $(u, \ldots, u^{m-1})$ is a local coordinate system of B at the point p. Let

$$(-1)^m du^1 \wedge \cdots \wedge du^{m-1} \tag{4.10}$$

specify the orientation of the boundary B in the coordinate neighborhood $U \cap B$ of the point p. We will prove that the orientations given in this way to the coordinate neighborhoods are consistent. Suppose $(V; v^i)$ is another coordinate neighborhood of a boundary point p consistent with the orientation of M. Then

$$\frac{\partial(v^1, \ldots, v^m)}{\partial(u^1, \ldots, u^m)} > 0. \tag{4.11}$$

Suppose $v^m = f^m(u^1, \ldots, u^m)$. Then for any fixed $u^i, \ldots, u^{m-1}$ the sign of the variable v^m is the same as that of u^m, and $v^m = 0$ when $u^m = 0$. Therefore, at the point p, $\partial v^m / \partial u^m > 0$. Without loss of generality, we may assume that $v^m = u^m$. Then (4.11) becomes

$$\frac{\partial(v^1, \ldots, v^{m-1})}{\partial(u^1, \ldots, u^{m-1})} > 0. \tag{4.12}$$

This shows that $(-1)^m du^1 \wedge \cdots \wedge du^{m-1}$ and $(-1)^m dv^1 \wedge \cdots \wedge dv^{m-1}$ give consistent orientations in $U \cap V \cap B$. Hence B is orientable. $\qquad\square$

The orientation of B given in (4.10) is called the **induced orientation** on the boundary B by an oriented manifold M. If D has the same orientation as M we denote the boundary B with the induced orientation by ∂D. It is easy to verify that the orientations of ∂D and $\partial \Sigma$ in the preceding four examples are induced in this way.

Theorem 4.2 (Stokes' Formula). *Suppose D is a region with boundary in an m-dimensional oriented manifold M, and ω is an exterior differential $(m - 1)$-form on M with compact support. Then*

$$\int_D d\omega = \int_{\partial D} \omega. \tag{4.13}$$

If $\partial D = \varnothing$, then the integral on the right hand side is zero.

Proof. Suppose $\{U_i\}$ is a coordinate covering consistent with the orientation of M, and $\{g_\alpha\}$ is a subordinate partition of unity. Then

$$\omega = \sum_\alpha g_\alpha \cdot \omega. \tag{4.14}$$

Since supp ω is compact, the right hand side of the above formula is a sum of finitely many terms. Therefore

$$\begin{aligned}
\int_D d\omega &= \sum_\alpha \int_D d(g_\alpha \cdot \omega), \\
\int_{\partial D} \omega &= \sum_\alpha \int_{\partial D} g_\alpha \cdot \omega.
\end{aligned} \tag{4.15}$$

This implies that we only need to prove

$$\int_D d(g_\alpha \cdot \omega) = \int_{\partial D} g_\alpha \cdot \omega. \tag{4.16}$$

for each α. We may assume that supp ω is contained in a coordinate neighborhood $(U; u^i)$ consistent with the orientation of M. Suppose ω can be expressed as

$$\omega = \sum_{j=1}^{m} (-1)^{j-1} a_j du^1 \wedge \cdots \wedge \widehat{du^j} \wedge \cdots \wedge du^m, \qquad (4.17)$$

where the a_j are smooth functions on U. Then

$$d\omega = \left(\sum_{j=1}^{m} \frac{\partial a_j}{\partial u^j} \right) du^1 \wedge \cdots \wedge du^m. \qquad (4.18)$$

There are two cases to consider.

1) If $U \cap \partial D = \varnothing$, the right hand side of (4.13) is zero. Then either U is contained in $M - D$ or in the interior of D. In the former case, the left hand side of (4.13) is obviously zero. In the latter, we have

$$\int_D d\omega = \sum_{j=1}^{m} \int_U \frac{\partial a_j}{\partial u^j} du^1 \cdots du^m. \qquad (4.19)$$

Consider a cube $C = \{u \in \mathbb{R}^m \mid |u^i| \leq K, 1 \leq i \leq m\}$ such that U is contained in C. Extend the functions a_j to C by letting them be zero outside U. Obviously the a_j are continuously differentiable in C. Hence

$$\int_U \frac{\partial a_j}{\partial u^j} du^1 \cdots du^m = \int_C \frac{\partial a^j}{\partial u^j} du^1 \cdots du^m$$

$$= \int_{\substack{|u^i| \leq K \\ i \neq j}} \left(\int_{-K}^{+K} \frac{\partial a_j}{\partial u^j} du^j \right) du^1 \cdots du^{j-1} du^{j+1} \cdots du^m$$

$$= 0. \qquad (4.20)$$

The last integral above vanishes since

$$\int_{-K}^{+K} \frac{\partial a_j}{\partial u^j} du^j = a_j(u^1, \ldots, u^{j-1}, K, u^{j+1}, \ldots, u^m)$$

$$-a_j(u^1, \ldots, u^{j-1}, -K, u^{j+1}, \ldots, u^m) \qquad (4.21)$$

$$= 0.$$

2) If $U \cap \partial D \neq \varnothing$, we may assume that U is an adapted coordinate neighborhood consistent with the orientation of M. Then we have

$$U \cap D = \{q \in U | u^m(q) \geq 0\}, \text{ and} \qquad (4.22)$$
$$U \cap \partial D = \{q \in U | u^m(q) = 0\}. \qquad (4.23)$$

Choose a cube $C = \{u \in \mathbb{R}^m \big| |u^i| \leq K, 1 \leq i \leq m - 1; 0 \leq u^m \leq K\}$. When K is sufficiently large, $U \cap D$ lies in the union of the interior of C and the boundary $u^m = 0$. Extend a_j as in 1). Then the right hand side of (4.13) becomes

$$
\begin{aligned}
\int_{\partial D} \omega &= \int_{U \cap \partial D} \omega \\
&= \sum_{j=1}^{m} (-1)^{j-1} \int_{U \cap \partial D} a_j \, du^1 \wedge \cdots \wedge du^{j-1} \wedge du^{j+1} \wedge \cdots du^m \\
&= (-1)^{m-1} \int_{U \cap \partial D} a_m \, du^1 \wedge \cdots \wedge du^{m-1} \\
&= - \int_{\substack{|u^i| \leq K \\ 1 \leq i \leq m-1}} a_m(u^1, \ldots, u^{m-1}, 0) du^1 \cdots du^{m-1},
\end{aligned}
$$

$$(4.24)$$

where the third equality holds since $du^m = 0$ on $U \cap \partial D$, while in the last equality we have used the induced orientation on ∂D by M.

The left-hand side of (4.13) is

$$
\begin{aligned}
\int_D d\omega &= \int_{D \cap U} d\omega \\
&= \sum_{j=1}^{m} \int_{D \cap U} \frac{\partial a_j}{\partial u^j} du^1 \wedge \cdots \wedge du^m.
\end{aligned}
$$

$$(4.25)$$

But for $1 \leq j \leq m-1$,

$$\int\limits_{D \cap U} \frac{\partial a_j}{\partial u^j} du^1 \wedge \cdots \wedge du^m$$

$$= \int\limits_{\substack{|u^i| \leq K, i \neq j, m \\ 0 \leq u^m \leq K}} \left(\int_{-K}^{+K} \frac{\partial a_j}{\partial u^j} du^j \right) du^1 \cdots du^{j-1} du^{j+1} \cdots du^m \qquad (4.26)$$

$$= 0.$$

Thus there is only one nonzero term in (4.25):

$$\int\limits_{D \cap U} \frac{\partial a_m}{\partial u^m} du^1 \wedge \cdots \wedge du^m$$

$$= \int\limits_{|u^i| \leq K, i \neq m} \left(a_m(u^1, \ldots, u^{m-1}, K) - a_m(u^1, \ldots, u^{m-1}, 0) \right) du^1 \cdots du^{m-1}$$

$$= - \int\limits_{|u^i| \leq K, i \neq m} a_m(u^1, \ldots, u^{m-1}, 0) du^1 \cdots du^{m-1}.$$

$$(4.27)$$

Hence (4.13) is true, and the theorem is proved. $\qquad \qquad \Box$

Remark. In most applications, the closed region D is compact. In such cases we need not assume that the exterior differential $(m-1)$-form has a compact support for the Stokes' formula to be valid.

The Stokes formula plays an important role in physics, mechanics, partial differential equations, and differential geometry. Integration is additive with respect to the domain of integration, hence we can further define the integral of exterior differential forms on **singular chains**. [e] Viewing the integral as a pairing between an exterior differential form and a domain of integration, every exterior differential form is equivalent to a **singular cochain** on the manifold M, and the Stokes' formula describes the duality between the boundary operator ∂ and the coboundary operator d. If we denote

$$(\partial D, \omega) = \int\limits_{\partial D} \omega,$$

$$(D, d\omega) = \int\limits_{D} d\omega, \qquad (4.28)$$

---

[e] See, for example Singer and Thorpe 1976.

then the Stokes' formula becomes

$$(\partial D, \omega) = (D, d\omega). \tag{4.29}$$

Now view $A^r(M)$ as a cochain group; $d : A^r(M) \longrightarrow A^{r+1}(M)$ is the coboundary operator, and $d \circ d = 0$ (Poincare's Lemma). Denote

$$\left\{ \begin{array}{rcl} Z^r(M, \mathbb{R}) & = & \{\omega \in A^r(M) | d\omega = 0\} \\ B^r(M, \mathbb{R}) & = & \{\omega \in A^r(M) | \omega = d\beta \text{ for some } \beta \in A^{r-1}(M)\}. \end{array} \right. \tag{4.30}$$

Then $Z^r(M, \mathbb{R})$ is the kernel of the homomorphism $d : A^r(M) \longrightarrow A^{r+1}(M)$ and $B^r(M, \mathbb{R})$ is the image of the homomorphism $d : A^{r-1}(M) \longrightarrow A^r(M)$. The elements of $Z^r(M, \mathbb{R})$ are called **closed** differential forms, and the elements of $B^{(}M, \mathbb{R})$ are called **exact** differential forms. Poincare's Lemma thus implies

$$B^r(M, \mathbb{R}) \subset Z^r(M, \mathbb{R}). \tag{4.31}$$

However, a closed differential form is not necessarily exact. The de Rham Theorem stated below indicates that M must have some specific topological properties in order for closed differential forms on M to be exact.

Definition 4.2. The quotient space

$$H^r(M, \mathbb{R}) = Z^r(M, \mathbb{R})/B^r(M, \mathbb{R}) \tag{4.32}$$

is called the r-th **de Rham cohomology group** of M.

Theorem 4.3 (de Rham Theorem). *Suppose M is a compact smooth manifold. Then the r-th de Rham cohomology group is isomorphic to the r-th cohomology group of M. If we denote*

$$\dim H^r(M, \mathbb{R}) = b_r,$$

then b_r is the r-th Betti number of M.

As a corollary, if the r-th Betti number of M is zero, then any closed differential r-form on M is exact. In particular, if all the Betti numbers ($r \geq 1$) of M are zero, then any arbitrary closed differential form on M is exact.

The de Rham group $H^r(M, \mathbb{R})$ derives from the differential structure of a manifold, while the Betti number is a purely topological invariant of the manifold. Therefore the de Rham Theorem establishes a connection between the local and the global properties of the manifold. The simplest proof of the de

Rham Theorem requires the concept of a **sheaf**. For details of an elementary proof, refer to Singer and Thorpe 1976.

Finally we look at Theorem 2.6 from the viewpoint of homology theory. Any smooth map $f : M \longrightarrow N$ induces a homomorphism

$$f^* : A^r(N) \longrightarrow A^r(M)$$

which commutes with the coboundary operator d. Such a map f^* is called a **chain map**. If $\omega \in Z^r(N, \mathbb{R})$, then

$$d(f^*\omega) = f^*(d\omega) = 0.$$

Therefore $f^*\omega \in Z^r(M, \mathbb{R})$, and f^* is a homomorphism from $Z^r(N, \mathbb{R})$ to $Z^r(M, \mathbb{R})$. Similarly, f^* provides a homomorphism from $B^r(N, \mathbb{R})$ to $B^r(M, \mathbb{R})$. Hence f^* induces a homomorphism between the de Rham groups:

$$f^* : H^r(N, \mathbb{R}) \longrightarrow H^r(M, \mathbb{R}).$$

As a corollary of Theorem 2.6, we conclude that if $f : M \longrightarrow N$ is a smooth map from M to N, then it induces a homomorphism f^* from the de Rham cohomology group $H^r(N, \mathbb{R})$ to the de Rham cohomology group $H^r(M, \mathbb{R})$.

Chapter 4

Connections

In order to "differentiate" sections of a vector bundle—or vector fields on a manifold, we need to introduce a structure called the connection on a vector bundle. For example, an affine connection is a structure attached to a differentiable manifold so that we can "differentiate" its tensor fields. We first introduce the general theory of connections on vector bundles.

§4–1 Connections on Vector Bundles

We have already described the concepts of a vector bundle and a section in §3–1. Suppose E is a q-dimensional real vector bundle on M, and $\Gamma(E)$ is the set of smooth sections of E on M. $\Gamma(E)$ is a real vector space; it is also a $C^\infty(M)$-module.

Definition 1.1. A **connection** on a vector bundle E is a map

$$D : \Gamma(E) \longrightarrow \Gamma(T^*(M) \otimes E), \tag{1.1}$$

which satisfies the following conditions:

1) For any $s_1, s_2 \in \Gamma(E)$,

$$D(s_1 + s_2) = Ds_1 + Ds_2.$$

2) For $s \in \Gamma(E)$ and any $\alpha \in C^\infty(M)$,

$$D(\alpha s) = d\alpha \otimes s + \alpha Ds.$$

Suppose X is a smooth tangent vector field on M and $s \in \Gamma(E)$. Let

$$D_X s = \langle X , Ds \rangle, \tag{1.2}$$

where $\langle\,,\,\rangle$ represents the pairing between $T(M)$ and $T^*(M)$. Then $D_X s$ is a section of E, called the **absolute differential quotient** or the **covariant derivative** of the section s along X.

Remark 1. Let $\alpha = -1$. Then by condition 2) above, we have

$$D(-s) = -Ds. \tag{1.3}$$

Hence D maps a zero section to a zero section. D is thus a linear operator from $\Gamma(E)$ to $\Gamma(T^*(M) \otimes E)$.

Remark 2. D is an operator on sections of E, but it also has local properties. If s_1 and s_2 are two sections of E, and the restrictions of s_1 and s_2 to an open set U of M are the same, then the restrictions of Ds_1 and Ds_2 to U are also the same. The proof of this resembles the proof of the local properties of d (Theorem 2.1 of Chapter 3). Using the linearity of D, we need only show that if the restriction of a section s to an open set $U \subset M$ is zero, then $Ds|_U = 0$.

To see this, choose any point $p \in U$. Then there exists an open set V such that $p \in V \subset \bar{V} \subset U$. By Lemma 3 of §1–3, there exists a smooth function h on M such that

$$h(p') = \begin{cases} 1, & p' \in V, \\ 0, & p' \notin U. \end{cases}$$

Hence hs is a zero section of E, and therefore, by Remark 1 and condition 2) of connections,

$$0 = D(hs) = dh \otimes s + hDs.$$

Since $dh = 0$ on V, we have

$$Ds(p) = 0.$$

Due to the arbitrariness of p in U, $Ds|_U = 0$.

Using the local property of D, we can define D as an operator on local sections. Suppose s is a section defined on an open set $U \in M$. By Lemma 3 of §1–3, for any point $p \in U$ there is a section $\tilde{s}$ of E such that $\tilde{s}$ and s are the same in a neighborhood of p. Let

$$Ds(p) = D\tilde{s}(p). \tag{1.4}$$

Since the value of $D\tilde{s}$ near p is independent of the choice of $\tilde{s}$, Ds is a well-defined section on U.

Remark 3. By (1.2), D as a bivalent map is an operator from $\Gamma(T(M)) \times \Gamma(E)$ to $\Gamma(E)$. It satisfies the following properties. Suppose X, Y are any two smooth tangent vector fields on M, s, s_1, s_2 are sections of E, and $\alpha \in C^\infty(M)$. Then

1) $D_{X+Y}s = D_X s + D_Y s$;
2) $D_{\alpha X}s = \alpha D_X s$;
3) $D_X(s_1 + s_2) = D_X s_1 + D_X s_2$;
4) $D_X(\alpha s) = (X\alpha)s + \alpha D_X s$.

These four conditions are direct consequences of the definition of the absolute differential quotient.

As described in Remark 2, absolute differential quotient operators have the following local properties:

1) If X_1, X_2 are two tangent vector fields with the same value at a point p in M, then for any section s of E, $D_{X_1}s$ and $D_{X_2}s$ also have the same value at p. From this, we can define the absolute differential quotient of a section of E with respect to a tangent vector of M at point p. For $X \in T_p(M)$, D_X is a map from $\Gamma(E)$ to E_p.
2) For a map $D_X : \Gamma(E) \longrightarrow E_p(X \in T_p(M))$, $D_X s_1 = D_X s_2$ if the values of the sections s_1, s_2 on a parametrized curve in M that is tangent to X are the same.

The proofs of these local properties are similar to those for Remark 2. The reader may fill in the details.

Locally, a connection is given by a set of differential 1-forms. Suppose U is a coordinate neighborhood of M with local coordinates u^i, $1 \leq i \leq m$. Choose q smooth sections $s_\alpha (1 \leq \alpha \leq q)$ of E on U such that they are linearly independent everywhere. Such a set of q sections is called a **local frame field** of E on U. It is obvious that at every point $p \in U$, $\{du^i \otimes s_\alpha, 1 \leq i \leq m, 1 \leq \alpha \leq q\}$ forms a basis for the tensor space $T_p^* \otimes E_p$.

Because Ds_α is a local section on U of the bundle $T^*(M) \otimes E$, we can write

$$Ds_\alpha = \sum_{1 \leq i \leq m, 1 \leq \beta \leq q} \Gamma^\beta_{\alpha i} du^i \otimes s_\beta, \tag{1.5}$$

where $\Gamma^\beta_{\alpha i}$ are smooth functions on U. Denote

$$\omega^\beta_\alpha = \sum_{1 \leq i \leq m} \Gamma^\beta_{\alpha i} du^i. \tag{1.6}$$

Then (1.5) becomes

$$Ds_\alpha = \sum_{\beta=1}^q \omega^\beta_\alpha \otimes s_\beta. \tag{1.7}$$

We will introduce the matrix notation to simplify calculations. Use S to denote the column matrix formed by the local frame field, and ω to denote the matrix with elements ω_α^β, that is,

$$
S = \begin{pmatrix} s_1 \\ \vdots \\ s_q \end{pmatrix},
$$

$$
\omega = \begin{pmatrix} \omega_1^1 & \cdots & \omega_1^q \\ \vdots & \ddots & \vdots \\ \omega_q^1 & \cdots & \omega_q^q \end{pmatrix}. \tag{1.8}
$$

Then (1.7) can be written as

$$
DS = \omega \otimes S. \tag{1.9}
$$

The matrix ω is called the **connection matrix**; it depends on the choice of the local frame field.

If $S' = {}^t(s'_1, \ldots, s'_q)$ is another local frame field on U, then we may assume that

$$
S' = A \cdot S, \tag{1.10}
$$

where

$$
A = \begin{pmatrix} a_1^1 & \cdots & a_1^q \\ \vdots & \ddots & \vdots \\ a_q^1 & \cdots & a_q^q \end{pmatrix}.
$$

Here the a_i^j are smooth functions on U, and $\det A \neq 0$. Suppose the matrix of the connection D with respect to the local frame field S' is ω'. Then by the conditions on connections in Definition 1.1, we have

$$
\begin{aligned}
DS' &= dA \otimes S + A \cdot DS \\
&= (dA + A \cdot \omega) \otimes S \tag{1.11} \\
&= (dA \cdot A^{-1} + A \cdot \omega \cdot A^{-1}) \otimes S'.
\end{aligned}
$$

Therefore

$$\omega' = dA \cdot A^{-1} + A \cdot \omega \cdot A^{-1}. \tag{1.12}$$

This is the transformation formula for a connection matrix under a change of the local frame field. It is a most important formula in differential geometry.

Conversely, suppose a coordinate covering $\{U, W, Z, \ldots\}$ is chosen for M. On each U fix a local frame field S_U of E and assign a $q \times q$ matrix ω_U of differential 1-forms which satisfies the transformation formula (1.12) when the corresponding coordinate neighborhoods intersect. In other words, on $U \cap W \neq \varnothing$, if

$$S_W = A_{WU} \cdot S_U \tag{1.13}$$

where A_{WU} is a $q \times q$ matrix of smooth functions on $U \cap W$, then

$$\omega_W = dA_{WU} \cdot A_{WU}^{-1} + A_{WU} \cdot \omega_U \cdot A_{WU}^{-1}. \tag{1.14}$$

Then there exists a connection D on E whose matrix representation on each member U of the coordinate covering is exactly ω_U. We prove this as follows.

Suppose s is an arbitrary section of E which can be expressed in U as

$$s = a_U \cdot S_U, \tag{1.15}$$

where $a_U = (a_U^1, \ldots, a_U^q)$ and the a_U^i are smooth functions on U. Let Ds be expressed on U by

$$Ds|_U = (da_U + a_U \cdot \omega_U) \otimes S_U. \tag{1.16}$$

We will show that if $U \cap W \neq \varnothing$, then on $U \cap W$,

$$(da_U + a_U \cdot \omega_U) \otimes S_U = (da_W + a_W \cdot \omega_W) \otimes S_W. \tag{1.17}$$

Thus the Ds defined in (1.16) is a section on M.

It follows from (1.13) that, on $U \cap W$,

$$\begin{cases} a_U &= a_W \cdot A_{WU}, \\ da_U &= da_W \cdot A_{WU} + a_W \cdot dA_{WU}. \end{cases} \tag{1.18}$$

Now using (1.14), we have, on $U \cap W$,

$$\begin{aligned} &(da_U + a_U \cdot \omega_U) \otimes S_U \\ &= (da_W \cdot A_{WU} + a_W \cdot dA_{WU} + a_W \cdot A_{WU} \cdot \omega_U) \otimes (A_{WU}^{-1} \cdot S_W) \\ &= (da_W + a_W \cdot dA_{WU} \cdot A_{WU}^{-1} + a_W \cdot \omega_W - a_W \cdot dA_{WU} \cdot A_{WU}^{-1}) \otimes S_W \\ &= (da_W + a_W \cdot \omega_W) \otimes S_W. \end{aligned} \tag{1.19}$$

It is easy to verify that the map $D : \Gamma(E) \longrightarrow \Gamma(T^*(M) \otimes E)$ defined in (1.16) satisfies the conditions in Definition 1.1. Hence D is a connection on E. Obviously the connection matrix of D on U is ω_U.

Theorem 1.1. *A connection always exists on a vector bundle.*

Proof. Choose a coordinate covering $\{U_\alpha\}_{\alpha \in \mathcal{A}}$ of M. Since vector bundles are trivial locally, we may assume that there is a local frame field S_α for any U_α. By the local structure of connections, we need only construct a $q \times q$ matrix ω_α on each U_α such that the matrices constructed satisfy (1.12) under a change of the local frame field.

As discussed in §3–3, we may assume that $\{U_\alpha\}$ is locally finite, and $\{g_\alpha\}$ is a corresponding subordinate partition of unity such that $\operatorname{supp} g_\alpha \subset U_\alpha$. When $U_\alpha \cap U_\beta \neq \varnothing$, there naturally exists a nondegenerate matrix $A_{\alpha\beta}$ of smooth functions on $U_\alpha \cap U_\beta$ such that

$$S_\alpha = A_{\alpha\beta} \cdot S_\beta, \qquad \det A_{\alpha\beta} \neq 0. \tag{1.20}$$

For every $\alpha \in \mathcal{A}$, choose an arbitrary $q \times q$ matrix φ_α of differential 1-forms on U_α. Let

$$\omega_\alpha = \sum_{\beta \in \mathcal{A}} g_\beta \cdot \left(dA_{\alpha\beta} \cdot A_{\alpha\beta}^{-1} + A_{\alpha\beta} \cdot \varphi_\beta \cdot A_{\alpha\beta}^{-1} \right), \tag{1.21}$$

where the terms in the sum over β with $U_\beta \cap U_\alpha = \varnothing$ are zero. Then ω_α is a matrix of differential 1-forms on U_α. We need only demonstrate the following transformation formula for $U_\alpha \cap U_\beta \neq \varnothing$:

$$\omega_\alpha = dA_{\alpha\beta} \cdot A_{\alpha\beta}^{-1} + A_{\alpha\beta} \cdot \omega_\beta \cdot A_{\alpha\beta}^{-1}. \tag{1.22}$$

This can be done by a direct calculation. First observe that when $U_\alpha \cap U_\beta \cap U_\gamma \neq \varnothing$, the following is true in the intersection:

$$A_{\alpha\beta} \cdot A_{\beta\gamma} = A_{\alpha\gamma}.$$

Thus on $U_\alpha \cap U_\beta \neq \varnothing$ we have

$$
\begin{aligned}
A_{\alpha\beta} \cdot & \omega_\beta \cdot A_{\alpha\beta}^{-1} \\
&= \sum_{\substack{\gamma \\ U_\gamma \cap U_\alpha \cap U_\beta \neq \varnothing}} g_\gamma \cdot A_{\alpha\beta} \cdot \left(dA_{\beta\alpha} \cdot A_{\beta\alpha}^{-1} + A_{\beta\gamma} \cdot \varphi_\gamma \cdot A_{\beta\gamma}^{-1} \right) \cdot A_{\alpha\beta}^{-1} \\
&= \omega_\alpha - dA_{\alpha\beta} \cdot A_{\alpha\beta}^{-1}.
\end{aligned}
$$

This is precisely (1.22). We see from the above that there is much freedom in the choice of a connection. $\qquad \square$

In particular, if we let $\varphi_\beta = 0$ in (1.21), then we obtain a connection D on E whose connection matrix on U_α is

$$\omega_\alpha = \sum_\beta g_\beta \cdot \left(dA_{\alpha\beta} \cdot A_{\alpha\beta}^{-1} \right).$$

By the transformation formula (1.12) for connection matrices, the vanishing of a connection matrix is not an invariant property. In fact, for an arbitrary connection, we can always find a local frame field with respect to which the connection matrix is zero at some point. This fact is useful in calculations involving connections.

Theorem 1.2. *Suppose D is a connection on a vector bundle E, and $p \in M$. Then there exists a local frame field S in a coordinate neighborhood of p such that the corresponding connection matrix ω is zero at p.*

Proof. Choose a coordinate neighborhood $(U; u^i)$ of p such that $u^i(p) = 0$, $1 \leq i \leq m$. Suppose S' is a local frame field on U with corresponding connection matrix $\omega' = (\omega'^\beta_\alpha)$, where

$$\omega'^\beta_\alpha = \sum_{i=1}^m \Gamma'^\beta_{\alpha i} u^i, \tag{1.23}$$

and the $\Gamma'^\beta_{\alpha i}$ are smooth functions on U. Let

$$a^\beta_\alpha = \delta^\beta_\alpha - \sum_{i=1}^m \Gamma'^\beta_{\alpha i}(p) \cdot u^i. \tag{1.24}$$

Then $A = (a^\beta_\alpha)$ is the identity matrix at p. Hence there exists a neighborhood $V \subset U$ of p such that A is nondegenerate in V. Thus

$$S = A \cdot S' \tag{1.25}$$

is a local frame field on V. Since

$$dA(p) = -\omega'(p),$$

we can obtain from (1.12) that

$$\begin{aligned} \omega(p) &= (dA \cdot A^{-1} + A \cdot \omega' \cdot A^{-1})(p) \\ &= -\omega'(p) + \omega'(p) = 0. \end{aligned} \tag{1.26}$$

Thus S is the desired local frame field. $\qquad\square$

On exteriorly differentiating (1.12) once, we have

$$d\omega' \cdot A - \omega' \wedge dA = dA \wedge \omega + A \cdot d\omega, \tag{1.27}$$

where the exterior product $\wedge$ between the matrices means that the products of matrix elements are exterior products when the two matrices are multiplied. On plugging $dA = \omega' \cdot A - A \cdot \omega$ [which follows directly from (1.12)] into (1.27), we have

$$(d\omega' - \omega' \wedge \omega') \cdot A = A \cdot (d\omega - \omega \wedge \omega). \tag{1.28}$$

Definition 1.2. $\Omega = d\omega - \omega \wedge \omega$ is called the **curvature matrix** of the connection D on U.

Thus (1.28) can be written as

$$\Omega' = A \cdot \Omega \cdot A^{-1}. \tag{1.29}$$

This is the transformation formula for the curvature matrix when the local frame field is changed. It is worth mentioning that the transformation formula for Ω is homogeneous, while that for the connection matrix ω is not. The curvature matrix Ω contains a wealth of information. In particular, with the help of Ω, we can construct differential forms defined globally on M (cf. §7–4).

Suppose X, Y are two arbitrary tangent vector fields on M. Then the curvature matrix Ω defines a linear transformation $R(X,Y)$ from $\Gamma(E)$ to $\Gamma(E)$. Choose any two tangent vectors $X, Y \in T_p(M)$, $p \in U$. Then we can define a linear transformation $R(X,Y)$ from the fiber $\pi^{-1}(p)$ to itself by the curvature matrix Ω as follows. Suppose $s \in \pi^{-1}(p)$. Using the local frame field $S_U = {}^t(s_1, \ldots, s_q)$ of the vector bundle E on U it can be expressed as

$$s = \sum_{\alpha=1}^{q} \lambda^\alpha s_\alpha|_p, \qquad \lambda^\alpha \in \mathbb{R}.$$

Then let

$$R(X,Y)s = \sum_{\alpha,\beta=1}^{q} \lambda^\alpha \left\langle X \wedge Y, \Omega_\alpha^\beta \right\rangle s_\beta|_p. \tag{1.30}$$

Because the curvature matrix $\Omega = (\Omega_\alpha^\beta)$ transforms according to formula (1.29) under a change of the local frame field, $\left\langle X \wedge Y, \Omega_\alpha^\beta \right\rangle$ is a $(1,1)$-type tensor on the linear space $\pi^{-1}(p)$. Therefore $R(X,Y)$, as defined by (1.30), is a linear transformation from $\pi^{-1}(p)$ to itself that is independent of the choice of local coordinates.

Now if X, Y are two smooth tangent vector fields on a smooth manifold M, then $R(X, Y)$ is a linear operator on $\Gamma(E)$: for any $s \in \Gamma(E)$, $p \in M$,

$$(R(X, Y)s)(p) = R(X_p, Y_p)s_p. \tag{1.31}$$

Obviously, $R(X, Y)$ has the following properties:

1) $R(X, Y) = -R(Y, X)$;
2) $R(fX, Y) = f \cdot R(X, Y)$;
3) $R(X, Y)(fs) = f \cdot (R(X, Y)s)$,

where $X, Y \in \Gamma(T(M))$, $f \in C^\infty(M)$, $s \in \Gamma(E)$. We call $R(X, Y)$ the **curvature operator** of the connection D.

Theorem 1.3. *Suppose X, Y are two arbitrary smooth tangent vector fields on the manifold M. Then*

$$R(X, Y) = D_X D_Y - D_Y D_X - D_{[X, Y]}. \tag{1.32}$$

Proof. Because the absolute differential quotient and the curvature operator are local operators, we need only consider the operations of both sides of (1.32) on a local section. Suppose $s \in \Gamma(E)$ has the local expression

$$s = \sum_{\alpha=1}^{q} \lambda^\alpha s_\alpha.$$

Then

$$D_X s = \sum_{\alpha=1}^{q} \left(X\lambda^\alpha + \sum_{\beta=1}^{q} \lambda^\beta \langle X, \omega_\beta^\alpha \rangle \right) s_\alpha, \tag{1.33}$$

and

$$D_Y D_X s = \sum_{\alpha=1}^{q} \left\{ Y(X\lambda^\alpha) + \sum_{\beta=1}^{q} \left(X\lambda^\beta \langle Y, \omega_\beta^\alpha \rangle + Y\lambda^\beta \langle X, \omega_\beta^\alpha \rangle \right) \right.$$
$$\left. + \sum_{\beta=1}^{q} \lambda^\beta \left(Y \langle X, \omega_\beta^\alpha \rangle + \sum_{\gamma=1}^{q} \langle X, \omega_\beta^\gamma \rangle \langle Y, \omega_\gamma^\alpha \rangle \right) \right\} s_\alpha.$$

Hence

$$D_X D_Y s - D_Y D_X s = \sum_{\alpha=1}^{q} \left\{ [X, Y]\lambda^\alpha + \sum_{\beta=1}^{q} \lambda^\beta \left(\langle [X, Y], \omega_\beta^\alpha \rangle \right. \right.$$
$$\left. \left. + \left\langle X \wedge Y, d\omega_\beta^\alpha - \sum_{\gamma=1}^{q} \omega_\beta^\gamma \wedge \omega_\gamma^\alpha \right\rangle \right) \right\} s_\alpha \tag{1.34}$$

$$= D_{[X, Y]}s + \sum_{\alpha, \beta=1}^{q} \lambda^\beta \langle X \wedge Y, \Omega_\beta^\alpha \rangle s_\alpha,$$

that is,

$$R(X,Y)s = D_X D_Y s - D_Y D_X s - D_{[X,Y]}s.$$

$\square$

Theorem 1.4. *The curvature matrix Ω satisfies the **Bianchi identity***

$$d\Omega = \omega \wedge \Omega - \Omega \wedge \omega. \tag{1.35}$$

Proof. Apply exterior differentiation to both sides of $\Omega = d\omega - \omega \wedge \omega$:

$$\begin{aligned}
d\Omega &= -d\omega \wedge \omega + \omega \wedge d\omega \\
&= -(\Omega + \omega \wedge \omega) \wedge \omega + \omega \wedge (\Omega + \omega \wedge \omega) \\
&= -\Omega \wedge \omega + \omega \wedge \Omega.
\end{aligned}$$

$\square$

If a section s of a vector bundle E satisfies the condition

$$Ds = 0, \tag{1.36}$$

then s is called a **parallel section**. The zero section is obviously a parallel section; yet in general nonzero parallel sections may not exist. If we express s with respect to the local frame field S by $s = \sum_{\alpha=1}^{q} \lambda^\alpha s_\alpha$, then equation (1.36) is equivalent to

$$d\lambda^\alpha + \sum_{\beta=1}^{q} \lambda^\beta \omega_\beta^\alpha = 0, \qquad 1 \le \alpha \le q. \tag{1.37}$$

This is a Pfaffian system of equations. If we let

$$\theta^\alpha = d\lambda^\alpha + \sum_{\beta=1}^{q} \lambda^\beta \omega_\beta^\alpha, \tag{1.38}$$

then

$$d\theta^\alpha = \sum_{\beta=1}^{q} \theta^\beta \wedge \omega_\beta^\alpha + \sum_{\beta=1}^{q} \lambda^\beta \Omega_\beta^\alpha. \tag{1.39}$$

Thus we see that if the curvature matrix of the connection D is zero, then

$$d\theta^\alpha \equiv 0 \quad \mod (\theta^1, \ldots, \theta^q),$$

that is, the system (1.37) is completely integrable. In this case there exist q linearly independent parallel sections. Similarly, for (1.37) to have nonzero solutions, we need to impose certain conditions on the connection.

Definition 1.3. Suppose C is a parametrized curve in M, and X is a tangent vector field along C. If a section s of the vector bundle E on C satisfies

$$D_X s = 0, \tag{1.40}$$

then we say s is **parallel** along the curve C.

Suppose the curve C is given in a local coordinate neighborhood U of M by

$$u^i = u^i(t), \qquad 1 \leq i \leq m. \tag{1.41}$$

Then the tangent vector field of C is

$$X = \sum_{i=1}^{m} \frac{du^i}{dt} \frac{\partial}{\partial u^i}.$$

Let S be a local frame field on U. Then $s = \sum_{\alpha=1}^{q} \lambda^\alpha s_\alpha$ is a parallel section along C if and only if it satisfies the following system of equations:

$$\langle X, Ds \rangle = \sum_{\alpha=1}^{q} \left(\frac{d\lambda^\alpha}{dt} + \sum_{\beta,i} \Gamma^\alpha_{\beta i} \frac{du^i}{dt} \lambda^\beta \right) s_\alpha = 0,$$

that is,

$$\frac{d\lambda^\alpha}{dt} + \sum_{\beta,i} \Gamma^\alpha_{\beta i} \frac{du^i}{dt} \lambda^\beta = 0, \qquad 1 \leq \alpha \leq q. \tag{1.42}$$

Since (1.42) is a system of ordinary differential equations, a unique solution exists for any given initial values. Thus we see that if any vector $v \in E_p$ is given at a point p on C, then it determines uniquely a vector field parallel along C, which is called the **parallel displacement** of v along C. Obviously, the parallel displacements along C introduce isomorphisms among the fibers of the vector bundle E at different points on C.

A connection D of the vector bundle E induces a connection (also denoted by D) on the dual bundle E^*. Suppose $s \in \Gamma(E)$, $s^* \in \Gamma(E^*)$, and the pairing $\langle s, s^* \rangle$ is a smooth function on M. Then the induced connection D on E^* is determined by the equation

$$d\langle s, s^* \rangle = \langle Ds, s^* \rangle + \langle s, Ds^* \rangle, \tag{1.43}$$

where the notation $\langle \, , \, \rangle$ on the right hand side still means the pairing between E and E^*.

Let us find the matrix of the induced connection on E^*. Suppose $s_\alpha (1 \le \alpha \le q)$ is a local frame field on E, and the dual local frame field on E^* is $s^{*\beta} (1 \le \beta \le q)$, i.e.

$$\langle s_\alpha, s^{*\beta} \rangle = \delta_\alpha^\beta. \tag{1.44}$$

Let

$$Ds^{*\beta} = \sum_{\gamma=1}^{q} \omega_\gamma^{*\beta} \otimes s^{*\gamma}. \tag{1.45}$$

Then from (1.43) we have

$$\begin{aligned} \omega_\alpha^\beta &= \langle Ds_\alpha, s^{*\beta} \rangle = - \langle s_\alpha, Ds^{*\beta} \rangle \\ &= -\omega_\alpha^{*\beta}. \end{aligned}$$

Thus

$$Ds^{*\beta} = - \sum_{\alpha=1}^{q} \omega_\alpha^\beta \otimes s^{*\alpha}. \tag{1.46}$$

If the section s^* of E^* is expressed locally as

$$s^* = \sum_{\alpha=1}^{q} x_\alpha s^{*\alpha},$$

then from (1.46) we obtain

$$Ds^* = \sum_{\alpha=1}^{q} \left(dx_\alpha - \sum_{\beta=1}^{q} x_\beta \omega_\alpha^\beta \right) \otimes s^{*\alpha}. \tag{1.47}$$

Suppose connections D are separately given on the vector bundles E_1 and E_2 (denoted by the same symbol). Suppose $s_1 \in \Gamma(E_1)$, $s_2 \in \Gamma(E_2)$. Then $s_1 \oplus s_2$ and $s_1 \otimes s_2$ are sections of $E_1 \oplus E_2$ and $E_1 \otimes E_2$, respectively. Let

$$\begin{aligned} D(s_1 \oplus s_2) &= Ds_1 \oplus Ss_2, &(1.48) \\ D(s_1 \otimes s_2) &= Ds_1 \otimes s_2 + s_1 \otimes Ds_2. &(1.49) \end{aligned}$$

Then these equations determine connections on $E_1 \oplus E_2$ and $E_1 \otimes E_2$, respectively called the **induced connections** on $E_1 \oplus E_2$ and $E_1 \otimes E_2$.

§4–2 Affine Connections

The tangent bundle $T(M)$ is an m-dimensional vector bundle determined intrinsically by the differentiable structure of an m-dimensional smooth manifold M. A connection on $T(M)$ is called an **affine connection** on M. By the general discussion of connections on vector bundles in §4–1, an affine connection necessarily exists on M. A manifold with a given affine connection is called an **affine connection space**.

Suppose M is an m-dimensional affine connection space with a given affine connection D. In this section we adopt the Einstein summation convention in which all indices assume integral values ranging between 1 and m.

Choose any coordinate system $(U; u^i)$ of M. Then the natural basis $\{s_i = \partial/\partial u^i, \ 1 \leq i \leq m\}$ forms a local frame field of the tangent bundle $T(M)$ on U. Thus we may assume that

$$Ds_i = \omega_i^j \otimes s_j = \Gamma_{ik}^j du^k \otimes s_j, \tag{2.1}$$

where Γ_{ik}^j are smooth functions on U, called the **coefficients** of the connection D with respect to the local coordinates u^i.

Now we consider the effect of a change of local coordinates on the coefficients of D. Suppose $(W; w^i)$ is another coordinate system of M. Let $s_i' = \partial/\partial w^i$. Then we have the following formula on $U \cap W \neq \varnothing$:

$$S' = J_{WU} \cdot S, \tag{2.2}$$

where

$$J_{WU} = \begin{pmatrix} \dfrac{\partial u^1}{\partial w^1} & \cdots & \dfrac{\partial u^m}{\partial w^1} \\ \vdots & \ddots & \vdots \\ \dfrac{\partial u^1}{\partial w^m} & \cdots & \dfrac{\partial u^m}{\partial w^m} \end{pmatrix}$$

is the Jacobian matrix of the change of local coordinates and $S = {}^t(s_1, \ldots, s_m)$.

From (1.12) we get

$$\omega' = dJ_{WU} \cdot J_{WU}^{-1} + J_{WU} \cdot \omega \cdot J_{WU}^{-1}, \tag{2.3}$$

that is,

$$\omega_i'^j = d\left(\frac{\partial u^p}{\partial w^i}\right) \frac{\partial w^j}{\partial u^p} + \frac{\partial u^p}{\partial w^i} \frac{\partial w^j}{\partial u^q} \omega_p^q, \tag{2.4}$$

where $\omega'^j_i = \Gamma'^j_{ik} dw^k$. Thus the transformation formula for the connection coefficients Γ^j_{ik} under a change of coordinates is

$$\Gamma'^j_{ik} = \Gamma^q_{pr} \frac{\partial w^j}{\partial u^q} \frac{\partial u^p}{\partial w^i} \frac{\partial u^r}{\partial w^k} + \frac{\partial^2 u^p}{\partial w^i \partial w^k} \cdot \frac{\partial w^j}{\partial u^p}. \qquad (2.5)$$

The above formula indicates that Γ^j_{ik} is not a tensor field on M, and therefore the vanishing of connection coefficients at a point in M is not an invariant property. In other words, there may exist a coordinate system such that the connection coefficients are zero at a given point (see Theorem 2.1 below). This fact is very useful for calculations involving connections.

The main purpose of introducing affine connections on M is to be able to differentiate tensor fields. We define the concept of the absolute differential of a tensor field as follows. Suppose X is a smooth vector field on M expressed in local coordinates as

$$X = x^i \frac{\partial}{\partial u^i}. \qquad (2.6)$$

By definition, we have

$$\begin{aligned} DX &= (dx^i + x^j \omega^i_j) \otimes \frac{\partial}{\partial u^i} \\ &= x^i{}_{,j} du^j \otimes \frac{\partial}{\partial u^i}, \end{aligned} \qquad (2.7)$$

where

$$x^i{}_{,j} = \frac{\partial x^i}{\partial u^j} + x^k \Gamma^i_{kj}. \qquad (2.8)$$

DX is a section of the vector bundle $T^*(M) \otimes T(M)$, thus it is a tensor field of type $(1,1)$ on M. We call DX the **absolute differential** of X. Equation (2.8) is precisely the formula for obtaining the absolute differential quotient of a contravariant vector with respect to u^j in classical tensor analysis.

We recall that the cotangent bundle is the dual bundle of the tangent bundle and the tensor bundle T^r_s is a tensor product of tangent and cotangent bundles:

$$T^r_s(M) = \underbrace{T(M) \otimes \cdots \otimes T(M)}_{r \text{ terms}} \otimes \underbrace{T^*(M) \otimes \cdots \otimes T^*(M)}_{s \text{ terms}}. \qquad (2.9)$$

By formulas (1.43) and (1.49), an affine connection D on M induces connections on the cotangent bundle $T^*(M)$ and tensor bundle T^r_s, respectively.

Under local coordinates u^i, the local coframe field of the cotangent bundle is

$$s^{*i} = du^i, \qquad 1 \leq i \leq m, \qquad (2.10)$$

which is dual to $\{s_i = \partial/\partial u^i, 1 \leq i \leq m\}$. From (1.46) we have

$$
\begin{aligned}
Ds^{*i} &= -\omega^i_j \otimes s^{*j} \\
&= -\Gamma^i_{jk} du^k \otimes du^j.
\end{aligned}
\qquad (2.11)
$$

If a cotangent vector field α on M is expressed in local coordinates as $\alpha = \alpha_i du^i$, then

$$
\begin{aligned}
D\alpha &= (d\alpha_i - \alpha_j \omega^j_i) \otimes du^i \\
&= \alpha_{i,j} du^j \otimes du^i,
\end{aligned}
\qquad (2.12)
$$

where

$$\alpha_{i,j} = \frac{\partial \alpha_i}{\partial u^j} - \alpha_k \Gamma^k_{ij}. \qquad (2.13)$$

$D\alpha$ is a $(0,2)$-type tensor field, called the **absolute differential** of the cotangent vector field α.

In general if t is an (r,s)-type tensor field, then the image of t under the induced connection D is an $(r, s+1)$-type tensor field Dt, called the **absolute differential** of t. We will use the case $r = 2$, $s = 1$ as an example to obtain the componentwise expression of the absolute differential.

Suppose t is a $(2,1)$-type tensor field expressed as

$$t = t^{ij}_k du^k \otimes \frac{\partial}{\partial u^i} \otimes \frac{\partial}{\partial u^j} \qquad (2.14)$$

under the local coordinates u^i. From (1.49) in we have

$$
\begin{aligned}
Dt &= dt^{ij}_k \otimes du^k \otimes \frac{\partial}{\partial u^i} \otimes \frac{\partial}{\partial u^j} + t^{ij}_k D(du^k) \otimes \frac{\partial}{\partial u^i} \otimes \frac{\partial}{\partial u^j} \\
&\quad + t^{ij}_k du^k \otimes D\left(\frac{\partial}{\partial u^i}\right) \otimes \frac{\partial}{\partial u^j} + t^{ij}_k du^k \otimes \frac{\partial}{\partial u^i} \otimes D\left(\frac{\partial}{\partial u^j}\right) \\
&= \left(dt^{ij}_k - t^{ij}_l \omega^l_k + t^{lj}_k \omega^i_l + t^{il}_k \omega^j_l\right) \otimes du^k \otimes \frac{\partial}{\partial u^i} \otimes \frac{\partial}{\partial u^j} \\
&= t^{ij}_{k,h} du^h \otimes du^k \otimes \frac{\partial}{\partial u^i} \otimes \frac{\partial}{\partial u^j},
\end{aligned}
\qquad (2.15)
$$

where

$$t^{ij}_{k,h} = \frac{\partial t^{ij}_k}{\partial u^h} - t^{ij}_l \Gamma^l_{kh} + t^{lj}_k \Gamma^i_{lh} + t^{il}_k \Gamma^j_{lh}. \tag{2.16}$$

The absolute differential of a scalar field is defined to be its ordinary differential.

Definition 2.1. Suppose $C : u^i = u^i(t)$ is a parametrized curve on M, and $X(t)$ is a tangent vector field defined on C given by

$$X(t) = x^i(t) \left(\frac{\partial}{\partial u^i} \right)_{C(t)}. \tag{2.17}$$

We say that $X(t)$ is **parallel** along C if its absolute differential along C is zero, i.e., if

$$\frac{DX}{dt} = 0. \tag{2.18}$$

If the tangent vectors of a curve C are parallel along C, then we call C a **self-parallel curve,** or a **geodesic.**

Equation (2.18) is equivalent to

$$\frac{dx^i}{dt} + x^j \Gamma^i_{jk} \frac{du^k}{dt} = 0. \tag{2.19}$$

This is a system of first-order ordinary differential equations. Thus a given tangent vector X at any point on C gives rise to a parallel tangent vector field, called the **parallel displacement** of X along the curve C. By the general discussion in §4–1, we see that a parallel displacement along C establishes an isomorphism between the tangent spaces at any two points on C.

If C is a geodesic, then its tangent vector

$$X(t) = \frac{du^i(t)}{dt} \left(\frac{\partial}{\partial u^i} \right)_{C(t)}$$

is parallel along C. Therefore a geodesic curve C should satisfy:

$$\frac{d^2 u^i}{dt^2} + \Gamma^i_{jk} \frac{du^j}{dt} \frac{du^k}{dt} = 0. \tag{2.20}$$

This is a system of second-order ordinary differential equations. Thus there exists a unique geodesic through a given point of M which is tangent to a given tangent vector at that point.

We now discuss the curvature matrix Ω of an affine connection. Since

$$\omega_i^j = \Gamma_{ik}^j du^k, \tag{2.21}$$

we have

$$
\begin{aligned}
d\omega_i^j - \omega_i^h \wedge \omega_h^j &= \frac{\partial \Gamma_{ik}^j}{\partial u^l} du^l \wedge du^k - \Gamma_{il}^h \Gamma_{hk}^j du^l \wedge du^k \\
&= \frac{1}{2}\left(\frac{\partial \Gamma_{il}^j}{\partial u^k} - \frac{\partial \Gamma_{ik}^j}{\partial u^l} + \Gamma_{il}^h \Gamma_{hk}^j - \Gamma_{ik}^h \Gamma_{hl}^j \right) du^k \wedge du^l.
\end{aligned}
$$

Therefore

$$\Omega_i^j = \frac{1}{2} R_{ikl}^j du^k \wedge du^l, \tag{2.22}$$

where

$$R_{ikl}^j = \frac{\partial \Gamma_{il}^j}{\partial u^k} - \frac{\partial \Gamma_{ik}^j}{\partial u^l} + \Gamma_{il}^h \Gamma_{hk}^j - \Gamma_{ik}^h \Gamma_{hl}^j. \tag{2.23}$$

If $(W; w^i)$ is another coordinate system of M, then the local frame field on W, $S' = {}^t\left(\frac{\partial}{\partial w^1}, \ldots, \frac{\partial}{\partial w^m} \right)$, is related to S on $U \cap W$ by (2.2). By (1.29) we have

$$\Omega' = J_{WU} \cdot \Omega \cdot J_{WU}^{-1}, \tag{2.24}$$

where Ω' is the curvature matrix of the connection D under the coordinate system $(W; w^i)$. Componentwise the above equation can be written

$$\Omega_i'^j = \Omega_p^q \frac{\partial u^p}{\partial w^i} \frac{\partial w^j}{\partial u^q}.$$

Thus

$$R'_{ikl}^j = R_{prs}^q \frac{\partial w^j}{\partial u^q} \frac{\partial u^p}{\partial w^i} \frac{\partial u^r}{\partial w^k} \frac{\partial u^s}{\partial w^l}, \tag{2.25}$$

where R'_{ikl}^j is determined by

$$\Omega_i'^j = \frac{1}{2} R'_{ikl}^j dw^k \wedge dw^l.$$

Comparing (2.25) with (2.9) of Chapter 2 we observe that R_{ikl}^j satisfies the transformation rule for the components of type-$(1,3)$ tensors. Therefore

$$R = R_{ikl}^j \frac{\partial}{\partial u^j} \otimes du^i \otimes du^k \otimes du^l \tag{2.26}$$

is independent of the choice of local coordinates, and is called the **curvature tensor** of the affine connection D.

For any two smooth tangent vector fields X, Y on M we have the curvature operator $R(X, Y)$ [see (1.30)] which maps a tangent vector field on M to another tangent vector field. By Theorem 1.3, $R(X, Y)$ can be written

$$R(X, Y) = D_X D_Y - D_Y D_X - D_{[X, Y]}. \tag{2.27}$$

Now we can express $R(X, Y)$ in terms of the curvature tensor. Suppose X, Y, Z are tangent vector fields with local expressions

$$X = X^i \frac{\partial}{\partial u^i}, \quad Y = Y^i \frac{\partial}{\partial u^i}, \quad Z = Z^i \frac{\partial}{\partial u^i}. \tag{2.28}$$

Then

$$\begin{aligned} R(X, Y)Z &= Z^i \left\langle X \wedge Y, \Omega_i^j \right\rangle \frac{\partial}{\partial u^j} \\ &= R_{ikl}^j Z^i X^k Y^l \frac{\partial}{\partial u^j}. \end{aligned} \tag{2.29}$$

Thus

$$R_{ikl}^j = \left\langle R\left(\frac{\partial}{\partial u^k}, \frac{\partial}{\partial u^l}\right) \frac{\partial}{\partial u^i}, du^j \right\rangle. \tag{2.30}$$

We know that the connection coefficients Γ_{ik}^j do not satisfy the transformation rule for tensors. But if we define

$$T_{ik}^j = \Gamma_{ki}^j - \Gamma_{ik}^j, \tag{2.31}$$

then (2.5) implies

$$T_{ik}'^j = T_{pr}^q \frac{\partial w^j}{\partial u^q} \frac{\partial u^p}{\partial w^i} \frac{\partial u^r}{\partial w^k}. \tag{2.32}$$

Hence T_{ik}^j satisfies the transformation rule for the components of $(1, 2)$-type tensors. Thus

$$T = T_{ik}^j \frac{\partial}{\partial u^j} \otimes du^i \otimes du^k \tag{2.33}$$

is a $(1, 2)$-type tensor, called the **torsion tensor** of the affine connection D. By (2.31) the components of the torsion tensor T are skew-symmetric with respect to the lower indices, that is,

$$T_{ik}^j = -T_{ki}^j. \tag{2.34}$$

Being a $(1,2)$-type tensor, T can be viewed as a map from $\Gamma(T(M)) \times \Gamma(T(M))$ to $\Gamma(T(M))$. Suppose X, Y are any two tangent vector fields on M. Then $T(X,Y)$ is a tangent vector field on M with local expression

$$T(X,Y) = T^k_{ij} X^i Y^j \frac{\partial}{\partial u^k}. \tag{2.35}$$

The reader should verify that

$$T(X,Y) = D_X Y - D_Y X - [X, Y]. \tag{2.36}$$

Definition 2.2. If the torsion tensor of an affine connection D is zero, then the connection is said to be **torsion-free**.

A torsion-free affine connection always exists. In fact, if the coefficients of a connection D are Γ^j_{ik}, then set

$$\tilde{\Gamma}^j_{ik} = \frac{1}{2}\left(\Gamma^j_{ik} + \Gamma^j_{ki}\right). \tag{2.37}$$

Obviously, $\tilde{\Gamma}^j_{ik}$ is symmetric with respect to the lower indices and satisfies (2.5) under a local change of coordinates. Therefore the $\tilde{\Gamma}^j_{ik}$ are the coefficients of some connection $\tilde{D}$, and $\tilde{D}$ is torsion-free.

Any connection can be decomposed into a sum of a multiple of its torsion tensor and a torsion-free connection. In fact, (2.31) and (2.37) give

$$\Gamma^j_{ik} = -\frac{1}{2}T^j_{ik} + \tilde{\Gamma}^j_{ik}, \tag{2.38}$$

that is,

$$D_X Z = \frac{1}{2}T(X,Z) + \tilde{D}_X Z. \tag{2.39}$$

The geodesic equation (2.20) is equivalent to

$$\frac{d^2 u^i}{dt^2} + \tilde{\Gamma}^i_{jk}\frac{du^j}{dt}\frac{du^k}{dt} = 0. \tag{2.40}$$

Thus a connection D and the corresponding torsion-free connection $\tilde{D}$ have the same geodesics.

The following two theorems indicate that torsion-free affine connections have relatively desirable properties.

Theorem 2.1. *Suppose D is a torsion-free affine connection on M. Then for any point $p \in M$ there exists a local coordinate system u^i such that the corresponding connection coefficients Γ^j_{ik} vanish at p.*

Proof. Suppose $(W; w^i)$ is a local coordinate system at p with connection coefficients Γ'^j_{ik}. Let

$$u^i = w^i + \frac{1}{2}\Gamma'^i_{jk}(p)\left(w^j - w^j(p)\right)\left(w^k - w^k(p)\right).\qquad(2.41)$$

Then

$$\left.\frac{\partial u^i}{\partial w^j}\right|_p = \delta^i_j, \qquad \left.\frac{\partial^2 u^i}{\partial w^j \partial w^k}\right|_p = \Gamma'^i_{jk}(p).\qquad(2.42)$$

Thus the matrix $\left(\dfrac{\partial u^i}{\partial w^j}\right)$ is nondegenerate near p, and (2.41) provides for a change of local coordinates in a neighborhood of p. From (2.5) we see that the connection coefficients Γ^j_{ik} in the new coordinate system u^i satisfy

$$\Gamma^j_{ik}(p) = 0, \qquad 1 \le i, j, k \le m.$$

$\square$

Theorem 2.2. *Suppose D is a torsion-free affine connection on M. Then we have the Bianchi identity:*

$$R^j_{ikl,h} + R^j_{ilh,k} + R^j_{ihk,l} = 0.\qquad(2.43)$$

Proof. From Theorem 1.4 we have

$$d\Omega^j_i = \omega^k_i \wedge \Omega^j_k - \Omega^k_i \wedge \omega^j_k,$$

that is,

$$\frac{\partial R^j_{ikl}}{\partial u^h} du^h \wedge du^k \wedge du^l = \left(\Gamma^p_{ih}R^j_{pkl} - \Gamma^j_{ph}R^p_{ikl}\right) du^h \wedge du^k \wedge du^l.$$

Therefore

$$
\begin{aligned}
R^j_{ikl,h} du^h \wedge du^k \wedge du^l &= -\left(\Gamma^p_{kh}R^j_{ipl} - \Gamma^p_{lh}R^j_{ikp}\right) du^h \wedge du^k \wedge du^l \\
&= 0,
\end{aligned}
$$

where in the last equality we have used the torsion-free property of the connection. Hence

$$\left(R^j_{ikl,h} + R^j_{ilh,k} + R^j_{ihk,l}\right) du^h \wedge du^k \wedge du^l = 0.\qquad(2.44)$$

Now since the coefficients in (2.44) are skew-symmetric with respect to k, l, h, we have

$$R^j_{ikl,h} + R^j_{ilh,k} + R^j_{ihk,l} = 0.$$

$\square$

From (2.27) we know that when the second-order absolute differential quotient of a tangent vector field is computed, the effect of interchanging the order of differentiation is measured by the curvature tensor. There is a similar result for tensor fields.

First we assume that f is a scalar field on M. Then

$$f_{,i} = \frac{\partial f}{\partial u^i}, \qquad f_{,ij} = \frac{\partial^2 f}{\partial u^i \partial u^j} - \Gamma_{ij}^k f_{,k}.$$

Thus

$$f_{,ij} - f_{,ji} = T_{ij}^k f_{,k}. \tag{2.45}$$

If X is a tangent vector field on M expressed locally by

$$X = X^i \frac{\partial}{\partial u^i},$$

then

$$X^i_{,p} = \frac{\partial X^i}{\partial u^p} + X^j \Gamma_{jp}^i,$$

$$X^i_{,pq} = \frac{\partial X^i_{,p}}{\partial u^q} + X^j_{,p} \Gamma_{jq}^i - X^i_{,l} \Gamma_{pq}^l.$$

Hence

$$X^i_{,pq} - X^i_{,qp} = -X^j R_{jpq}^i + X^i_{,l} T_{pq}^l. \tag{2.46}$$

Similarly, for a $(2, 1)$-type tensor t, we have

$$t_k^{ij}{}_{,pq} - t_k^{ij}{}_{,qp} = -t_k^{lj} R_{lpq}^i - t_k^{il} R_{lpq}^j + t_l^{ij} R_{kpq}^l + t_k^{ij}{}_{,l} T_{pq}^l. \tag{2.47}$$

It is worthwhile to point out that the commutation formula for absolute differentials is completely determined by the curvature tensor and the torsion tensor.

§4–3 Connections on Frame Bundles

There is a close relationship between frame bundles and tangent bundles on differentiable manifolds.

Suppose M is an m-dimensional differentiable manifold. A **frame** refers to a combination of the form $(p; e_1, \ldots, e_m)$, where p is a point in M and $e_1, \ldots, e_m$ are m linearly independent tangent vectors at p. The set of all frames on M is denoted by P. We now introduce a differentiable structure on P so that it becomes a smooth manifold, and the natural projection

$$\pi(p; e_1, \ldots, e_m) = p \tag{3.1}$$

is a smooth map from P to M. (P, M, π) is called the **frame bundle** on M.

The procedure to introduce a differentiable structure on P is similar to that for tensor bundles. Suppose $(U; u^i)$ is any coordinate neighborhood of M. Then there is a natural frame field $\left(\dfrac{\partial}{\partial u^1}, \ldots, \dfrac{\partial}{\partial u^m} \right)$ on U. Hence any frame $(p; e_1, \ldots, e_m)$ on U can be written as

$$e_i = X_i^k \left(\frac{\partial}{\partial u^k} \right)_p, \qquad 1 \leq i \leq m, \tag{3.2}$$

where (X_i^k) is a nondegenerate $m \times m$ matrix, and therefore an element of $GL(m; \mathbb{R})$. Thus we can define a map $\varphi_U : U \times GL(m; \mathbb{R}) \longrightarrow \pi^{-1}(U)$ such that for any $p \in U$ and $(X_i^k) \in GL(m; \mathbb{R})$ we have

$$\varphi_U(p, X_i^k) = (p; e_1, \ldots, e_m), \tag{3.3}$$

where e_i is given by (3.2). Obviously φ_U is one to one.

Now choose a coordinate covering $\{U, W, Z, \ldots\}$ of M with the corresponding maps defined by (3.3) denoted by $\varphi_U, \varphi_W, \varphi_Z, \ldots$. The images of all the open sets in the topological products $U \times GL(m; \mathbb{R})$ under φ_U form a topological basis for P. With respect to this topological structure of P, the map

$$\varphi_U : U \times GL(m; \mathbb{R}) \longrightarrow \pi^{-1}(U)$$

is a homeomorphism.

Through the map φ_U, $\pi^{-1}(U)$ becomes a coordinate neighborhood in P, with the local coordinate system (u^i, X_i^k). If $U \cap W \neq \varnothing$, then M has the local change of coordinates on $U \cap W$:

$$w^i = w^i(u^1, \ldots, u^m), \qquad 1 \leq i \leq m. \tag{3.4}$$

The corresponding natural bases have the following relationship:

$$\frac{\partial}{\partial u^i} = \frac{\partial w^j}{\partial u^i} \cdot \frac{\partial}{\partial w^j}. \tag{3.5}$$

If $(p; e_1, \ldots, e_m)$ is a frame on $U \cap W$, then its coordinates (u^i, X_i^k) and (w^i, Y_i^k) under the two coordinate systems satisfy the following relation

$$\varphi_U(u^i, X_i^k) = \varphi_W(w^i, Y_i^k), \tag{3.6}$$

that is, w^i and u^i are related by (3.4), and

$$X_i^k \frac{\partial}{\partial u^k} = Y_i^k \frac{\partial}{\partial w^k},$$

or

$$Y_i^k = X_i^j \frac{\partial w^k}{\partial u^j}. \tag{3.7}$$

Thus (3.4) and (3.7) together constitute the coordinate transformation formulas for the manifold P. Obviously w^i and Y_i^k are all smooth functions of u^i and X_i^k; hence the coordinate neighborhoods $\pi^{-1}(U)$ and $\pi^{-1}(W)$ are C^∞-compatible. Therefore P becomes an $(m + m^2)$-dimensional smooth manifold, and the natural projection $\pi : P \longrightarrow M$ is a smooth surjective map.

Equation (3.3) indicates that the map φ_U provides a local product structure of the manifold P, which means $\pi^{-1}(U)$ is diffeomorphic to the direct product $U \times \mathrm{GL}(m; \mathbb{R})$. For any $p \in U$, let

$$\varphi_{U,p}(X) = \varphi_U(p, X), \qquad X \in \mathrm{GL}(m; \mathbb{R}). \tag{3.8}$$

Then $\varphi_{U,p} : \mathrm{GL}(m; \mathbb{R}) \longrightarrow \pi^{-1}(p)$ is a homeomorphism.

If $U \cap W \neq \varnothing$, for $p \in U \cap W$, the map $\varphi_{W,p}^{-1} \circ \varphi_{U,p}$ is a homeomorphism from $\mathrm{GL}(m; \mathbb{R})$ to itself. By (3.7) we know that $\varphi_{W,p}^{-1} \circ \varphi_{U,p}$ is precisely the right translation of the Jacobian matrix $J_{UW} = (\partial w^k / \partial u^j)$ on $\mathrm{GL}(m; \mathbb{R})$. Thus $\{J_{UW}\}$ forms a family of transition functions on the frame bundle. Therefore the frame bundle P is the **principal bundle** associated with the tangent bundle $T(M)$ of M; its typical fiber and structure group are both $\mathrm{GL}(m; \mathbb{R})$ (see Steenrod 1951). The frame bundle is a fiber bundle that is not a vector bundle.

Note that the structure group $\mathrm{GL}(m; \mathbb{R})$ acts on the frame bundle P in a natural way and forms a group of homeomorphisms on P. Suppose $a = (a_i^j) \in \mathrm{GL}(m; \mathbb{R})$. Then $\det a \neq 0$. The action L_a of a on P is defined by

$$L_a(p; e_1, \ldots, e_m) = (p; e_1', \ldots, e_m'), \tag{3.9}$$

where

$$e_i' = a_i^j e_j. \tag{3.10}$$

It is obvious that every L_a is a homeomorphism from P to itself, and preserves the fibers, that is,

$$\pi \circ L_a = \pi : P \longrightarrow M. \tag{3.11}$$

We call L_a the **left translation** of the element $a \in \mathrm{GL}(m; \mathbb{R})$ on P. If $a, b \in \mathrm{GL}(m; \mathbb{R})$, then it is obvious that

$$L_{ab} = L_a \circ L_b. \tag{3.12}$$

Suppose $(U; u^i)$ and $(W; w^i)$ are two coordinate systems on M with the corresponding coordinate systems (u^i, X_i^k) and (w^i, Y_i^k) on P. Use (X_i^{*k}) and (Y_i^{*k}) to denote the inverse matrices of (X_i^k) and (Y_i^k), respectively, that is,

$$
\begin{aligned}
X_i^k X_k^{*j} &= X_i^{*k} X_k^j = \delta_i^j, \\
Y_i^k Y_k^{*j} &= Y_i^{*k} Y_k^j = \delta_i^j.
\end{aligned}
$$

If $U \cap W \neq \varnothing$, then on $U \cap W$ we have

$$dw^i = \frac{\partial w^i}{\partial u^j} du^j. \tag{3.13}$$

By (3.7), we have, on the other hand,

$$X_i^{*j} = \frac{\partial w^k}{\partial u^i} Y_k^{*j}. \tag{3.14}$$

Hence

$$X_i^{*j} du^i = Y_i^{*j} dw^i. \tag{3.15}$$

This implies that the differential 1-form

$$\theta^i = X_j^{*i} du^j \tag{3.16}$$

is independent of the choice of local coordinates of P. Hence θ^i is a differential 1-form on P.

The Pfaffian system of equations

$$\theta^i = 0, \qquad 1 \leq i \leq m \tag{3.17}$$

defines an m^2-dimensional tangent subspace field V on P and determines at each point an m^2-dimensional tangent subspace called the **vertical space**. From (3.16) we obtain

$$du^i = X_j^i \theta^j.$$

Thus the system of equations (3.17) is equivalent to

$$du^i = 0, \qquad 1 \leq i \leq m, \tag{3.18}$$

in every coordinate neighborhood $\pi^{-1}(U)$, and the Pfaffian system of equations (3.17) is completely integrable. The maximal integral manifold of (3.17) is

$$u^i = \text{const}, \qquad 1 \le i \le m, \qquad (3.19)$$

which is the fiber $\pi^{-1}(p)$, $p \in M$, of P. Hence the vertical space is the tangent space of each fiber.

Now suppose M is an m-dimensional affine connection space with connection D. Suppose the connection matrix of D under the local coordinate system $(U; u^i)$ is $\omega = (\omega_i^j)$. Then the absolute differential of the vector field $e_i = X_i^k \frac{\partial}{\partial u^k}$ is

$$De_i = (dX_i^k + X_i^j \omega_j^k) \otimes \frac{\partial}{\partial u^k}.$$

If we view the X_i^k as independent variables, then

$$DX_i^k \equiv dX_i^k + X_i^j \omega_j^k \qquad (3.20)$$

is a differential 1-form on the coordinate neighborhood $\pi^{-1}(U)$ of P.

Our purpose is to find a set of differential 1-forms determined by the connection D and defined on the whole manifold P. Suppose $(W; w^i)$ is another local coordinate system of M. If $U \cap W \ne \varnothing$, then we have, on $U \cap W$,

$$Y_i^k = X_i^j \frac{\partial w^k}{\partial u^j}.$$

Thus from (2.4)

$$
\begin{aligned}
dY_i^k + Y_i^j \omega'^k_j &= dX_i^j \frac{\partial w^k}{\partial u^i} + X_i^j d\left(\frac{\partial w^k}{\partial u^j}\right) \\
&\quad + Y_i^j \left(d\left(\frac{\partial u^h}{\partial w^j}\right)\frac{\partial w^k}{\partial u^h} + \frac{\partial u^h}{\partial w^j}\omega_h^p \frac{\partial w^k}{\partial u^p}\right) \\
&= \left(dX_i^j + X_i^l \omega_l^j\right)\frac{\partial w^k}{\partial u^j},
\end{aligned}
\qquad (3.21)
$$

or

$$DY_i^k = DX_i^j \frac{\partial w^k}{\partial u^j}. \qquad (3.22)$$

By (3.14) we then have

$$Y_k^{*j} DY_i^k = X_k^{*j} DX_i^k. \qquad (3.23)$$

Hence the differential 1-form

$$\theta_i^j = X_k^{*j} D X_i^k = X_k^{*j} \left(dX_i^k + X_i^l \omega_l^k \right) \qquad (3.24)$$

is independent of the choice of the local coordinate system, and is therefore a differential 1-form on P.

Because (u^i, X_i^k) is a local coordinate system on P, (du^i, dX_i^k) are coordinates of the tangent space at a point in P. Therefore $(u^i, X_i^k; du^i, dX_i^k)$ is a local coordinate system of the tangent bundle of P. Now θ^i along with θ_i^k are $m + m^2$ differential 1-forms defined on P. They can be written as linear combinations of du^i, dX_i^k in any coordinate neighborhood $\pi^{-1}(U)$, and vice versa. Thus θ^i and θ_i^k are linearly independent everywhere, that is, $\{\theta^i, \theta_i^k\}$ forms a coframe field on the whole of P whose dual is a global frame field on P.

Generally speaking, there may not exist a global frame field on M. Since there always exists an affine connection on M, there always exists a global frame field on the frame bundle P. In this sense, the manifold P is simpler than M.

Under the local coordinate system $(U; u^i)$, we obtain from (3.16) and (3.24) that

$$du^i = X_j^i \theta^j, \qquad (3.25)$$

$$dX_i^j = -X_i^k \omega_k^j + X_k^j \theta_i^k. \qquad (3.26)$$

Exteriorly differentiating (3.25) we have

$$0 = dX_j^i \wedge \theta^j + X_j^i d\theta^j$$

$$= X_j^i \left(d\theta^j - \theta^k \wedge \theta_k^j \right) - X_k^p X_l^q \Gamma_{pq}^i \theta^l \wedge \theta^k.$$

Hence

$$d\theta^j - \theta^k \wedge \theta_k^j = X_r^{*j} X_k^p X_l^q \Gamma_{pq}^r \theta^l \wedge \theta^k$$

$$= \frac{1}{2} X_r^{*j} X_k^p X_l^q T_{pq}^r \theta^k \wedge \theta^l. \qquad (3.27)$$

Exteriorly differentiating (3.26) we get

$$0 = -dX_i^k \wedge \omega_k^j - X_i^k d\omega_k^j + dX_k^j \wedge \theta_i^k + X_k^j d\theta_i^k$$

$$= -X_i^k \Omega_k^j + X_k^j \left(d\theta_i^k - \theta_i^l \wedge \theta_l^k \right).$$

Hence

$$d\theta_i^j - \theta_i^l \wedge \theta_l^j = X_h^{*j} X_i^k \Omega_k^h$$

$$= \frac{1}{2} X_q^{*j} X_i^p X_k^r X_l^s R_{prs}^q \theta^k \wedge \theta^l. \qquad (3.28)$$

Here T^r_{pq} and R^q_{prs} are, respectively, the torsion tensor and curvature tensor defined in (2.31) and (2.23). Let

$$\begin{cases} P^j_{kl} = X^{*j}_r X^p_k X^q_l T^r_{pq}, \\ S^j_{ikl} = X^{*j}_q X^p_i X^r_k X^s_l R^q_{prs}. \end{cases} \tag{3.29}$$

Then (3.27) and (3.28) become

$$\begin{cases} d\theta^j - \theta^k \wedge \theta^j_k = \dfrac{1}{2} P^j_{kl} \theta^k \wedge \theta^l, \\ d\theta^j_i - \theta^k_i \wedge \theta^j_k = \dfrac{1}{2} S^j_{ikl} \theta^k \wedge \theta^l. \end{cases} \tag{3.30}$$

Obviously P^j_{kl} and S^j_{ikl} are independent of the choice of local coordinates. Therefore (3.30) is valid on the whole frame bundle P, and comprises the so-called **structure equations** of the connection.

For the natural frame $\left\{ \dfrac{\partial}{\partial u^i} \right\}$ we have

$$X^k_i = X^{*k}_i = \delta^k_i.$$

Therefore, (3.30) becomes

$$-du^k \wedge \omega^j_k = \frac{1}{2} T^j_{kl} du^k \wedge du^l,$$

$$d\omega^j_i - \omega^k_i \wedge \omega^j_k = \frac{1}{2} R^j_{ikl} du^k \wedge du^l,$$

restricted to the natural frame. This returns us to the original definitions of the torsion tensor and the curvature tensor.

If we denote

$$\begin{cases} \Theta^j = \dfrac{1}{2} P^j_{kl} \theta^k \wedge \theta^l, \\ \Theta^j_i = \dfrac{1}{2} S^j_{ikl} \theta^k \wedge \theta^l, \end{cases} \tag{3.31}$$

then the structure equations become

$$\begin{cases} d\theta^j - \theta^k \wedge \theta^j_k = \Theta^j, \\ d\theta^j_i - \theta^k_i \wedge \theta^j_k = \Theta^j_i. \end{cases} \tag{3.32}$$

Exteriorly differentiating once more, we obtain

$$\begin{cases} d\Theta^j + \Theta^k \wedge \theta^j_k - \theta^k \wedge \Theta^j_k = 0, \\ d\Theta^j_i + \Theta^k_i \wedge \theta^j_k - \theta^k_i \wedge \Theta^j_k = 0. \end{cases} \tag{3.33}$$

Equation (3.33) is also called the **Bianchi identity**.

We know that the differential forms θ^i are determined by the differentiable structure of M. The importance of the structure equations (3.30) is that collectively they give a sufficient condition for the m^2 differential forms θ^k_i to define an affine connection on M.

Theorem 3.1. *Suppose θ^j_i ($1 \leq i, j \leq m$) are m^2 differential 1-forms on the frame bundle P. If they and the θ^i satisfy the structure equations*

$$\begin{cases} d\theta^i - \theta^j \wedge \theta^i_j = \dfrac{1}{2}P^i_{kl}\theta^k \wedge \theta^l, \\[2mm] d\theta^j_i - \theta^k_i \wedge \theta^j_k = \dfrac{1}{2}S^j_{ikl}\theta^k \wedge \theta^l, \end{cases} \tag{3.34}$$

where P^i_{kl} and S^j_{ikl} are certain functions defined on P, then there exists an affine connection D such that θ^j_i and D are related as in (3.24).

Proof. Choose a local coordinate system (u^i, X^k_i) in P. Then

$$\theta^i = X^{*i}_k du^k, \tag{3.35}$$

where (X^{*i}_k) is the inverse matrix of (X^k_i). Therefore

$$\begin{aligned} d\theta^i &= dX^{*i}_k \wedge du^k \\ &= \left(dX^{*i}_k X^k_j\right) \wedge \theta^j \\ &= -X^{*i}_k dX^k_j \wedge \theta^j. \end{aligned}$$

Plugging this into the first equation in (3.34) we have

$$\theta^j \wedge \left(\theta^i_j + \frac{1}{2}P^i_{jk}\theta^k - X^{*i}_k dX^k_j\right) = 0. \tag{3.36}$$

Since the θ^j are linearly independent, by Cartan's Lemma, $\theta^i_j - X^{*i}_k DX^k_j$ are linear combinations of the θ^l. Thus we may assume

$$X^k_j \theta^j_i - dX^k_i = \omega^k_j X^j_i, \tag{3.37}$$

where ω^k_j are linear combinations of θ^l, and hence of du^i. Let

$$\omega^k_j = \Gamma^k_{ji}du^i, \tag{3.38}$$

where Γ^k_{ji} are functions on P. If we can show that the Γ^k_{ji} are functions of u^i only, and independent of X^j_i, then Γ^k_{ji} are the coefficients of some connection under the local coordinates u^i, and the theorem will be proved.

Exteriorly differentiating (3.37), we obtain

$$dX_j^k \wedge \theta_i^j + X_j^k d\theta_i^j = d\omega_j^k \cdot X_i^j - \omega_j^k \wedge dX_i^j.$$

Using (3.34), the above equation can be simplified to

$$X_i^j \left(d\omega_j^k - \omega_j^l \wedge \omega_l^k \right) = \frac{1}{2} X_j^k S_{ilh}^j \theta^l \wedge \theta^h.$$

Since the right hand side contains only the differentials du^i, and $\omega_j^l \wedge \omega_l^k$ also contains only the differentials du^i, $d\omega_j^k$ should contain only the differentials du^i. From (3.38), we have

$$d\omega_j^k = \sum_{i,l} \frac{\partial \Gamma_{ji}^k}{\partial u^l} du^l \wedge du^i + \sum_{i,l,h} \frac{\partial \Gamma_{ji}^k}{\partial X_l^h} dX_l^h \wedge du^i.$$

Hence

$$\frac{\partial \Gamma_{ji}^k}{\partial X_l^h} = 0. \tag{3.39}$$

Suppose $(W; w^i)$ is another coordinate neighborhood of M. Then (w^i, Y_i^k) is the local coordinate system of P in $\pi^{-1}(W)$. If $U \cap W \neq \varnothing$, then on $U \cap W$ we have

$$
\begin{aligned}
\theta_i^j &= X_k^{*j} dX_i^k + X_k^{*j} \omega_l^k X_i^l \\
&= Y_k^{*j} dY_i^k + Y_k^{*j} \omega_l'^k Y_i^l,
\end{aligned}
$$

where $\omega_l'^k = \Gamma_{lj}'^k dw^j$ and the $\Gamma_{lj}'^k$ are functions of w^j only. Plugging (3.7) into the above equation, we get

$$\omega_i'^j = d \left(\frac{\partial u^p}{\partial w^i} \right) \frac{\partial w^j}{\partial u^p} + \frac{\partial u^p}{\partial w^i} \frac{\partial w^j}{\partial u^q} \omega_p^q. \tag{3.40}$$

We see from this that (ω_i^j) indeed defines an affine connection D on M, such that (ω_i^j) is the connection matrix of D under the local coordinate system $(U; u^i)$. □

We have stated previously that the Pfaffian system of equations

$$\theta^i = 0, \qquad 1 \le i \le m$$

defines the vertical space field V on P. We call the m-dimensional tangent subspace $H(x)$ determined by the Pfaffian system of equations

$$\theta_j^i = 0, \qquad 1 \le i, j \le m \tag{3.41}$$

at each point $x \in P$ the **horizontal space**. The m-dimensional distribution H determined by (3.41) is called a **field of horizontal spaces**. Obviously the field of horizontal spaces H determined by the affine connection D on the frame bundle P has the following properties:

1) At any point $x \in P$, the tangent space $T_x(P)$ can be decomposed into the direct sum:

$$T_x(P) = V(x) \oplus H(x), \tag{3.42}$$

and the image of the horizontal space $H(x)$ under the projection π is isomorphic to the tangent space $T_p(M)$ $(p = \pi(x))$.

2) H is invariant under the left translation L_a $[a \in \mathrm{GL}(m; \mathbb{R})]$ on P, that is, for any point $x \in P$, we have

$$(L_a)_* H(x) = H(L_a(x)). \tag{3.43}$$

Since the sum of the dimensions of $V(x)$ and $H(x)$ is exactly the dimension of $T_x(P)$, $m^2 + m$, we only need to show $V(x) \cap H(x) = 0$ in order to prove (3.42). Suppose $X \in V(x) \cap H(x)$. From the definitions of vertical and horizontal spaces we have

$$\theta^i(X) = 0, \quad \theta^i_j(X) = 0, \quad 1 \le i, j \le m.$$

Since $\{\theta^i, \theta^i_j\}$ forms a coframe field on P, $X = 0$, which proves (3.42). Since the map $\pi : P \longrightarrow M$ is smooth and surjective, $\pi_* : T_x(P) \longrightarrow T_p(M)$ $(p = \pi(x))$ is a surjective homomorphism; and since $\pi_*(V(x)) = 0$, $\pi_* : H(x) \longrightarrow T_p(M)$ is an isomorphism. Hence property 1) follows.

To show property 2) we only need to express the left translation L_a in the local coordinates of the frame bundle. Suppose U is a coordinate neighborhood of M with local coordinates u^i. Then the local coordinates of the frame bundle P in the coordinate neighborhood $\pi^{-1}(U)$ are (u^i, X^i_j) [see (3.3)]. Suppose the frame $(p; e'_i)$ is the image of $(p; e_i)$ under L_a, where e'_i is given by (3.10). Let

$$e'_i = X'^j_i \frac{\partial}{\partial u^j}.$$

Then it is obvious that

$$X'^j_i = a^k_i X^j_k, \quad X'^{*j}_i = X^{*k}_i (a^{-1})^j_k, \tag{3.44}$$

where (X'^{*j}_i) is the inverse matrix of (X'^j_i). Thus

$$X'^{*j}_k D X'^k_i = a^p_i (X^{*q}_k D X^k_p)(a^{-1})^j_q,$$

that is,

$$(L_a)^* \theta_i^j = a_i^p (a^{-1})_q^j \theta_p^q. \tag{3.45}$$

Furthermore the horizontal space H is the annihilator subspace of θ_i^j ($1 \leq i, j \leq m$). Therefore (3.45) implies that the horizontal space field H is invariant under the left translation L_a.

Conversely, if we are given an m-dimensional tangent subspace H of the frame bundle P that satisfies the above two properties, then there exists a connection D on M such that H is the field of horizontal spaces of the frame bundle P with respect to the connection D (see Kobayashi and Nomizu 1963 and 1969). Therefore, from the viewpoint of the frame bundle, an affine connection is equivalent to an m-dimensional tangent subspace that satisfies the above properties.

Chapter 5

Riemannian Geometry

§5–1 The Fundamental Theorem of Riemannian Geometry

Suppose M is an m-dimensional smooth manifold, and G is a symmetric covariant tensor field of rank 2 on M. If $(U; u^i)$ is a local coordinate system on M, then the tensor field G can be expressed as

$$G = g_{ij} du^i \otimes du^j, \qquad (1.1)$$

on U, where $g_{ij} = g_{ji}$ is a smooth function on U. G provides a bilinear function on $T_p(M)$ at every point $p \in M$. Suppose $X = X^i \frac{\partial}{\partial u^i}, Y = Y^i \frac{\partial}{\partial u^i}$, then

$$G(X, Y) = g_{ij} X^i Y^j. \qquad (1.2)$$

We say that the tensor G is **nondegenerate** at the point p if, whenever $X \in T_p(M)$ and

$$G(X, Y) = 0$$

for all $Y \in T_p(M)$, it must be true that $X = 0$. This implies that G is nondegenerate at p if and only if the system of linear equations

$$g_{ij}(p) X^i = 0, \qquad 1 \le j \le m$$

has zero as its only solution, i.e., $\det(g_{ij}(p)) \ne 0$.

If for all $X \in T_p(M)$ we have

$$G(X, X) \ge 0, \qquad (1.3)$$

Refer to Chapter 8 (§8–3.2) to see the relationship between Riemannian geometry and Finsler geometry.

133

and the equality holds only if $X = 0$, then we say G is **positive definite** at p. From linear algebra a necessary and sufficient condition for G to be positive definite is that the matrix (g_{ij}) is positive definite. Thus a positive definite tensor G is necessarily nondegenerate.

Definition 1.1. If an m-dimensional smooth manifold M is given a smooth, everywhere nondegenerate symmetric covariant tensor field of rank 2, G, then M is called a **generalized Riemannian manifold,** and G is called a **fundamental tensor** or **metric tensor** of M. If G is positive definite, then M is called a **Riemannian manifold**

For a generalized Riemannian manifold M, (1.2) specifies an inner product on the tangent space $T_p(M)$ at every point $p \in M$. For any $X, Y \in T_p(M)$, let

$$X \cdot Y = G(X, Y) = g_{ij}(p) X^i Y^j. \qquad (1.4)$$

When G is positive definite, it is meaningful to define the length of a tangent vector and the angle between two tangent vectors at the same point, i.e.,

$$|X| = \sqrt{g_{ij} X^i X^j}, \qquad (1.5)$$

$$\cos \angle(X, Y) = \frac{X \cdot Y}{|X| \cdot |Y|}. \qquad (1.6)$$

Thus a Riemannian manifold is a differentiable manifold which has a positive definite inner product on the tangent space at every point. The inner product is required to be smooth: if X, Y are smooth tangent vector fields, then $X \cdot Y$ is a smooth function on M.

The differential 2-form

$$ds^2 = g_{ij} du^i du^j \qquad (1.7)$$

is independent of the choice of the local coordinate system u^i and is usually called the **metric form** or **Riemannian metric**. ds is precisely the length of an infinitesimal tangent vector, and is called the **element of arc length.** Suppose $C : u^i = u^i(t)$, $t_0 \leq t \leq t_1$, is a continuous and piecewise smooth parametrized curve on M. Then the arc length of C is defined to be

$$s = \int_{t_0}^{t_1} \sqrt{g_{ij} \frac{du^i}{dt} \frac{du^j}{dt}} dt. \qquad (1.8)$$

Theorem 1.1. *There exists a Riemannian metric on any m-dimensional smooth manifold M.*

Proof. Choose a locally finite coordinate covering $\{(U_\alpha; u^i_\alpha)\}$ of M. Suppose $\{h_\alpha\}$ is the corresponding partition of unity so that supp $h_\alpha \subset U_\alpha$. Let

$$ds^2_\alpha = \sum_{i=1}^m (du^i_\alpha)^2, \tag{1.9}$$

$$ds^2 = \sum_\alpha h_\alpha \cdot ds^2_\alpha, \tag{1.10}$$

where the $h_\alpha \cdot ds^2_\alpha$ are defined by

$$(h_\alpha \cdot ds^2_\alpha)(p) = \begin{cases} h_\alpha(p)ds^2_\alpha, & p \in U_\alpha, \\ \\ 0, & p \notin U_\alpha. \end{cases} \tag{1.11}$$

Equations (1.9) and (1.10) define smooth differential 2-forms on M. Since the right hand side of (1.10) is a sum of finitely many terms at every point $p \in M$, the formula is meaningful. In fact, if we choose a coordinate neighborhood $(U; u^i)$ such that $\overline{U}$ is compact, then U intersects only finitely many $U_{\alpha_1}, \dots, U_{\alpha_r}$ because $\{U_\alpha\}$ is locally finite. Therefore the restriction of (1.10) to U is

$$ds^2 = \sum_{\lambda=1}^r h_{\alpha_\lambda} \cdot ds^2_{\alpha_\lambda} = g_{ij}du^i du^j,$$

where

$$g_{ij} = \sum_{\lambda=1}^r \sum_{k=1}^m h_{\alpha_\lambda} \frac{\partial u^k_{\alpha_\lambda}}{\partial u^i} \frac{\partial u^k_{\alpha_\lambda}}{\partial u^j}. \tag{1.12}$$

Since $0 \leq h_\alpha \leq 1$ and $\sum_\alpha h_\alpha = 1$, there exists an index β such that $h_\beta(p) > 0$. Hence

$$ds^2(p) \geq h_\beta \cdot ds^2_\beta.$$

Thus ds^2 is positive definite everywhere on M.   □

Remark. The existence of a Riemannian metric on a smooth manifold is an extraordinary result. In general, there may not exist a non-positive definite Riemannian metric on M (which is harder to prove). In the context of fiber bundles, the existence of a Riemannian metric on M implies the existence of a positive definite smooth section of the bundle of symmetric covariant tensors of order 2 on M. However, for arbitrary vector bundles, there may not exist a smooth section which is nonzero everywhere.

Let us assume that M is a generalized Riemannian manifold in the following discussion. When the local coordinate system is changed, the transformation formula for the components of a fundamental tensor G is

$$g'_{ij} = g_{kl}\frac{\partial u^k}{\partial u'^i}\frac{\partial u^l}{\partial u'^j}.$$

Since the matrix (g_{ij}) is nondegenerate, we may denote its inverse by (g^{ij}), i.e.,

$$g^{ik}g_{kj} = g_{jk}g^{ki} = \delta^i_j. \tag{1.13}$$

Then it is easy to show that the transformation formula for g^{ij} under a change of coordinates is given by

$$g'^{ij} = g^{kl}\frac{\partial u'^i}{\partial u^k}\frac{\partial u'^j}{\partial u^l}. \tag{1.14}$$

Hence (g^{ij}) is a symmetric contravariant tensor of rank 2. (The reader should verify this).

With the help of a fundamental tensor, we may identify a tangent space with a cotangent space, and hence a contravariant vector and a covariant vector can be viewed as different expressions of the same vector. In fact, if $X \in T_p(M)$, let

$$\alpha_X(Y) = G(X,Y), \quad Y \in T_p(M). \tag{1.15}$$

Then α_X is a linear functional on $T_p(M)$, i.e., $\alpha_X \in T^*_p(M)$. Conversely, since G is nondegenerate, any element of $T^*_p(M)$ can be expressed in the form α_X. Thus α establishes an isomorphism between $T_p(M)$ and $T^*_p(M)$. Componentwise, if

$$X = X^i\frac{\partial}{\partial u^i}, \quad \alpha_X = X_i du^i,$$

then we obtain from (1.15) that

$$X_i = g_{ij}X^j, \quad X^j = g^{ij}X_i. \tag{1.16}$$

Moreover we can verify directly that if (X^i) is a contravariant vector, then the (X_i) defined in (1.16) obey the transformation rule for covariant vectors.

In general, if $(t^i{}_{jk})$ is a (1,2)-type tensor, then

$$t_{ijk} = g_{il}t^l{}_{jk}, \quad t^{ij}{}_k = g^{jl}t^i{}_{lk} \tag{1.17}$$

are $(0,3)$-type and $(2,1)$-type tensors, respectively. The operations in (1.17) are usually called the **lowering** and **raising** of tensorial indices, respectively.

Definition 1.2. Suppose (M, G) is an m-dimensional generalized Riemannian manifold, and D is an affine connection on M. If

$$DG = 0, \tag{1.18}$$

then D is called a **metric-compatible connection** on (M, G).

Condition (1.18) means that the fundamental tensor G is parallel with respect to metric-compatible connections. If the connection matrix of D under the local coordinates u^i is $\omega = (\omega_i^j)$, then

$$DG = (dg_{ij} - \omega_i^k g_{kj} - \omega_j^k g_{ik}) \otimes du^i \otimes du^j.$$

Thus (1.18) is equivalent to

$$dg_{ij} = \omega_i^k g_{kj} + \omega_j^k g_{ik}, \tag{1.19}$$

or, in matrix notation,

$$dG = \omega \cdot G + G \cdot {}^t\omega, \tag{1.20}$$

where G represents the matrix

$$G = \begin{pmatrix} g_{11} & \cdots & g_{1m} \\ \vdots & \ddots & \vdots \\ g_{m1} & \cdots & g_{mm} \end{pmatrix}, \tag{1.21}$$

and

$$\omega = \begin{pmatrix} \omega_1^1 & \cdots & \omega_1^m \\ \vdots & \ddots & \vdots \\ \omega_m^1 & \cdots & \omega_m^m \end{pmatrix}. \tag{1.22}$$

The geometric meaning of metric-compatible connections is that parallel translations preserve the metric. In particular, on a Riemannian manifold, the length of a tangent vector and the angle between two tangent vectors are invariant under parallel translations. In fact, if $X(t), Y(t)$ are parallel vector fields along the curve $C : u^i = u^i(t)$ $(1 \leq i \leq m)$ with respect to a metric-compatible connection, then

$$\begin{cases} \dfrac{dX^i}{dt} + \Gamma_{jk}^i X^j \dfrac{du^k}{dt} = 0, \\ \dfrac{dY^i}{dt} + \Gamma_{jk}^i Y^j \dfrac{du^k}{dt} = 0. \end{cases} \tag{1.23}$$

Hence

$$
\begin{aligned}
\frac{d}{dt}(g_{ij}X^iY^j) &= \frac{dg_{ij}}{dt}X^iY^j + g_{ij}\frac{dX^i}{dt}Y^j + g_{ij}X^i\frac{dY^j}{dt} \\
&= \left(\frac{dg_{ij}}{dt} - g_{ik}\Gamma^k_{jh}\frac{du^h}{dt} - g_{jk}\Gamma^k_{ih}\frac{du^h}{dt}\right)X^iY^j.
\end{aligned}
\tag{1.24}
$$

By (1.19), the right hand side of the last equality above is zero. Therefore, along C, we have

$$
g_{ij}X^iY^j = \text{const.}
\tag{1.25}
$$

Theorem 1.2 (Fundamental Theorem of Riemannian Geometry).
*Suppose M is an m-dimensional generalized Riemannian manifold. Then there exists a unique torsion-free and metric-compatible connection on M, called the **Levi-Civita connection** of M, or the **Riemannian connection** of M.*

Proof. Suppose D is a torsion-free and metric-compatible connection on M. Denote the connection matrix of D under the local coordinates u^i by $\omega = (\omega^j_i)$, where

$$
\omega^j_i = \Gamma^j_{ik}du^k.
\tag{1.26}
$$

Then we have

$$
dg_{ij} = \omega^k_i g_{kj} + \omega^k_j g_{ki},
\tag{1.27}
$$
$$
\Gamma^j_{ik} = \Gamma^j_{ki}.
\tag{1.28}
$$

Denote

$$
\Gamma_{ijk} = g_{lj}\Gamma^l_{ik}, \qquad \omega_{ik} = g_{lk}\omega^l_i.
\tag{1.29}
$$

Then it follows from (1.27) and (1.28) that

$$
\frac{\partial g_{ij}}{\partial u^k} = \Gamma_{ijk} + \Gamma_{jik},
\tag{1.30}
$$
$$
\Gamma_{ijk} = \Gamma_{kji}.
\tag{1.31}
$$

Cycling the indices in (1.30), we get

$$
\frac{\partial g_{ik}}{\partial u^j} = \Gamma_{ikj} + \Gamma_{kij},
\tag{1.32}
$$
$$
\frac{\partial g_{jk}}{\partial u^i} = \Gamma_{jki} + \Gamma_{kji}.
\tag{1.33}
$$

Calculating (1.32)+(1.33)−(1.30) and using (1.31), we then obtain

$$\Gamma_{ikj} = \frac{1}{2}\left(\frac{\partial g_{ik}}{\partial u^j} + \frac{\partial g_{jk}}{\partial u^i} - \frac{\partial g_{ij}}{\partial u^k}\right), \tag{1.34}$$

$$\Gamma_{ij}^k = \frac{1}{2}g^{kl}\left(\frac{\partial g_{il}}{\partial u^j} + \frac{\partial g_{jl}}{\partial u^i} - \frac{\partial g_{ij}}{\partial u^l}\right). \tag{1.35}$$

Thus we see that the torsion-free and metric-compatible connection is determined uniquely by the metric tensor.

Conversely, the Γ_{ij}^k defined in (1.35) indeed satisfy the transformation equation for connection coefficients under a change of local coordinates [(2.5) in Chapter 4]. Hence they define an affine connection D on M. Calculations also verify that they satisfy (1.30) and (1.31). Thus D is a torsion-free and metric-compatible connection on M. □

The existence and uniqueness of the Levi-Civita connection is a most important result in Riemannian Geometry. The Γ_{ikj} and Γ_{ij}^k defined in (1.34) and (1.35) are called **Christoffel symbols** of the first kind and second kind, respectively.

Remark. It is more convenient to use an arbitrary frame field instead of the natural frame field in a neighborhood of a Riemannian manifold. A local frame field on a manifold is a local section of the frame bundle. Suppose $(e_1, \ldots, e_m)$ is a local frame field with coframe field $(\theta^1, \ldots, \theta^m)$. Let

$$De_i = \theta_i^j e_j, \tag{1.36}$$

where $\theta = (\theta_i^j)$ is the connection matrix of the connection D with respect to the frame field $(e_1, \ldots, e_m)$. Here the θ^i, θ_i^j are none other than the forms obtained by pulling the differential 1-forms θ^i and θ_i^j on the frame bundle P back to local sections (we use the same notation here). Hence by the structure equations of connections we know that the statement that D is a torsion-free connection is equivalent to the statement that the θ_j^i satisfy the equations

$$d\theta^i - \theta^j \wedge \theta_j^i = 0. \tag{1.37}$$

If we still denote

$$g_{ij} = G(e_i, e_j), \tag{1.38}$$

then the metric form is

$$ds^2 = g_{ij}\theta^i\theta^j.$$

Since $G = g_{ij}\theta^i \otimes \theta^j$, we have

$$DG = (dg_{ij} - g_{ik}\theta_j^k - g_{kj}\theta_i^k) \otimes \theta^i \otimes \theta^j.$$

Therefore the condition for D to be a metric-compatible connection is still

$$dg_{ij} = \theta_i^k g_{kj} + \theta_j^k g_{ik}. \tag{1.39}$$

Now Theorem 1.2 can be restated as follows.

Theorem 1.3. *Suppose* (M, G) *is a generalized Riemannian manifold, and* $\{\theta^i, 1 \leq i \leq m\}$ *is a set of differential 1-forms on a neighborhood* $U \subset M$ *which is linearly independent everywhere. Then there exists a unique set of* m^2 *differential 1-forms* θ_j^k *on* U *such that*

$$d\theta^i = \theta^j \wedge \theta_j^i = 0,$$
$$dg_{ij} = \theta_i^k g_{kj} + \theta_j^k g_{ik}, \tag{1.40}$$

where the g_{ij} *are the components of* G *with respect to the local coframe field* $\{\theta^i\}$, *i.e.,*

$$G = g_{ij}\theta^i \otimes \theta^j. \tag{1.41}$$

Remark 1. Suppose M is a Riemannian manifold, and G is positive definite. Then we may choose an orthogonal frame field $\{e_i, 1 \leq i \leq m\}$ in U with $g_{ij} = \delta_{ij}$, that is,

$$ds^2 = \sum_{i=1}^{m} (\theta^i)^2.$$

The second formula in (1.40) then becomes

$$\theta_i^j + \theta_j^i = 0, \tag{1.42}$$

which implies that the connection matrix $\theta = (\theta_i^j)$ is skew-symmetric.

Remark 2. Recall that equation (1.30) can be written as

$$\frac{\partial g_{ij}}{\partial u^k} = g_{il}\Gamma_{jk}^l + g_{jl}\Gamma_{ik}^l,$$

or

$$g_{ij,k} = 0. \tag{1.43}$$

This is another form of condition (1.18). It implies that for a metric-compatible connection, the metric tensor g_{ij} behaves like a constant under absolute differentiation.

By definition, the curvature matrix of the Levi-Civita connection ω is

$$\Omega = d\omega - \omega \wedge \omega.$$

Exterior differentiation of (1.20) yields

$$
\begin{aligned}
d\omega \cdot G - \omega \wedge dG + dG \wedge {}^t\omega + G \cdot {}^t(d\omega) &= 0, \\
(d\omega - \omega \wedge \omega) \cdot G + G \cdot {}^t(d\omega - \omega \wedge \omega) &= 0,
\end{aligned}
$$

i.e.,

$$\Omega \cdot G + {}^t(\Omega \cdot G) = 0. \tag{1.44}$$

Let

$$\Omega_{ij} = \Omega_i^k g_{kj}. \tag{1.45}$$

Then $\Omega \cdot G = (\Omega_{ij})$, and (1.44) becomes

$$\Omega_{ij} + \Omega_{ji} = 0, \tag{1.46}$$

that is, Ω_{ij} is skew-symmetric with respect to the lower indices. By a direct calculation we get

$$\Omega_{ij} = d\omega_{ij} + \omega_i^l \wedge \omega_{jl}. \tag{1.47}$$

Also, by (2.22) of Chapter 4, we have

$$\Omega_i^j = \frac{1}{2} R_{ikl}^j du^k \wedge du^l, \tag{1.48}$$

where

$$R_{ikl}^j = \frac{\partial \Gamma_{il}^j}{\partial u^k} - \frac{\partial \Gamma_{ik}^j}{\partial u^l} + \Gamma_{il}^h \Gamma_{hk}^j - \Gamma_{ik}^h \Gamma_{hl}^j. \tag{1.49}$$

If we now let

$$R_{ijkl} = R_{ikl}^h g_{hj}, \tag{1.50}$$

then

$$\Omega_{ij} = \frac{1}{2} R_{ijkl} du^k \wedge du^l, \tag{1.51}$$

and

$$R_{ijkl} = \frac{\partial \Gamma_{ijl}}{\partial u^k} - \frac{\partial \Gamma_{ijk}}{\partial u^l} + \Gamma_{ik}^h \Gamma_{jhl} - \Gamma_{il}^h \Gamma_{jhk}. \tag{1.52}$$

R_{ijkl} is a covariant tensor of rank 4. It is determined completely by a given generalized Riemannian metric on M, and is called the **curvature tensor** of the generalized Riemannian manifold M. A curvature tensor has very special properties.

Theorem 1.4. *The curvature tensor R_{ijkl} of a generalized Riemannian manifold satisfies the following properties:*

   1) $R_{ijkl} = -R_{jikl} = -R_{ijlk};$
   2) $R_{ijkl} + R_{iklj} + R_{iljk} = 0;$
   3) $R_{ijkl} = R_{klij}.$

Proof. 1) This is a direct corollary of (1.46) and (1.52).
2) From the torsion-free property of the Levi-Civita connection we have

$$du^i \wedge \omega_{ij} = 0. \tag{1.53}$$

Exteriorly differentiating and using (1.47) we then have

$$du^i \wedge (\Omega_{ij} - \omega_i^l \wedge \omega_{jl}) = 0,$$
$$du^i \wedge \Omega_{ij} = 0.$$

Plugging (1.51) into the above equation we get

$$R_{jikl}du^i \wedge du^k \wedge du^l = 0,$$
$$(R_{jikl} + R_{jkli} + R_{jlik}) du^i \wedge du^k \wedge du^l = 0. \tag{1.54}$$

Since R_{ijkl} is skew-symmetric in the last two indices, the coefficients in the above equation are skew-symmetric in the last three indices. Thus

$$R_{jikl} + R_{jkli} + R_{jlik} = 0. \tag{1.55}$$

3) From (1.55) we get

$$R_{ijkl} + R_{iklj} + R_{iljk} = 0.$$

Subtracting (1.55) from the above equation we then have

$$2R_{ijkl} + R_{iklj} + R_{ljik} + R_{iljk} + R_{jkil} = 0.$$

Similarly we have

$$2R_{klij} + R_{kijl} + R_{jlki} + R_{kjli} + R_{likj} = 0.$$

Due to the skew-symmetry property 1), we finally have

$$R_{ijkl} = R_{klij}. \tag{1.56}$$

$\square$

As a corollary, under the same conditions as in Theorem 1.4, we have

$$R^i_{jkl} + R^i_{klj} + R^i_{ljk} = 0. \tag{1.57}$$

Further, from (1.43), we obtain

$$
\begin{aligned}
R_{ijkl,h} &= (g_{jp}R^p_{ikl})_{,h} \\
&= g_{jp}R^p_{ikl,h}.
\end{aligned}
$$

Thus, by using Theorem 2.2 in Chapter 4, we have

$$R_{ijkl,h} + R_{ijlh,k} + R_{ijhk,l} = 0. \tag{1.58}$$

This is also called the **Bianchi identity**.

Remark. The concept of a Riemannian manifold can be generalized to that of a Riemannian vector bundle. Suppose (E, M, π) is a real vector bundle. If for every $p \in M$, we are given a nondegenerate symmetric bilinear function G on the fiber $\pi^{-1}(p)$, and $G(X, Y)$ is a smooth function on M for any two smooth sections X, Y of E, then E is called a **generalized Riemannian vector bundle**. If G is positive definite, then E is called a **Riemannian vector bundle**, and G is called a **Riemannian structure** on E. As in Theorem 1.1 there always exists a Riemannian structure on any real vector bundle. Similarly we can define the concept of metric-compatible connections on generalized Riemannian vector bundles.

A tensor bundle on a Riemannian manifold can naturally be viewed as a Riemannian vector bundle. Take the tensor bundle $T^1_2(M)$, for example. If $a, b \in T^1_2(M)$, then the inner product of a and b is

$$a \cdot b = \sum_{\substack{i,j,k \\ l,r,s}} g^{ij} g^{kl} g_{rs} a^r_{ik} b^s_{jl}. \tag{1.59}$$

§5–2 Geodesic Normal Coordinates

Definition 2.1. Suppose M is an m-dimensional Riemannian manifold. If a parametrized curve C is a geodesic curve in M with respect to the Levi-Civita connection, then C is called a **geodesic** of the Riemannian manifold M.

Suppose the coefficients of the Levi-Civita connection D under the local coordinates u^i are Γ^i_{jk}. Then the curve $C : u^i = u^i(t)$ $(1 \le i \le m)$ is a geodesic if it satisfies the following second order differential equation:

$$\frac{d^2 u^i}{dt^2} + \Gamma^i_{jk} \frac{du^j}{dt} \frac{du^k}{dt} = 0. \tag{2.1}$$

Here $X^i = du^i/dt$ is a tangent vector of C. By definition, the tangent vector of a geodesic is parallel along the curve itself with respect to the Levi-Civita connection, which also preserves metric properties under parallel displacement. Therefore the length of the tangent vector X^i of a geodesic is constant, that is,

$$g_{ij}\frac{du^i}{dt}\frac{du^j}{dt} = \text{const.},$$

or

$$\frac{ds}{dt} = \text{const.} \tag{2.2}$$

Hence we see that the parameter for a geodesic curve in a Riemannian manifold must be a linear function of the arc length s, i.e.,

$$t = \lambda s + \mu, \tag{2.3}$$

where $\lambda\,(\neq 0)$ and μ are constants.

We now consider a special coordinate system near a point such that the coordinates of any geodesic starting from that point are linear functions of the arc length. First we discuss this under the general assumption that M is an affine connection space.

Suppose the equation of a geodesic under the coordinates system $(U; u^i)$ is

$$\frac{d^2u^i}{dt^2} + \Gamma^i_{jk}\frac{du^j}{dt}\frac{du^k}{dt} = 0. \tag{2.4}$$

By the theory of ordinary differential equations, there exist for any point $x_0 \in U$ a neighborhood $W \subset U$ of x_0 and positive numbers r, δ such that for any initial value $x \in W$ and $\alpha \in \mathbb{R}^m$ satisfying $\|\alpha\| < r$ (see footnote [a]) the system of equations (2.4) has a unique solution[b] in U:

$$u^i = f^i(t, x^k, \alpha^k), \qquad |t| < \delta, \tag{2.5}$$

that satisfies the initial conditions

$$\begin{cases} u^i(0) &= f^i(0, x^k, \alpha^k) = x^i, \\ \dfrac{du^i}{dt}(0) &= \dfrac{\partial f^i(t, x^k, \alpha^k)}{\partial t}\bigg|_{t=0} = \alpha^i. \end{cases} \tag{2.6}$$

Furthermore the functions f^i depend smoothly on the independent variable t and the initial values x^k, α^k.

---

[a] Here, $\|\alpha\|$ means $\sqrt{\sum_{i=1}^m (\alpha^i)^2}$.
[b] See Hurewicz 1966.

If we choose a non-zero constant c, then the functions $f^i(ct, x^k, \alpha^k)$, ($x \in W, \|\alpha\| < r, t < \delta/|c|$) still satisfy (2.4), and

$$\begin{cases} f^i(ct, x^k, \alpha^k)|_{t=0} = x^i, \\ \dfrac{\partial f^i(ct, x^k, \alpha^k)}{\partial t}\bigg|_{t=0} = c\alpha^i. \end{cases} \tag{2.7}$$

By the uniqueness property of the solution of (2.4), whenever $\|\alpha\|, \|c\alpha\| < r$ and $|t|, |ct| < \delta$, we have

$$f^i(ct, x^k, \alpha^k) = f^i(t, x^k, c\alpha^k). \tag{2.8}$$

But the left hand side of the above equation is always well-defined when $x \in W$, $\|\alpha\| < r$, $|t| < \delta/|c|$. Hence we can use it to define the right hand side. Thus the function $f^i(t, x^k, \alpha^k)$ is always defined for $x \in W$, $|t| < \delta/|c|$, and $\|\alpha\| < |c|r$. In particular, we can choose $|c| < \delta$, so that $f^i(t, x^k, \alpha^k)$ is defined for $x \in W$, $|t| \leq 1$, and $\|\alpha\| < |c|r$. Let

$$u^i = f^i(1, x^k, \alpha^k). \tag{2.9}$$

Then

$$f^i(1, x^k, 0) = f^i(0, x^k, \alpha^k) = x^i. \tag{2.10}$$

Thus for a fixed $x \in W$, (2.9) provides a smooth map from a neighborhood of the origin in the tangent space $T_x(M)$ ($= \mathbb{R}^m$) to a neighborhood of x in the manifold M. Because

$$\frac{\partial f^i(1, x^k, t\alpha^k)}{\partial t}\bigg|_{t=0} = \frac{\partial f^i(1, x^k, \alpha^k)}{\partial \alpha^j}\bigg|_{\alpha=0} \cdot \alpha^j,$$

and, on the other hand,

$$\frac{\partial f^i(1, x^k, t\alpha^k)}{\partial t}\bigg|_{t=0} = \frac{\partial f^i(t, x^k, \alpha^k)}{\partial t}\bigg|_{t=0} = \alpha^i,$$

we have

$$\left(\frac{\partial u^i}{\partial \alpha^j}\right)_{\alpha=0} = \delta^i_j, \tag{2.11}$$

that is, the map (2.9) is regular at the origin $\alpha = 0$. Hence the α^i can be chosen to be local coordinates of x in M, called the **geodesic normal coordinates** of x, or simply **normal coordinates**. Since a tangent space is a linear space, and any two coordinate systems on it are the same up to a nondegenerate linear transformation, a normal coordinate system of a point in M is determined up to a nondegenerate linear transformation.

Fix $\alpha^k = \alpha_0^k$. As t changes, $t\alpha_0^k$ describes a straight line in $T_x(M)$ starting from the origin, and traces a geodesic curve on the manifold starting from x and tangent to the tangent vector (α_0^k). Therefore the equation for this geodesic curve under the normal coordinate system α^i is

$$\alpha^k = t\alpha_0^k, \tag{2.12}$$

where α_0^k is a constant.

Theorem 2.1. *If M is a torsion-free affine connection space, then with respect to a normal coordinate system α^i at the point x, the connection coefficients Γ_{ik}^j are zero at x.*

Proof. Since the geodesic curve $\alpha^i = t\alpha_0^i$ satisfies (2.4) under the normal coordinate system α^i, we have, for any α_0^k,

$$\Gamma_{jk}^i(0)\alpha_0^j\alpha_0^k = 0. \tag{2.13}$$

Since Γ_{jk}^i is symmetric in the lower indices for torsion-free connections, we have

$$\Gamma_{jk}^i(0) = 0, \qquad 1 \le i, j, k \le m. \tag{2.14}$$

$\square$

Theorem 2.2. *For any point x_0 in an affine connection space M, there exists a neighborhood W of x_0 such that every point in W has a normal coordinate neighborhood that contains W.*

Proof. Suppose $(U; u^i)$ is a normal coordinate system at a point x_0. Let

$$U(x_0; \rho) = \left\{ x \in U \mid \sum_{i=1}^m (u^i(x))^2 < \rho^2 \right\}. \tag{2.15}$$

By the above discussion on the solution of equation (2.4), there exists a neighborhood $W = U(x_0; r)$ of x_0 and a positive number δ such that for any $x \in W$ and $\alpha \in \mathbb{R}^m$, $\|\alpha\| < \delta$, there is a unique geodesic curve

$$u^i = f^i(t, x^k, \alpha^k), \qquad |t| < 2, \tag{2.16}$$

with initial condition (x, α^k). We use $B(0; \delta)$ to denote the set $\{\alpha \in \mathbb{R}^m \mid \|\alpha\| < \delta\}$. Then (2.16) gives a map $\varphi : W \times B(0; \delta) \longrightarrow W \times U$ such that

$$\varphi(x, \alpha) = (x^k, f^k(1, x^i, \alpha^i)), \tag{2.17}$$

where $x \in W$, $\alpha \in B(0; \delta)$. Since the function f^k depends on x and α smoothly, the map φ is smooth. By (2.11) we have

$$\frac{\partial(x^k, f^k)}{\partial(x^i, \alpha^i)}\bigg|_{(x_0, 0)} = 1.$$

Thus the Jacobian matrix of the map φ is nondegenerate near the point $(x_0, 0) \in W \times B(0; \delta)$. By the Inverse Function Theorem there exists a neighborhood V of the point $(x_0, 0) \in W \times B(0; \delta)$ and a positive number $a < \delta$ such that $\varphi : V \longrightarrow U(x_0; a) \times U(x_0; a)$ is a diffeomorphism. For any $x \in U(x_0; a)$, let

$$B_x = \{\alpha \in B(0; a)| \text{ such that } (x, \alpha) \in V\}. \tag{2.18}$$

Then the map

$$u^i = f^i(1, x^k, \alpha^k), \qquad \alpha \in B_x, \tag{2.19}$$

given by (2.16) is a diffeomorphism from B_x to $U(x_0; a)$. Choose $W' = U(x_0; a)$. Then the above formula shows that for any point in W' there exists a normal coordinate neighborhood that contains W'. □

Corollary. *For every point x_0 in an affine connection space M, there exists a neighborhood W of x_0 such that any two points in W can be connected by a geodesic curve.*

Remark. A more delicate argument will show that this geodesic curve is contained in the neighborhood W. Then W is called a **geodesic convex neighborhood**. Theorem 2.7 below proves the existence of geodesic convex neighborhoods under the assumptions for Riemannian manifolds. With some adjustment the proof can be applied to an affine connection space.

Theorem 2.3. *A torsion-free affine connection is completely determined locally by the curvature tensor.*

Proof. Consider a normal coordinate system α^i at a fixed point O. Choose a natural frame at O, and parallelly displace the frame along the geodesic curves starting from O. Thus we get a frame field $\{e_i, 1 \le i \le m\}$ in a neighborhood of O. Let θ^i be the dual differential 1-forms of e_j, and denote the restriction of the everywhere linearly independent m^2 differential 1-forms θ_i^j of the frame bundle to the frame field described as above by the same notation. Then θ^i, θ_i^j are differential 1-forms of t, α^k. When the α^k are constants, θ^i, θ_i^j are restricted to the geodesic curve $\alpha^i t$. Since the frame field is parallel along the geodesic curve $\alpha^i t$, we have

$$\begin{cases} \theta^i \equiv \alpha^i dt (\bmod d\alpha^k), \\ \theta_i^j \equiv 0 (\bmod d\alpha^k), \end{cases} \tag{2.20}$$

or

$$\theta^i = \alpha^i dt + \bar{\theta}^i, \qquad \theta_i^j = \bar{\theta}_i^j, \qquad (2.21)$$

where $\bar{\theta}^i$ and $\bar{\theta}_i^j$ are the parts of θ^i and θ_i^j without dt. Plugging (2.21) into the structure equations (§4–3)

$$\begin{cases} d\theta^i - \theta^j \wedge \theta_j^i &= 0, \\ d\theta_i^j - \theta_i^k \wedge \theta_k^j &= \dfrac{1}{2} S_{ikl}^j \theta^k \wedge \theta^l, \end{cases}$$

and comparing the terms with dt, we get

$$\left(d\alpha^i - \frac{\partial \bar{\theta}^i}{\partial t} + \alpha^j \bar{\theta}_j^i \right) \wedge dt = 0,$$

$$\left(\frac{\partial \bar{\theta}_i^j}{\partial t} - \alpha^k S_{ikl}^j \bar{\theta}^l \right) \wedge dt = 0,$$

where $\partial \bar{\theta}^i / \partial t$, $\partial \bar{\theta}_i^j / \partial t$ represent the differential 1-forms obtained by taking the partial derivatives of the coefficients of $\bar{\theta}^i$, $\bar{\theta}_i^j$, respectively, with respect to t. Since dt does not occur inside the parentheses, we have

$$\begin{cases} \dfrac{\partial \bar{\theta}^i}{\partial t} &= d\alpha^i + \alpha^j \bar{\theta}_j^i, \\ \dfrac{\partial \bar{\theta}_i^j}{\partial t} &= \alpha^k S_{ikl}^j \bar{\theta}^l. \end{cases} \qquad (2.22)$$

This is a system of ordinary differential equations with t as the independent variable. Differentiating the first formula with respect to t again, we obtain

$$\frac{\partial^2 \bar{\theta}^i}{\partial t^2} = \alpha^j \frac{\partial \bar{\theta}_j^i}{\partial t} = \alpha^j \alpha^k S_{jkl}^i \bar{\theta}^l. \qquad (2.23)$$

Since the frame field e_i is parallel along any direction at the point O, we have

$$\bar{\theta}_i^j \Big|_{t=0} = 0. \qquad (2.24)$$

Moreover, by definition, we have

$$\theta^i \big|_{t=0} = \alpha^i dt. \qquad (2.25)$$

Thus

$$\bar{\theta}^i\big|_{t=0} = 0. \tag{2.26}$$

Hence, from (2.24) and the first formula of (2.22), we obtain

$$\frac{\partial \bar{\theta}^i}{\partial t}\bigg|_{t=0} = d\alpha^i. \tag{2.27}$$

For a given curvature tensor, the system of second-order ordinary differential equations (2.23) has a unique solution for $\bar{\theta}^i$ under the initial conditions (2.26) and (2.27), and $\bar{\theta}^j_i$ is completely determined by the first formula in (2.22). Hence the curvature tensor completely determines the torsion-free affine connection locally. □

Now assume M is am m-dimensional Riemannian manifold. Suppose $x_0 \in M$, and choose a fixed orthogonal frame F_0 in the tangent space $T_{x_0}(M)$. Then the normal coordinate system u^i at x_0 can be expressed as

$$u^i = \alpha^i s, \tag{2.28}$$

where (α^i) is a unit vector in $T_{x_0}(M)$ and s is the arc length of the geodesic curve starting from x_0. Displace the frame F_0 parallelly along geodesic curves originating from x_0 to obtain an orthogonal frame field in a neighborhood of x_0. By the proof of Theorem 2.3, we can write

$$\begin{cases} \theta^i &= \alpha^i ds + \bar{\theta}^i, \\ \theta^j_i &= \bar{\theta}^j_i, \end{cases} \tag{2.29}$$

where $\bar{\theta}^i$, $\bar{\theta}^j_i$ do not contain the differential ds, and satisfy the equations

$$\begin{cases} \dfrac{\partial \bar{\theta}^i}{\partial s} &= d\alpha^i + \alpha^j \bar{\theta}^i_j, \\[2mm] \dfrac{\partial \bar{\theta}^j_i}{\partial s} &= \alpha^k S^j_{ikl} \bar{\theta}^l, \\[2mm] \bar{\theta}^j_i + \bar{\theta}^i_j &= 0, \end{cases} \tag{2.30}$$

with initial conditions

$$\bar{\theta}^i\big|_{s=0} = 0, \quad \bar{\theta}^j_i\big|_{s=0} = 0, \quad \frac{\partial \bar{\theta}^i}{\partial s}\bigg|_{s=0} = d\alpha^i. \tag{2.31}$$

If we write

$$\bar{\theta}^i = s\,d\alpha^i + A^i_j d\alpha^j, \tag{2.32}$$

then the A^i_j satisfy the initial conditions

$$A^i_j|_{s=0} = 0, \qquad \frac{\partial A^i_j}{\partial s}\bigg|_{s=0} = 0. \tag{2.33}$$

Thus the arc length element near the point O can be expressed by

$$
\begin{aligned}
d\sigma^2 &= \sum_{i=1}^m (\theta^i)^2 \\
&= ds^2 + 2ds \sum_{i=1}^m \alpha^i \bar{\theta}^i + \sum_{i=1}^m (\bar{\theta}^i)^2.
\end{aligned}
\tag{2.34}
$$

Since $\sum_{i=1}^m \alpha^i d\alpha^i = 0$, $\bar{\theta}^i_j + \bar{\theta}^j_i = 0$, it is easy to see that

$$\frac{\partial}{\partial s}\left(\sum_{i=1}^m \alpha^i \bar{\theta}^i\right) = \sum_{j=1}^m \alpha^i \left(d\alpha^i + \sum_{i=1}^m \alpha^j \bar{\theta}^i_j\right) = 0,$$

and

$$\sum_{i=1}^m \alpha^i \bar{\theta}^i\bigg|_{s=0} = 0.$$

Thus

$$\sum_{i=1}^m \alpha^i \bar{\theta}^i = 0. \tag{2.35}$$

Hence, from (2.34), the arc length element near O is

$$d\sigma^2 = ds^2 + \sum_{i=1}^m (\bar{\theta}^i)^2. \tag{2.36}$$

Corollary. *A hypersurface $s = $ constant is orthogonal to geodesic curves starting from the point O.*

Theorem 2.4. *For every point O in a Riemannian manifold M, there exists a normal coordinate neighborhood W such that*

1) *Every point in W has a normal coordinate neighborhood that contains W.*

2) *The geodesic curve that connects O and $p \in W$ is the unique shortest curve in W connecting these two points.*

Proof. Apply Theorem 2.2 to the Levi-Civita connection of M, and 1) follows. Now assume that u^i is the normal coordinate system of the point O given by (2.28). A normal coordinate neighborhood W as required in 1) is

$$W = \left\{ p \in M \left| \sum_{i=1}^{m} (u^i(p))^2 < \epsilon^2 \right. \right\},$$

where ϵ is a sufficiently small positive number. Because W is a normal coordinate neighborhood, for any point $p \in W$ there exists a unique geodesic curve γ in W that connects O and p. Suppose the length of γ is s_0.

First, we show that γ is the shortest path in W that connects O and p. Suppose C is any piecewise smooth curve in W that connects O and p. We may assume that the parametrized equation for C is $u^i = u^i(s)$, where s is the arc length parameter of γ. Then the arc length of C is

$$\int_0^{s_0} d\sigma = \int_0^{s_0} \sqrt{ds^2 + \sum_{i=1}^{m} (\theta^i)^2} \geq \int_0^{s_0} ds = s_0. \tag{2.37}$$

If C is the shortest path in W connecting O and p, then the equality in (2.37) holds. Hence we must have, along the curve C,

$$\bar{\theta}^i = 0.$$

By (2.32) we have

$$d\alpha^i + \sum_{j=1}^{m} A_j^i \frac{d\alpha^j}{s} = 0. \tag{2.38}$$

Since the A_j^i satisfy the initial conditions (2.33), $A_j^i = o(s)$. Letting $s \to 0$ in (2.38), we then have

$$d\alpha^i = 0, \qquad \alpha^i = \text{const.},$$

that is, C is a geodesic curve connecting O and p. Hence $C = \gamma$. $\square$

Theorem 2.5. *Suppose U is a normal coordinate neighborhood of the point O. Then there exists a positive number ϵ such that, for any $0 < \delta < \epsilon$, the hypersphere*

$$\Sigma_\delta = \left\{ p \in U \left| \sum_{i=1}^{m} (u^i(p))^2 = \delta^2 \right. \right\}$$

has the following properties:

1) *Every point on Σ_δ can be connected to O by a unique shortest geodesic curve in U.*

2) *Any geodesic curve tangent to Σ_δ is strictly outside Σ_δ in a neighborhood of the tangent point.*

Proof. Choose W to be a normal coordinate neighborhood as required in Theorem 2.4. We may assume that W is a spherical neighborhood with radius ϵ:

$$W = \left\{ p \in U \,\bigg|\, \sum_{i=1}^{m} (u^i(p))^2 < \epsilon^2 \right\}.$$

When $0 < \delta < \epsilon$, since $\Sigma_\delta \subset W \subset U$ and U is a normal coordinate neighborhood, property 1) is a corollary of Theorem 2.4. Now we decrease ϵ so that 2) is true.

Since $(U; u^i)$ is a normal coordinate system, it follows from Theorem 2.1 that

$$\Gamma^j_{ik}(0) = 0. \tag{2.39}$$

The equation for the hypersphere Σ_δ can be written as

$$F(u^1, \dots, u^m) = \frac{1}{2}[(u^1)^2 + \cdots + (u^m)^2 - \delta^2] = 0. \tag{2.40}$$

Suppose γ is a geodesic curve tangent to Σ_δ at p, and its equation is

$$u^i = u^i(\sigma), \tag{2.41}$$

where σ is the arc length of γ measured from the point p. Then

$$F(u^i(\sigma))|_{\sigma=0} = 0. \tag{2.42}$$

Since the hypersphere Σ_δ is orthogonal to geodesic curves starting from the point O, the geodesic curve γ tangent to Σ_δ at the point p should be orthogonal to the geodesic curve connecting O and p. Therefore

$$\sum_{i=1}^{m} u^i(\sigma) \frac{du^i}{d\sigma}\bigg|_{\sigma=0} = 0. \tag{2.43}$$

From direct calculation we obtain

$$\frac{d}{d\sigma} F(u^i(\sigma))\bigg|_{\sigma=0} = \sum_{i=1}^{m} u^i(\sigma) \frac{du^i}{d\sigma}\bigg|_{\sigma=0} = 0, \tag{2.44}$$

$$\frac{d^2}{d\sigma^2}F(u^i(\sigma))\Big|_{\sigma=0} = \sum_{i,j=1}^{m}\left[\delta_{ij} - \sum_{k=1}^{m}u^k(p)\Gamma_{ij}^k(p)\right]\left(\frac{du^i}{d\sigma}\right)_0\left(\frac{du^j}{d\sigma}\right)_0. \quad (2.45)$$

Hence

$$F(u^i(\sigma)) = \frac{1}{2}\sum_{i,j=1}^{m}\left[\delta_{ij} - \sum_{k=1}^{m}u^k(p)\Gamma_{ij}^k(p)\right]\left(\frac{du^i}{d\sigma}\right)_0\left(\frac{du^j}{d\sigma}\right)_0 \cdot \sigma^2 + o(\sigma^2).$$
$$(2.46)$$

Because of (2.39), we can make the values of Γ_{ij}^k arbitrarily small in a neighborhood of O. Hence we can choose a sufficiently small ϵ such that when $0 < \delta < \epsilon$, (2.45) is always positive. Thus the geodesic curve (2.41) lies strictly outside Σ_δ near p, and has only one point in common with Σ_δ, namely p. $\quad\square$

When studying the geometry of a Riemannian manifold, an effective technique is to introduce a distance function on the manifold so that it becomes a metric space.

Definition 2.2. Suppose M is a connected Riemannian manifold, and p, q are two arbitrary points in M. Let

$$\rho(p,q) = \inf \widehat{pq}, \quad (2.47)$$

where $\widehat{pq}$ denotes the arc length of a curve connecting p and q with measurable arc length. Then $\rho(p,q)$ is called the **distance** between points p and q.

Because M is connected, there always exists a curve connecting p and q with measurable arc length. Therefore (2.47) is always meaningful, and defines a real function on $M \times M$.

Theorem 2.6. *The function* $\rho : M \times M \longrightarrow \mathbb{R}$ *has the following properties:*

1) *for any* $p, q \in M$, $\rho(p,q) \geq 0$, *and the equality holds only when* $p = q$;
2) $\rho(p,q) = \rho(q,p)$;
3) *for any three points* $p, q, r \in M$ *we have*

$$\rho(p,q) + \rho(q,r) \geq \rho(p,r).$$

Therefore ρ becomes a distance function on M and makes M a metric space. The topology of M as a metric space and the original topology of M as a manifold are equivalent.

Proof. According to definition (2.47), the above properties are obvious. We need only show that $\rho(p, q) > 0$ whenever $p \neq q$.

Suppose p, q are any two points in M, $p \neq q$. Since M is a Hausdorff space, there exists a neighborhood U of p such that $q \notin U$. By Theorem 2.4, there must exist a normal coordinate neighborhood $W \subset U$ of p such that its normal coordinates are $u^i = \alpha^i s$, where $\sum_{i=1}^{m} (\alpha^i)^2 = 1$ and $0 \leq s \leq s_0$. Choose δ such that $0 < \delta < s_0$. Then the hypersurface $\Sigma_\delta \subset W$. Suppose γ is a measurable curve connecting p and q. Then the length of γ is at least δ, that is

$$\rho(p, q) \geq \delta > 0.$$

By Theorem 2.5, the interior of Σ_δ is precisely the set

$$\{q \in M | \rho(p, q) < \delta\},$$

that is, the interior of Σ_δ is a δ-ball neighborhood of p when M is viewed as a metric space. Thus the topology of M viewed as a metric space and the original topology of M are equivalent. $\square$

We note that if W is a ball-shaped normal coordinate neighborhood at the point O constructed as in Theorem 2.4, then for any point $p \in W$ the unique geodesic curve connecting O and p in W has length $\rho(O, p)$.

Theorem 2.7. *There exists a η-ball neighborhood W at any point p in a Riemannian manifold M, where η is a sufficiently small positive number, such that any two points in W can be connected by a unique geodesic curve.*

Any neighborhood satisfying the above property is called a **geodesic convex neighborhood**. Thus the theorem states that there exists a geodesic convex neighborhood at every point in a Riemannian manifold.

Proof. Suppose $p \in M$. By Theorem 2.4 there exists a ball-shaped normal coordinate neighborhood U of p with radius ϵ such that for any point q in U there is a normal coordinate neighborhood V_q that contains U. We may assume that ϵ also satisfies the requirements of Theorem 2.5. Choose a positive number $\eta \leq \frac{1}{4}\epsilon$. Then the η-ball neighborhood W of p is a geodesic convex neighborhood of p.

Choose any $q_1, q_2 \in W$. Then

$$\rho(q_1, q_2) \leq \rho(p, q_1) + \rho(p, q_2) < 2\eta \leq \frac{\epsilon}{2}. \tag{2.48}$$

Suppose $U(q_1; \epsilon/2)$ is an $\epsilon/2$-ball neighborhood of q_1. Then the above formula indicates that $q_2 \in U(q_1; \epsilon/2)$. For any $q \in U(q_1; \epsilon/2)$ we have

$$\rho(p, q) \leq \rho(p, q_1) + \rho(q_1, q) < \frac{3\epsilon}{4}.$$

Hence

$$U\left(q_1; \frac{\epsilon}{2}\right) \subset U \subset V_{q_1}, \tag{2.49}$$

that is, the $\epsilon/2$-ball neighborhood of q_1 is contained in the normal coordinate neighborhood of q_1. By Theorem 2.4 and the statement immediately following the proof of Theorem 2.6, there exists a unique geodesic curve γ in $U(q_1; \epsilon/2)$ connecting q_1 and q_2, whose length is precisely $\rho(q_1, q_2)$. In particular, if $r \in \gamma$, then

$$\rho(q_1, r) \leq \rho(q_1, q_2). \tag{2.50}$$

Finally we prove that the geodesic curve γ lies inside W. Since $\gamma \subset U(q_1; \epsilon/2) \subset U$, the function $\rho(p, q)$ $(q \in \gamma)$ is bounded. If γ does not lie inside W completely, and $q_1, q_2 \in W$, then the function $\rho(p, q)$ $(q \in \gamma)$ must attain its maximum at an interior point q_0 of γ. Let $\delta = \rho(p, q_0)$. Then $\delta < \epsilon$, and the hypersphere Σ_δ is tangent to γ at q_0. By Theorem 2.5, γ lies completely outside Σ_δ near q_0, which contradicts the fact that $\rho(p, q)$ $(q \in \gamma)$ attains its maximum at q_0. Hence $\gamma \subset W$. $\qquad\square$

§5–3 Sectional Curvature

Suppose M is an m-dimensional Riemannian manifold whose curvature tensor R is a covariant tensor of rank 4, and u^i is a local coordinate system in M. Then R can be expressed as

$$R = R_{ijkl} du^i \otimes du^j \otimes du^k \otimes du^l, \tag{3.1}$$

where R_{ijkl} is defined as in (1.50). A covariant tensor of rank 4 can be viewed as a linear function on the space of contravariant tensors of rank 4 (see §2–2), so at every point $p \in M$ we have a multilinear function $R : T_p(M) \times T_p(M) \times T_p(M) \times T_p(M) \longrightarrow R$, defined by

$$R(X, Y, Z, W) = \langle X \otimes Y \otimes Z \otimes W, R \rangle, \tag{3.2}$$

where the notation $\langle \, , \, \rangle$ is defined as in (2.17) of Chapter 2. If we let

$$X = X^i \frac{\partial}{\partial u^i}, \quad Y = Y^i \frac{\partial}{\partial u^i}, \quad Z = Z^i \frac{\partial}{\partial u^i}, \quad W = W^i \frac{\partial}{\partial u^i}, \tag{3.3}$$

then

$$R(X, Y, Z, W) = R_{ijkl} X^i Y^j Z^k W^l. \tag{3.4}$$

In particular,

$$R_{ijkl} = R\left(\frac{\partial}{\partial u^i}, \frac{\partial}{\partial u^j}, \frac{\partial}{\partial u^k}, \frac{\partial}{\partial u^l}\right). \tag{3.5}$$

In §4–2, we have already interpreted the curvature tensor of a connection D as a curvature operator: for any given $Z, W \in T_p(M)$, $R(Z, W)$ is a linear map from $T_p(M)$ to $T_p(M)$ defined by

$$R(Z, W)X = R^j_{ikl} X^i Z^k W^l \frac{\partial}{\partial u^j}. \tag{3.6}$$

If D is the Levi-Civita connection of a Riemannian manifold M, then we have

$$R(X, Y, Z, W) = (R(Z, W)X) \cdot Y, \tag{3.7}$$

where the notation "$\cdot$" on the right hand side is the inner product defined by (1.4).

By Theorem 1.4, the 4-linear function $R(X, Y, Z, W)$ has the following properties:

1) $R(X, Y, Z, W) = -R(X, Y, W, Z) = -R(Y, X, Z, W)$;
2) $R(X, Y, Z, W) + R(X, Z, W, Y) + R(X, W, Y, Z) = 0$;
3) $R(X, Y, Z, W) = R(Z, W, X, Y)$.

Using the fundamental tensor G of M, we can also define a 4-linear function as follows:

$$G(X, Y, Z, W) = G(X, Z)G(Y, W) - G(X, W)G(Y, Z). \tag{3.8}$$

Obviously the function defined above is linear with respect to every variable, and also has the same properties 1)—3) as $R(X, Y, Z, W)$.

If $X, Y \in T_p(M)$, then

$$G(X, Y, X, Y) = |X|^2 \cdot |Y|^2 - (X \cdot Y)^2 = |X|^2 \cdot |Y|^2 \cdot \sin^2 \angle(X, Y). \tag{3.9}$$

Therefore, when X, Y are linearly independent, $G(X, Y, X, Y)$ is precisely the square of the area of the parallelogram determined by the tangent vectors X and Y. Hence $G(X, Y, X, Y) \neq 0$.

Suppose X', Y' are another two linearly independent tangent vectors at the point p, and that they span the same 2-dimensional tangent subspace E as that spanned by X and Y. Then we may assume that

$$X' = aX + bY, \qquad Y' = cX + dY,$$

where $ad - bc \neq 0$. By properties 1)—3) we have

$$\begin{aligned}
R(X',Y',X',Y') &= (ad - bc)^2 R(X,Y,X,Y), \\
G(X',Y',X',Y') &= (ad - bc)^2 G(X,Y,X,Y).
\end{aligned}$$

Thus

$$\frac{R(X',Y',X',Y')}{G(X',Y',X',Y')} = \frac{R(X,Y,X,Y)}{G(X,Y,X,Y)}.$$

This implies that the above expression is a function of the 2-dimensional subspace E of $T_p(M)$, and is independent of the choice of X and Y.

Definition 3.1. Suppose E is a 2-dimensional subspace of $T_p(M)$, and X, Y are any two linearly independent vectors in E. Then

$$K(E) = -\frac{R(X,Y,X,Y)}{G(X,Y,X,Y)} \tag{3.10}$$

is a function of E independent of the choice of X and Y in E. We call it the **Riemannian curvature**, or **sectional curvature**, of M at (p, E).

We know that the product of the two principal curvatures at a point on a surface in 3-dimensional Euclidean space is called the **total curvature**, or **Gauss curvature**, of the surface at that point. A result of Gauss which he proclaimed as being "amazing" (the **Theorema Egregium**) is this: even though the total curvature of a surface at a point is defined extrinsically (i.e., the definition uses not only the first fundamental form of the surface, but also its second fundamental form), it depends only on the first fundamental form of the surface, that is, the total curvature K is

$$K = -\frac{R_{1212}}{g}, \tag{3.11}$$

where $g = g_{11}g_{22} - g_{12}^2$, and R_{1212} is defined in (1.52):

$$R_{1212} = \frac{\partial \Gamma_{122}}{\partial u^1} - \frac{\partial \Gamma_{121}}{\partial u^2} + \Gamma_{11}^h \Gamma_{2h2} - \Gamma_{12}^h \Gamma_{2h1}. \tag{3.12}$$

This fact gives a geometric explanation of the sectional curvature. Suppose $m \geq 3$ and E is a 2-dimensional subspace of $T_p(M)$. Choose an orthogonal frame $\{e_i\}$ at p such that E is spanned by $\{e_1, e_2\}$. Suppose u^i is the geodesic normal coordinate system determined by this frame near p. Now consider the 2-dimensional submanifold S of all geodesic curves starting from p and tangent to E. Obviously the equation for S is

$$u^r = 0, \qquad 3 \leq r \leq m, \tag{3.13}$$

and (u^1, u^2) are the normal coordinates of S at p. S is called the geodesic submanifold at p tangent to E. W will prove that the sectional curvature $K(E)$ of M at (p, E) is exactly the total curvature of the surface S (with Riemannian metric induced from M) at p.

Suppose the Riemannian metric of M near p is

$$ds^2 = g_{ij} du^i du^j. \tag{3.14}$$

Then its induced metric on S is

$$d\bar{s}^2 = \bar{g}_{\alpha\beta} du^\alpha du^\beta, \qquad 1 \le \alpha, \beta \le 2, \tag{3.15}$$

where

$$\bar{g}_{\alpha\beta}(u^1, u^2) = g_{\alpha\beta}(u^1, u^2, 0, \dots, 0). \tag{3.16}$$

Therefore

$$
\begin{aligned}
\Gamma_{\alpha\beta\gamma}|_S &= \frac{1}{2}\left(\frac{\partial g_{\beta\gamma}}{\partial u^\alpha} + \frac{\partial g_{\alpha\beta}}{\partial u^\gamma} - \frac{\partial g_{\alpha\gamma}}{\partial u^\beta}\right)\bigg|_S \\
&= \frac{1}{2}\left(\frac{\partial \bar{g}_{\beta\gamma}}{\partial u^\alpha} + \frac{\partial \bar{g}_{\alpha\beta}}{\partial u^\gamma} - \frac{\partial \bar{g}_{\alpha\gamma}}{\partial u^\beta}\right) \\
&= \bar{\Gamma}_{\alpha\beta\gamma}.
\end{aligned}
\tag{3.17}
$$

Since (u^i) and (u^α) are normal coordinate systems of M and S, respectively, at p, it follows from Theorem 2.1 that

$$\bar{\Gamma}_{\alpha\beta\gamma}(p) = \Gamma_{ijk}(p) = 0. \tag{3.18}$$

Hence

$$
\begin{aligned}
R_{1212}(p) &= \left(\frac{\partial \Gamma_{122}}{\partial u^1} - \frac{\partial \Gamma_{121}}{\partial u^2} + \Gamma^i_{11}\Gamma_{2i2} - \Gamma^i_{12}\Gamma_{2i1}\right)_p \\
&= \left(\frac{\partial \bar{\Gamma}_{122}}{\partial u^1} - \frac{\partial \bar{\Gamma}_{121}}{\partial u^2}\right) = \bar{R}_{1212}(p).
\end{aligned}
\tag{3.19}
$$

From this, the sectional curvature of M at (p, E) is

$$
\begin{aligned}
K(E) &= -\frac{R(e_1, e_2, e_1, e_2)}{G(e_1, e_2, e_1, e_2)} \\
&= -\frac{R_{1212}}{g_{11}g_{22} - g_{12}^2}\bigg|_p \\
&= -\frac{\bar{R}_{1212}}{\bar{g}_{11}\bar{g}_{22} - \bar{g}_{12}^2}\bigg|_p = \bar{K}(p).
\end{aligned}
$$

The right hand side is precisely the total curvature of the surface S at p.

The importance of the sectional curvature lies in the following:

Theorem 3.1. *The curvature tensor of a Riemannian manifold M at a point p is uniquely determined by the sectional curvatures of all the 2-dimensional tangent subspaces at p.*

Proof. Suppose there is a 4-linear function $\bar{R}(X,Y,Z,W)$ satisfying all the properties 1)—3) of the curvature tensor $R(X,Y,Z,W)$, and that for any two linearly independent tangent vectors X,Y at p,

$$\frac{\bar{R}(X,Y,X,Y)}{G(X,Y,X,Y)} = \frac{R(X,Y,X,Y)}{G(X,Y,X,Y)}. \tag{3.20}$$

We will show that for any $X,Y,Z,W \in T_p(M)$,

$$\bar{R}(X,Y,Z,W) = R(X,Y,Z,W). \tag{3.21}$$

If we let

$$S(X,Y,Z,W) = \bar{R}(X,Y,Z,W) - R(X,Y,Z,W), \tag{3.22}$$

then S is also a 4-linear function satisfying the properties 1)—3); and, because of (3.20), we have, for any $X,Y \in T_p(M)$,

$$S(X,Y,X,Y) = 0. \tag{3.23}$$

Thus (3.21) is equivalent to the statement that S is the zero function.

From (3.23) we obtain

$$S(X+Z,Y,X+Z,Y) = 0.$$

Expanding and using the properties of the function S, we have

$$S(X,Y,Z,Y) = 0, \tag{3.24}$$

where X,Y,Z are any three elements of $T_p(M)$. Thus

$$S(X,Y+W,Z,Y+W) = 0,$$

and by expanding we obtain

$$S(X,Y,Z,W) + S(X,W,Z,Y) = 0. \tag{3.25}$$

From property 1) we then have

$$\begin{aligned}
S(X,Y,Z,W) &= -S(X,W,Z,Y) = S(X,W,Y,Z) \\
&= -S(X,Z,Y,W) = S(X,Z,W,Y).
\end{aligned} \tag{3.26}$$

On the other hand, property 2) implies

$$S(X,Y,Z,W) + S(X,Z,W,Y) + S(X,W,Y,Z) = 0.$$

Thus

$$3S(X,Y,Z,W) = 0, \tag{3.27}$$

which completes the proof.   □

Definition 3.2. Suppose M is a Riemannian manifold. If the sectional curvature $K(E)$ at the point p is a constant (i.e., independent of E), then we say that M is **wandering** at p.

If M is wandering at p, then the sectional curvature of M at p can be denoted by $K(p)$. Hence, for any $X, Y \in T_p(M)$, we have

$$R(X,Y,X,Y) = -K(p)G(X,Y,X,Y). \tag{3.28}$$

According to the proof of Theorem 3.1, for any $X, Y, Z, W \in T_p(M)$, we have, on the other hand,

$$R(X,Y,Z,W) = -K(p)G(X,Y,Z,W). \tag{3.29}$$

Thus the condition for a Riemannian manifold to be wandering at p is

$$R_{ijkl}(p) = -K(p)(g_{ik}g_{jl} - g_{il}g_{jk})(p),$$

or

$$\Omega_{ij}(p) = \frac{1}{2}R_{ijkl}(p)du^k \wedge du^l = -K(p)\theta_i \wedge \theta_j(p), \tag{3.30}$$

where

$$\theta_i = g_{ij}du^j \tag{3.31}$$

Definition 3.3. If M is a Riemannian manifold which is wandering at every point and the sectional curvature $K(p)$ is a constant function on M, then M is called a **constant curvature space**.

Spheres, planes and pseudo-spheres are all surfaces in the 3-dimensional Euclidean space whose total curvatures are constant. Hence they are all 2-dimensional constant curvature Riemannian spaces.

Theorem 3.2 (F. Schur's Theorem). *Suppose M is a connected m-dimensional Riemannian manifold that is everywhere wandering. If $m \geq 3$, then M is a constant curvature space.*

Proof. Since M is wandering everywhere, we obtain from (3.30) that

$$\Omega_{ij} = -K\theta_i \wedge \theta_j, \tag{3.32}$$

where K is a smooth function on M, and θ_i is given in (3.31). Exteriorly differentiating (3.32) we get

$$d\Omega_{ij} = -dK \wedge \theta_i \wedge \theta_j - K d\theta_i \wedge \theta_j + K\theta_i \wedge d\theta_j. \tag{3.33}$$

But

$$
\begin{aligned}
d\theta_i &= dg_{ij} \wedge du^j \\
&= (g_{ik}\omega_j^k + g_{kj}\omega_i^k) \wedge du^j \\
&= (\omega_{ji} + \omega_{ij}) \wedge du^j,
\end{aligned}
$$

where

$$\omega_{ij} = g_{jk}\omega_i^k = \Gamma_{ijk}du^k.$$

Since the Levi-Civita connection is torsion-free, we have

$$\omega_{ji} \wedge du^j = \Gamma_{jik}du^k \wedge du^j = 0.$$

Hence

$$d\theta_i = \omega_{ij} \wedge du^j = \omega_i^j \wedge \theta_j. \tag{3.34}$$

On the other hand, by the Bianchi identity (Theorem 2.2 in Chapter 4), we have

$$
\begin{aligned}
d\Omega_{ij} &= d(\Omega_i^l \cdot g_{lj}) \\
&= d\Omega_i^l \cdot g_{lj} + \Omega_i^l \wedge dg_{lj} \\
&= (\omega_i^k \wedge \Omega_k^l - \Omega_i^k \wedge \omega_k^l) \cdot g_{lj} + \Omega_i^l \wedge (\omega_{lj} + \omega_{jl}) \\
&= \omega_i^k \wedge \Omega_{kj} + \Omega_{ik} \wedge \omega_j^k.
\end{aligned}
\tag{3.35}
$$

Thus

$$
\begin{aligned}
d\Omega_{ij} &= -K\omega_i^k \wedge \theta_k \wedge \theta_j - K\theta_i \wedge \theta_k \wedge \omega_j^k \\
&= -K d\theta_i \wedge \theta_j + K\theta_i \wedge d\theta_j.
\end{aligned}
\tag{3.36}
$$

Comparing (3.36) with (3.33), we obtain

$$dK \wedge \theta_i \wedge \theta_j = 0. \tag{3.37}$$

Since $\{\theta_i\}$ and $\{du^i\}$ are both local coframes, we may assume that

$$dK = \sum_{i=1}^{m} a^i \theta_i.$$

Since $m \geq 3$, for any three indices $1 \leq i < j < k \leq m$, we have

$$dK \wedge \theta_i \wedge \theta_j = dK \wedge \theta_j \wedge \theta_k = dK \wedge \theta_i \wedge \theta_k = 0.$$

Hence $a^i = 0$ $(1 \leq i \leq m)$, i.e.,

$$dK = 0. \tag{3.38}$$

Since M is a connected manifold, K is a constant function on M.   □

Example. Suppose

$$ds^2 = \frac{(du^1)^2 + \cdots + (du^m)^2}{[1 + \frac{K}{4}((u^1)^2 + \cdots + (u^m)^2)]^2}, \tag{3.39}$$

where K is a real number. Then the Riemannian space with ds^2 as its metric form is a constant curvature space, and its sectional curvature is K (the proof is left to the reader).

The formula (3.39) was given by B. Riemann in 1854 in his inaugural speech, "On the Basic Assumptions of Geometry," at Gottingen University in Germany. This speech founded what is now known as "Riemannian Geometry."

§5–4   The Gauss–Bonnet Theorem

The Gauss–Bonnet Theorem is a classical theorem in global differential geometry. It establishes a connection between local and global properties of Riemannian manifolds. In this section we will only prove the Theorem for 2-dimensional Riemannian manifolds.

Suppose M is an oriented 2-dimensional Riemannian manifold. If we choose a smooth frame field $\{e_1, e_2\}$ in a coordinate neighborhood U whose orientation is consistent with that of M, with coframe field $\{\theta^1, \theta^2\}$, then the Riemannian metric is

$$ds^2 = g_{ij}\theta^i\theta^j, \qquad 1 \leq i, j \leq 2, \tag{4.1}$$

where $g_{ij} = G(e_i, e_j)$. By the Fundamental Theorem of Riemannian Geometry, there exists a unique set of differential 1-forms θ_i^j such that

$$
\begin{cases}
d\theta^i - \theta^j \wedge \theta_j^i = 0, \\[2mm]
dg_{ij} = g_{ik}\theta_j^k + g_{kj}\theta_i^k.
\end{cases}
\tag{4.2}
$$

The θ_i^j define the Levi-Civita connection on M:

$$
De_i = \theta_i^j e_j.
\tag{4.3}
$$

The curvature form for the connection is

$$
\Omega_i^j = d\theta_i^j - \theta_i^k \wedge \theta_k^j.
\tag{4.4}
$$

Let $\Omega_{ij} = \Omega_i^k g_{kj}$. Then by (1.46), Ω_{ij} is skew-symmetric. Now the indices i, j only take the values 1 and 2, so the only nonzero element in the curvature form Ω_{ij} is Ω_{12}.

We will consider the transformation rule for Ω_{12} under a change of the local frame field. Let Ω denote the curvature matrix (Ω_i^j), and write

$$
G = \begin{pmatrix} g_{11} & g_{12} \\[2mm] g_{21} & g_{22} \end{pmatrix}.
\tag{4.5}
$$

If (e_1', e_2') is another local frame field in a coordinate neighborhood $W \subset M$ with orientation consistent with that of M, then in $U \cap W$, when $U \cap W \neq \varnothing$,

$$
\begin{pmatrix} e_1' \\[2mm] e_2' \end{pmatrix} = A \cdot \begin{pmatrix} e_1 \\[2mm] e_2 \end{pmatrix},
\tag{4.6}
$$

where

$$
A = \begin{pmatrix} a_1^1 & a_1^2 \\[2mm] a_2^1 & a_2^2 \end{pmatrix}, \qquad \det A > 0.
$$

Let G' and Ω' denote the corresponding quantities with respect to the frame field (e_1', e_2'). Then

$$
G' = A \cdot G \cdot {}^t\!A, \qquad \Omega' = A \cdot \Omega \cdot A^{-1},
\tag{4.7}
$$

where the second formula is equation (1.29) in Chapter 4. Therefore

$$\Omega' \cdot G' = A \cdot (\Omega \cdot G) \cdot {}^t A, \tag{4.8}$$

i.e.,

$$\begin{pmatrix} 0 & \Omega'_{12} \\ -\Omega'_{12} & 0 \end{pmatrix} = A \cdot \begin{pmatrix} 0 & \Omega_{12} \\ -\Omega_{12} & 0 \end{pmatrix} \cdot {}^t A.$$

Thus

$$\Omega'_{12} = (\det A) \cdot \Omega_{12}. \tag{4.9}$$

We also obtain from (4.7) that

$$\begin{aligned} g' &= \det G' = (\det A)^2 \cdot \det G \\ &= (\det A)^2 \cdot g. \end{aligned} \tag{4.10}$$

Hence

$$\frac{\Omega'_{12}}{\sqrt{g'}} = \frac{\Omega_{12}}{\sqrt{g}}. \tag{4.11}$$

In other words, $\Omega_{12}/\sqrt{g}$ is independent of the choice of the orientation-consistent local frame field, and is therefore an exterior differential 2-form defined on the whole manifold. If we choose a local coordinate system u^i with the same orientation as M, and $\{e_1, e_2\}$ is the natural basis, then

$$\begin{aligned} \Omega_{12} &= \frac{1}{2} R_{12kl} du^k \wedge du^l \\ &= R_{1212} du^1 \wedge du^2. \end{aligned}$$

Thus

$$\frac{\Omega_{12}}{\sqrt{g}} = \frac{R_{1212}}{g} \cdot \sqrt{g} du^1 \wedge du^2 = -K d\sigma, \tag{4.12}$$

where K is the Gauss curvature of M, and $d\sigma = \sqrt{g} du^1 \wedge du^2$ is the oriented area element of M. The Gauss–Bonnet Theorem is concerned with the integral of the exterior differential form $K d\sigma$ on M.

If $\{e_1, e_2\}$ is an orthogonal local frame field with an orientation consistent with that of M, then $g = g_{11} g_{22} - g_{12}^2 = 1$. Thus

$$K d\sigma = -\Omega_{12}. \tag{4.13}$$

On the other hand, it follows from (1.47) that

$$\Omega_{12} = d\theta_{12} + \theta_1^i \wedge \theta_{2i}.$$

Since θ_i^j is skew-symmetric (see Remark 1 after Theorem 1.3), we have

$$\Omega_{12} = d\theta_{12}, \tag{4.14}$$

where $\theta_{12} = De_1 \cdot e_2$. It follows from (4.13) and (4.14) that

$$K d\sigma = -d\theta_{12}. \tag{4.15}$$

We need to point out that as long as there exists a smooth orthogonal frame field $\{e_1, e_2\}$ with an orientation consistent with that of M in an open set $U \subset M$, then there exists a connection form θ_{12} on U, and hence (4.15) holds.

On an oriented 2-dimensional Riemannian manifold, a smooth orthogonal frame field with an orientation consistent with that of M corresponds to a tangent vector field that is never zero. In fact, the tangent vector e_2 in the frame $\{e_1, e_2\}$ is obtained by rotating e_1 by 90° according to the orientation of M. Therefore an orthogonal frame field $\{e_1, e_2\}$ with an orientation consistent with that of M is equivalent to the unit tangent vector field e_1.

A null point of a tangent vector field is called a **singular point**. We first describe the concept of the index of a tangent vector field at a singular point. Assume that there is a smooth vector field X on U that has exactly one singular point p, i.e., when $q \in U - \{p\}$, $X_q \neq 0$. Then there is smooth unit tangent vector field

$$a_1 = \frac{X}{|X|} \tag{4.16}$$

which determines an orthogonal frame field $\{a_1, a_2\}$ with an orientation consistent with that of M in $U - \{p\}$. Therefore, if $\{c_1, e_2\}$ is a given orthogonal frame field on U that is also orientation-consistent with M, then we may assume that

$$\begin{cases} a_1 &=& e_1 \cos\alpha + e_2 \sin\alpha, \\ a_2 &=& -e_1 \sin\alpha + e_2 \cos\alpha, \end{cases} \tag{4.17}$$

where $\alpha = \angle(e_1, a_1)$ is the oriented angle from e_1 to a_1. Obviously α is a multi-valued function, but at every point the difference between two values of α is an integer multiple of 2π. Thus by the differentiability of frame fields and vector fields there always exists a continuous branch of α in a neighborhood of any point. The single-valued function obtained from this branch is smooth in the neighborhood; and the difference between any two continuous branches of α is an integer multiple of 2π. Let

$$\omega_{12} = Da_1 \cdot a_2. \tag{4.18}$$

Then by (4.17), (4.18) and the fact that $De_1 \cdot e_2 = \theta_{12}$, we have

$$\omega_{12} = d\alpha + \theta_{12}. \tag{4.19}$$

Suppose D is a simply connected domain containing the point p whose boundary is a smooth simple closed curve $C = \partial D$. According to the discussion in §3–4, it has an induced orientation from M. Suppose the arc length parameter of C is s, $0 \leq s \leq L$, and the direction along the curve as s increases is the same as the induced direction of C. So $C(0) = C(L)$. Since C is compact, it can be covered by finitely many neighborhoods, and there exists a continuous branch of α in each neighborhood. Therefore, there exists a continuous function $\alpha = \alpha(s)$, $0 \leq s \leq L$, on C. But in general, $\alpha(0) \neq \alpha(L)$, and the difference between any two function branches is an integer multiple of 2π. By the fundamental theorem of calculus, we have

$$\alpha(L) - \alpha(0) = \int_0^L d\alpha. \tag{4.20}$$

But $\alpha(L)$ and $\alpha(0)$ are the angles between the tangent vectors a_1 and e_1 at the same point $C(0)$. Therefore the left hand side of (4.20) is an integer multiple of 2π, and is independent of the choice of the continuous branch of $\alpha(s)$. It is also independent of the choice of the frame field $\{e_1, e_2\}$.

We will show that the value of formula (4.20) does not depend on the choice of the simple closed curve C surrounding the point p. Suppose there is another simply connected domain $D_1 \subset \overset{\circ}{D}$ containing p. Let $C_1 = \partial D_1$. Then $D - \overset{\circ}{D}_1$ is a domain with boundary in M, and its boundary with induced orientation is $C - C_1$.[c] By (4.19) and the Stokes formula, we have

$$\int_{C-C_1} d\alpha = \int_{C-C_1} \omega_{12} - \int_{C-C_1} \theta_{12}$$

$$= \int_{C-C_1} \omega_{12} - \int_{D-D_1} d\theta_{12} \tag{4.21}$$

$$= \int_{C-C_1} \omega_{12} + \int_{D-D_1} K \, d\sigma.$$

---

[c] We use the "chain" notation from topology. In fact $C - C_1$ means the union of C and the negatively oriented C_1, c.f. Singer and Thorpe 1976.

The right hand side is obviously independent of the choice of the frame field $\{e_1, e_2\}$ on $D - D_1$. Hence we may assume that $e_i = a_i$, $i = 1, 2$. Then $\bar{\alpha} = \angle(e_1, a_1) = 0$, and (4.21) still holds. Therefore

$$\int_{C-C_1} \omega_{12} + \int_{D-D_1} K \, d\sigma = \int_{C-C_1} d\bar{\alpha} = 0.$$

By plugging this into (4.21) we have

$$\int_{C-C_1} d\alpha = 0, \qquad \int_C d\alpha = \int_{C_1} d\alpha.$$

Definition 4.1. Suppose X is a smooth tangent vector field with an isolated singular point p, and U is a coordinate neighborhood of p such that p is the only singular point of X in U. Then the integer

$$I_p = \frac{1}{2\pi}[\alpha(L) - \alpha(0)] = \frac{1}{2\pi} \int_C d\alpha, \qquad (4.22)$$

obtained by the above construction is independent of the choice of the simple closed curve C surrounding p, and the choice of the frame field $\{e_1, e_2\}$ on U (with an orientation consistent with that of M). It is called the **index** of the tangent vector field X at the point p.

Intuitively, the index I_p represents the number of times the tangent vector field X loops around the singular point p.

Integrating (4.19) over C, we get

$$\frac{1}{2\pi} \int_C \omega_{12} = \frac{1}{2\pi} \int_C d\alpha - \frac{1}{2\pi} \int_D K \, d\sigma.$$

Since the Gauss curvature K is continuous at p, when D is shrunk to a point, the integral

$$\frac{1}{2\pi} \int_D K \, d\sigma \longrightarrow 0.$$

But $\dfrac{1}{2\pi} \displaystyle\int_C d\alpha$ is the constant I_p, hence

$$I_p = \frac{1}{2\pi} \lim_{C \to p} \int_C \omega_{12}. \qquad (4.23)$$

Theorem 4.1 (Gauss–Bonnet Theorem). *Suppose M is a compact oriented 2-dimensional Riemannian manifold. Then*

$$\frac{1}{2\pi} \int_M K \, d\sigma = \chi(M), \qquad (4.24)$$

*where $\chi(M)$ is the **Euler characteristic** of M.*

Proof. Choose a smooth tangent vector field X on M with only finitely many isolated singular points p_i, $1 \le i \le r$. For each p_i, we choose an ϵ-ball neighborhood D_i, where ϵ is a sufficiently small positive number such that each D_i contains no singular points of X other that p_i. Let $C_i = \partial D_i$, then C_i is a simple closed curve with induced orientation from M on D_i. Thus the tangent vector field X determines a smooth orthogonal frame field $\{e_1, e_2\}$ on $M - \bigcup_i D_i$ that is orientation consistent, with $e_1 = X/|X|$. Suppose

$$\theta_{12} = De_1 \cdot e_2. \qquad (4.25)$$

By (4.15), we have, on $M - \bigcup_i D_i$,

$$d\theta_{12} = \Omega_{12} = -K \, d\sigma.$$

Also, by the Stokes formula,

$$
\begin{aligned}
\int_{M - \bigcup_i D_i} K \, d\sigma &= -\int_{M - \bigcup_i D_i} d\theta_{12} \\
&= \sum_{i=1}^{r} \int_{\partial D_i} \theta_{12} \\
&= \sum_{i=1}^{r} \int_{C_i} \theta_{12}.
\end{aligned}
\qquad (4.26)
$$

We note that the boundary of $M - \bigcup\limits_{1 \le i \le r} D_i$ is identical to the boundary of $\bigcup\limits_{1 \le i \le r} D_i$ in set theory, but the former has an induced orientation on its boundary which is exactly the opposite of the orientation of $\sum_i \partial D_i = \sum_i C_i$. In the second equality above we have made use of this fact.

Since the orthogonal frame field $\{e_1, e_2\}$ is in fact well-defined on $M - \bigcup_i \{p_i\}$, Equation (4.26) still holds as $\epsilon \to 0$. Also, since K is a continuously differentiable function defined on the whole of M, we have

$$\lim_{\epsilon \to 0} \int_{M - \bigcup_i D_i} K \, d\sigma = \int_M K \, d\sigma,$$

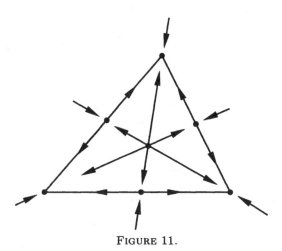

FIGURE 11.

and the final term in (4.26) is precisely $2\pi \sum_{i=1}^{r} I_{p_i}$ when $\epsilon \to 0$ [see (4.23)]. Hence

$$\frac{1}{2\pi} \int_M K \, d\sigma = \sum_{i=1}^{r} I_{p_i}. \tag{4.27}$$

The left hand side of the above equation is independent of the tangent vector field X. We construct a special tangent vector field on M as follows. Choose a triangulation of M (since it is compact, it is triangulable, cf. Singer and Thorpe 1976). Construct a smooth tangent vector field X such that the center of mass of each face in each dimension is a singular point; and its indices at the centers of mass of any 2-dimensional face, 1-dimensional face, and 0-dimensional face are $+1$, -1, and $+1$, respectively (see Figure 11). Thus

$$\sum I_{p_i} = f - e + v = \chi(M), \tag{4.28}$$

where f, e, v represent the numbers of 2-dimensional faces, 1-dimensional faces, and 0-dimensional faces of the triangulation of M, respectively. Hence

$$\frac{1}{2\pi} \int_M K \, d\sigma = \chi(M).$$

$\square$

From the above proof we have already obtained **Hopf's Index Theorem:**

Corollary. *Suppose there is a smooth tangent vector field on a compact oriented 2-dimensional Riemannian manifold with finitely many singular points. Then the sum of its indices at the various singular points is equal to the Euler characteristic of the manifold.*

Remark. We see from the above corollary that a necessary condition for a compact oriented 2-dimensional smooth manifold to have a nonvanishing smooth tangent vector field is that the Euler characteristic of the manifold is zero. Conversely, from the classification of 2-dimensional manifolds, if the Euler characteristic of a compact oriented 2-dimensional smooth manifold is zero, then it must be diffeomorphic to a torus. Hence there must exist a nowhere vanishing smooth tangent vector field on a torus.

The Gauss–Bonnet formula can be generalized to domains with boundaries. Suppose C is a smooth curve on M, and a_1 is a unit tangent vector to C. Choose a unit normal vector a_2 to C such that the orientation determined by $\{a_1, a_2\}$ is consistent with that of M. Since Da_1 is collinear with a_2, we may assume

$$\kappa_g = \frac{Da_1}{ds} \cdot a_2. \tag{4.29}$$

κ_g is called the **geodesic curvature** of C. Obviously a necessary and sufficient condition for C to be a geodesic curve is

$$\kappa_g \equiv 0.$$

Suppose D is a compact domain with boundary in an oriented 2-dimensional Riemannian manifold M whose boundary ∂D is composed of finitely many piecewise smooth simple closed curves with induced orientation from D. Suppose the interior angle of ∂D at each vertex p_i is α_i, $1 \leq i \leq l$. Then we have the Gauss–Bonnet formula [see S. S. Chern 1967(a)]

$$\sum_{i=1}^{l}(\pi - \alpha_i) + \int_{\partial D} \kappa_g ds + \iint_D K\, d\sigma = 2\pi\chi(D), \tag{4.30}$$

where κ_g is the geodesic curvature along ∂D, and $\chi(D)$ is the Euler characteristic of D.

If D is a geodesic triangle in M, and ∂D is a closed curve composed of three geodesic segments, then $\chi(D) = 1$ and (4.30) therefore becomes

$$\alpha_1 + \alpha_2 + \alpha_3 - \pi = \iint_D K\, d\sigma. \tag{4.31}$$

This generalizes the theorem that "the sum of the internal angles of a planar triangle is 180°." Formula (4.31) for the case of a spherical surface was obtained by Gauss.

Analyzing in detail the proof of the Gauss–Bonnet theorem, it is not difficult to see that the key is to represent the exterior differential 2-form $K\, d\sigma$ on M by $-d\theta_{12}$, where the latter is viewed as an object on the frame bundle of M, that is, on the sphere bundle of M (a fiber bundle composed of unit tangent vectors of M). Thus an integral on M is transformed into an integral on a section of the sphere bundle. But a section of the sphere bundle, that is, a unit tangent vector field, may not exist. Hence, via Stokes's theorem, the integral becomes one surrounding singular points, and thus involves indices of the vector field at the singular points. This idea is used to the fullest extent in the proof of the Gauss–Bonnet theorem in higher-dimensional manifolds.

Let M be a compact $2n$-dimensional oriented Riemannian manifold. Suppose $\{e_i, 1 \leq i \leq 2n\}$ is a local orthogonal frame field whose coframe field is $\{\theta_i, 1 \leq i \leq 2n\}$; and the connection form and curvature form are ω_{ij} and Ω_{ij}, respectively. Consider the exterior differential $2n$-form

$$\Omega = (-1)^n \frac{1}{2^{2n}\pi^n n!} \delta_{1\cdots 2n}^{i_1 \cdots i_{2n}} \Omega_{i_1 i_2} \wedge \cdots \wedge \Omega_{i_{2n-1} i_{2n}}. \tag{4.32}$$

Ω is independent of the choice of the local orthogonal frame field, and hence is an exterior differential $2n$-form defined globally on M. Ω can be denoted by

$$\Omega = K\, d\sigma, \tag{4.33}$$

where $d\sigma = \theta_1 \wedge \cdots \wedge \theta_{2n}$ is the volume element of M, and K is the **Lipschitz– Killing** curvature of M given by

$$K = \frac{1}{2^{2n}(2\pi)^n n!} \delta_{j_1 \cdots j_{2n}}^{i_1 \cdots i_{2n}} R_{i_1 i_2 j_1 j_2} \cdots R_{i_{2n-1} i_{2n} j_{2n-1} j_{2n}}. \tag{4.34}$$

The Gauss–Bonnet theorem states that

$$\int_M K\, d\sigma = \chi(M). \tag{4.35}$$

The key to proving this theorem is to represent the exterior differential form Ω on the sphere bundle of M (the fiber bundle formed by unit tangent vectors of M) by the exterior differential of an exterior differential $(2n-1)$-form Π:

$$\Omega = d\Pi. \tag{4.36}$$

In gerneral, when a characteristic form Ω on M is pulled back in a bundle $\pi : E \longrightarrow M$, $\pi^*\Omega$ my become an exact form. This process is called **transgression**, and is of great importance. For details see S. S. Chern, 1944(b).

Chapter 6

Lie Groups and Moving Frames

§6–1 Lie Groups

The real number field is richly endowed with structures. On the one hand, it has an algebraic structure which permits addition, subtraction, multiplication and division; on the other, it has topological and differentiable structures which allow us to discuss continuity and differentiability. Our object of discussion in this section, the Lie group, is a composite containing group and differentiable structures. Here we only introduce briefly some basic concepts in Lie group and Lie algebra theory.

Definition 1.1. Let G be a nonempty set. If

1) G is a group (whose operation is denoted by multiplication);
2) G is an r-dimensional smooth manifold; and
3) the inverse map $\tau : G \longrightarrow G$ such that $\tau(g) = g^{-1}$ and the multiplication map $\varphi : G \times G \longrightarrow G$ such that $\varphi(g_1, g_2) = g_1 \cdot g_2$ are both smooth maps,

then G is called an r-dimensional **Lie group**.

Because $\tau^2 = \text{id} : G \longrightarrow G$ [i.e., $\tau^2(g) = g$], τ is a diffeomorphism from G to itself. Moreover G possesses another two sets of diffeomorphisms—the right and left translations—defined as follows. For $g \in G$, the **right translation** by g on G is $R_g : G \longrightarrow G$ such that

$$R_g(x) = \varphi(x, g) = x \cdot g, \tag{1.1}$$

and the **left translation** is $L_g : G \longrightarrow G$ such that

$$L_g(x) = \varphi(g, x) = g \cdot x. \tag{1.2}$$

Since the inverse of L_g is $L_{g^{-1}}$, and the inverse of R_g is $R_{g^{-1}}$, L_g and R_g are both diffeomorphisms from G to itself.

Example 1. $\mathbb{R}^n$ is an n-dimensional Lie group with respect to the addition of vectors.

Example 2 (The n-dimensional torus group $\mathbb{T}^n$). Choose n linearly independent vectors, e_i, $1 \leq i \leq n$, in $\mathbb{R}^n$. The lattice formed by them is

$$L = \left\{ \sum_{i=1}^{n} \alpha_i e_i, \alpha_i \in \mathbb{Z} \right\} = \mathbb{Z}^n, \tag{1.3}$$

which is a group under addition and is a subgroup of $\mathbb{R}^n$. The torus group $\mathbb{T}^n$ is the quotient group $\mathbb{R}^n / L$. Topologically it is an n-dimensional torus. Therefore it is an n-dimensional compact Lie group.

If G_1, G_2 are two Lie groups, then we can define an operation on the product manifold $G_1 \times G_2$ as follows. Suppose $(a_1, a_2), (b_1, b_2) \in G_1 \times G_2$, and let

$$(a_1, a_2) \cdot (b_1, b_2) = (a_1 \cdot b_1, a_2 \cdot b_2). \tag{1.4}$$

Then $G_1 \times G_2$ becomes a Lie group with respect to the operation in (1.4), called the **direct product** of the Lie groups G_1 and G_2.

Naturally, the 1-dimensional torus $\mathbb{T}^1$ can be viewed as a circle $S^1 = \{e^{i2\pi t}\}_{t \in \mathbb{R}} \cong \mathbb{R}^1 / \mathbb{Z}^1$. Therefore the n-dimensional torus is the direct product of n circles:

$$\mathbb{T}^n \cong S^1 \times \cdots \times S^1 \ (\ n \text{ factors}). \tag{1.5}$$

Example 3 (General linear groups $\mathrm{GL}(n; \mathbb{R})$ and $\mathrm{GL}(n; \mathbb{C})$).
$\mathrm{GL}(n; \mathbb{R})$ is the set of nondegenerate $n \times n$ real matrices with matrix multiplication for its group operation. Since $\mathrm{GL}(n; \mathbb{R})$ is an open subset of $\mathbb{R}^{n^2}$, it has the differentiable structure induced from $\mathbb{R}^{n^2}$. Suppose

$$A = \left(A_i^j \right), B = \left(B_i^j \right) \in \mathrm{GL}(n; \mathbb{R}).$$

Then

$$(A \cdot B)_i^j = \sum_{k=1}^{n} A_i^k B_k^j. \tag{1.6}$$

The right hand side is a polynomial of the elements of the matrices A and B. Hence the map

$$\varphi(A, B) = A \cdot B \tag{1.7}$$

is smooth. Moreover, since the elements of A^{-1} are rational functions of the elements A_i^j, the inverse map is also smooth. Hence GL $(n; \mathbb{R})$ is a Lie group.

Similarly, the multiplicative group GL $(n; \mathbb{C})$ of nondegenerate $n \times n$ complex matrices is a $2n^2$-dimensional Lie group.

Example 4. GL $(1; \mathbb{C})$ is the multiplicative group of nonzero complex numbers, also denoted $\mathbb{C}^*$. Topologically $\mathbb{C}^*$ is identical to $\mathbb{R}^2 - \{0\}$, and the coordinates of its elements $x + iy$ are (x, y). Therefore $\mathbb{C}^*$ is a 2-dimensional smooth manifold.

Suppose $z_\alpha = x_\alpha + iy_\alpha$, $\alpha = 1, 2$, $z_\alpha \neq 0$. Then the product of the z_α in terms of coordinates is

$$(x_1, y_1) \cdot (x_2, y_2) = (x_1 x_2 - y_1 y_2, x_1 y_2 + x_2 y_1), \tag{1.8}$$

and the inverse of $z = x + iy$ is

$$(x, y)^{-1} = \left(\frac{x}{x^2 + y^2}, -\frac{y}{x^2 + y^2} \right). \tag{1.9}$$

Obviously the multiplication and inverse maps are both smooth. Thus C^* is a 2-dimensional Lie group.

Example 5. Suppose G is a Lie group, and H is a subgroup of G. If H is a regular submanifold of G, then it can be shown that the restrictions of the mappings in Definition 1.1, namely

$$\psi|_{H \times H} \quad : \quad H \times H \longrightarrow H \subset G,$$
$$\tau|_H \quad : \quad H \longrightarrow H \subset G,$$

are both smooth (the reader should verify this, cf. Boothby 1975, p. 83).

Suppose

$$\begin{aligned} \mathrm{SL}\,(n; \mathbb{R}) &= \{A \in \mathrm{GL}\,(n; \mathbb{R}) \,|\, \det A = 1\}, \\ \mathrm{O}\,(n; \mathbb{R}) &= \{A \in \mathrm{GL}\,(n; \mathbb{R}) \,|\, A \cdot {}^tA = I\}, \end{aligned}$$

where I is the identity matrix, that is, the identity element of GL $(n; \mathbb{R})$. Then SL $(n; \mathbb{R})$ and O $(n; \mathbb{R})$ are both subgroups, and, in fact, regular submanifolds, of GL $(n; \mathbb{R})$. Therefore they are Lie groups. SL $(n; \mathbb{R})$ and O $(n; \mathbb{R})$ are called the **special linear group** and the **real orthogonal group**, respectively.

Suppose G is an r-dimensional Lie group with identity element e. Since for every $a \in G$ the map $R_{a^{-1}}$ is a diffeomorphism from G to itself and $R_{a^{-1}}(a) = e$, the tangent map $(R_{a^{-1}})_* : G_a \longrightarrow G_e$ is a linear isomorphism, where G_a represents the tangent space of G at a. Suppose $X \in G_a$. Let

$$\omega(X) = (R_{a^{-1}})_* X. \tag{1.10}$$

Then ω is a differential 1-form defined on G with values in G_e, called the **right fundamental differential form** or **Maurer–Cartan form** of the Lie group G. If we choose a basis δ_i, $1 \le i \le r$, for G_e, then we may write

$$\omega = \sum_{i=1}^{r} \omega^i \delta_i, \tag{1.11}$$

where the $\omega^i (1 \le i \le r)$ are r differential 1-forms on G that are linearly independent everywhere.

Now we derive the expression for ω^i in coordinates. Choose local coordinate systems $(U; x^i)$ and $(W; y^i)$ at points e and a respectively. Due to the continuity of φ, when U is sufficiently small, there must exist a neighborhood $W_1 \subset W$ of a such that $\varphi(U \times W_1) \subset W$. Choose $\delta_i = \left.\dfrac{\partial}{\partial x^i}\right|_e$ and let

$$\varphi^i(x, y) = y^i \circ \varphi(x, y), \qquad (x, y) \in U \times W_1.$$

Then the linear isomorphism $(R_a)_* : G_e \longrightarrow G_a$ is given as follows:

$$(R_a)_* \delta_i = \sum_{j=1}^{r} \left(\left.\frac{\partial \varphi^j(x, a)}{\partial x^i}\right|_{x=e} \cdot \left.\frac{\partial}{\partial y^j}\right|_a \right). \tag{1.12}$$

Because $(R_{a^{-1}})_* \circ (R_a)_* = \text{id} : G_e \longrightarrow G_e$, we have

$$(R_{a^{-1}})_* \left.\frac{\partial}{\partial y^i}\right|_a = \sum_{j=1}^{r} \Lambda_i^j(a) \delta_j,$$

where $(\Lambda_i^j(a))$ is the inverse matrix of $\left(\left.\dfrac{\partial \varphi^i(x, a)}{\partial x^j}\right|_{x=e} \right)$. Therefore

$$\omega^i = \sum_{j=1}^{r} \Lambda_j^i(a) dy^j. \tag{1.13}$$

From this expression we realize that ω^i is a smooth differential 1-form.

Theorem 1.1. *Suppose $\sigma : G \longrightarrow G$ is a smooth map. Then a necessary and sufficient condition for σ to be a right translation of the Lie group G is that it preserves the right fundamental differential form, i.e.,*

$$\sigma^* \omega^i = \omega^i, \qquad 1 \leq i \leq r.$$

Proof. Suppose σ is the right translation R_x, $x \in G$. Then for any $X \in G_a$ we have

$$
\begin{aligned}
(R_x)^* \omega(X) &= \omega((R_x)_* X) \\
&= (R_{(ax)^{-1}})_* \circ (R_x)_* X \\
&= (R_{a^{-1}})_* X \\
&= \omega(X).
\end{aligned}
$$

Hence

$$(R_x)^* \omega = \omega. \tag{1.14}$$

Conversely, suppose σ preserves the right fundamental differential form. Since the Pfaffian system of equations,

$$\omega^i(a) = \omega^i(b), \quad (a,b) \in G \times G, \quad 1 \leq i \leq r, \tag{1.15}$$

determines an r-dimensional planar distribution on the manifold $G \times G$, its r-dimensional integral manifold is unique. Now the map $\sigma : G \longrightarrow G$ preserves the right fundamental differential form. Therefore

$$\omega^i(X) = \omega^i(\sigma_* X),$$

where $X \in G_a$, $\sigma_* X \in G_{\sigma(a)}$. This indicates that σ gives the r-dimensional integral manifold of (1.15) passing through $(e, \sigma(e))$. Moreover the right translation $R_{\sigma(e)}$ also gives an r-dimensional integral manifold satisfying the same initial condition. So by uniqueness, we have

$$\sigma = R_{\sigma(e)}.$$

$\square$

Because $d \circ \sigma^* = \sigma^* \circ d$ (Theorem 2.6 in Chapter 3), $d\omega^i$ is still invariant under right translations. Let

$$
\left\{
\begin{array}{l}
d\omega^i = -\dfrac{1}{2} \displaystyle\sum_{j,k=1}^{r} c^i_{jk} \omega^j \wedge \omega^k, \\[2mm]
c^i_{jk} + c^i_{kj} = 0.
\end{array}
\right. \tag{1.16}
$$

Because ω^i and $d\omega^i$ are both right-invariant, the c^i_{jk} are constants, called the **structure constants** of the Lie group G. Equation (1.16) is called the **structure equation** or the **Maurer–Cartan equation** of the Lie group G.

Theorem 1.2. *The structure constants c^i_{jk} satisfy the Jacobi identity*

$$\sum_{j=1}^{r}(c^i_{jk}c^j_{hl} + c^i_{jh}c^j_{lk} + c^i_{jl}c^j_{kh}) = 0. \tag{1.17}$$

Proof. Exteriorly differentiating (1.16) we get

$$
\begin{aligned}
0 &= -\frac{1}{2}\sum_{j,k} c^i_{jk}(d\omega^j \wedge \omega^k - \omega^j \wedge d\omega^k)\\
&= \frac{1}{2}\sum_{j,k,h,l} c^i_{jk}c^j_{hl}\omega^k \wedge \omega^h \wedge \omega^l\\
&= \frac{1}{6}\sum_{k,h,l=1}^{r}\sum_{j=1}^{r}\left(c^i_{jk}c^j_{hl} + c^i_{jh}c^j_{lk} + c^i_{jl}c^j_{kh}\right)\omega^k \wedge \omega^h \wedge \omega^l.
\end{aligned}
$$

The terms inside the parentheses are skew-symmetric with respect to k, h, l. Hence (1.17) follows. $\square$

Remark. The importance of the structure constants is that they determine the structure of a local Lie group. By a local Lie group V we mean a smooth manifold such that for some element $e \in V$, there is a smooth map from a neighborhood W of (e, e) in $V \times V$ to V, denoted by $(x, y) \longmapsto x \cdot y$, that satisfies the following conditions.

1) If $(e, y) \in W$, then $e \cdot y = y$; if $(x, e) \in W$, then $x \cdot e = x$;
2) If $(x, y), (y, z), (x \cdot y, z), (x, y \cdot z) \in W$, then $(x \cdot y) \cdot z = x \cdot (y \cdot z)$.

The element e is called the identity element of the local Lie group.

The difference between a local Lie group and a Lie group is that, for the former, multiplication is only defined near the identity element. Obviously a Lie group is also a local Lie group. We can similarly define the Maurer–Cartan form and structure constants on a local Lie group. The following theorem holds:

If r^3 constants c^i_{jk} satisfy the conditions

$$\left\{\begin{aligned} &c^i_{jk} + c^i_{kj} = 0\\ &\sum_{j=1}^{r}\left(c^i_{jk}c^j_{hl} + c^i_{jh}c^j_{lk} + c^i_{jl}c^j_{kh}\right) = 0,\end{aligned}\right. \tag{1.18}$$

then there exists an r-dimensional local Lie group V with c^i_{jk} as its structure constants, and any two such local Lie groups are isomorphic (see Chevalley 1946).

Definition 1.2. Suppose X is a smooth tangent vector field on a Lie group G. If, for any $a \in G$,

$$(R_a)_* X = X, \tag{1.19}$$

then we say that the vector field X is a **right-invariant vector field** on G.

Choose an arbitrary tangent vector $X_e \in G_e$, and let

$$X_a = (R_a)_* X_e. \tag{1.20}$$

Then we obtain a smooth tangent vector field X on G. Obviously, the value of the right fundamental differential form ω on X is constant, i.e.,

$$\omega(X) = X_e. \tag{1.21}$$

By Theorem 1.1 we have

$$\omega(X) = (R_a)^* \omega(X) = \omega((R_a)_* X).$$

Therefore

$$(R_a)_* X = X,$$

that is, the tangent vector field defined in (1.20) is right-invariant.

Let X_i denote the right-invariant vector field obtained by the right translation of $\delta_i \in G_e$. Then the $X_i (1 \leq i \leq r)$ are tangent vector fields which are everywhere linearly independent on G, and any right-invariant vector field on G is a linear combination of the X_i with constant coefficients. Hence the set of right-invariant vector fields on G forms an r-dimensional vector space, denoted by $\mathcal{G}$, and is isomorphic to G_e.

By (1.21) we have

$$\omega(X_i) = \delta_i,$$

that is,

$$\omega^i(X_j) = \langle X_j, \omega^i \rangle = \delta^i_j. \tag{1.22}$$

Thus the fundamental differential forms $\omega^i (1 \leq i \leq r)$ and the right-invariant vector fields $X_j (1 \leq j \leq r)$ constitute sets of mutually dual coframe fields and frame fields, respectively, on the Lie group G. Therefore a necessary and sufficient condition for a tangent vector field X on G to be right-invariant is that the value of the right fundamental differential form on X is constant.

Theorem 1.3. *If X, Y are right-invariant vector fields on G, then $[X, Y]$ is also a right-invariant vector field on G.*

Proof. By Theorem 2.3 in Chapter 3,

$$\langle X \wedge Y, d\omega^i \rangle = X \langle Y, \omega^i \rangle - Y \langle X, \omega^i \rangle - \langle [X, Y], \omega^i \rangle. \tag{1.23}$$

From the structure equation (1.16) we obtain

$$\begin{aligned}
\langle X \wedge Y, d\omega^i \rangle &= -\frac{1}{2} \sum_{j,k=1}^{r} c_{jk}^i \langle X \wedge Y, \omega^j \wedge \omega^k \rangle \\
&= -\sum_{j,k=1}^{r} c_{jk}^i \omega^j(X)\omega^k(Y).
\end{aligned}$$

Since X, Y are right-invariant vector fields, we have

$$\omega^i(X) = \text{const}, \qquad \omega^i(Y) = \text{const}.$$

Therefore we obtain from (1.23) that

$$\omega^i([X, Y]) = \sum_{j,k=1}^{r} c_{jk}^i \omega^j(X)\omega^k(Y) = \text{const}. \tag{1.24}$$

This implies that $[X, Y]$ is right-invariant. $\square$

Theorem 1.3 shows that the Poisson bracket product for smooth tangent vector fields is closed in $\mathcal{G}$. Hence the Poisson bracket defines a multiplication operation on $\mathcal{G}$, which satisfies the following conditions (see §1–4):

1) Distributive Law

$$[a_1 X_1 + a_2 X_2, Y] = a_1 [X_1, Y] + a_2 [X_2, Y];$$

2) Skew-symmetric Law

$$[X, Y] = -[Y, X];$$

3) Jacobi Identity

$$[X, [Y, Z]] + [Y, [Z, X]] + [Z, [X, Y]] = 0.$$

If an n-dimensional real vector space has a multiplication operation that satisfies the distributive law, the skew-symmetric law and the Jacobi identity, then we call it an n-dimensional **Lie algebra**. For example, the 3-dimensional Euclidean space becomes a 3-dimensional Lie algebra with respect to the cross product for vectors; and the set of all smooth tangent vector fields on a manifold M becomes an infinite-dimensional Lie algebra with respect to the Poisson bracket product. Thus, Theorem 1.3 tells us that the vector space $\mathcal{G}$ of all right-invariant vector fields on a Lie group G is an r-dimensional Lie algebra. This Lie algebra $\mathcal{G}$ is called the **Lie algebra** of the Lie group G.

The structure constants of a Lie group provide the multiplication table for its Lie algebra $\mathcal{G}$. In fact, from (1.24) we have

$$\omega^i([X_j , X_k]) = c^i_{jk}.$$

Therefore

$$[X_j , X_k] = \sum_{i=1}^{r} c^i_{jk} X_i. \tag{1.25}$$

The skew-symmetry of the structure constants c^i_{jk} with respect to the lower indices and the Jacobi identity satisfied by these constants correspond to the skew-symmetry of the bracket and its Jacobi identity. Thus if we let

$$[\delta_j , \delta_k] = \sum_{i=1}^{r} c^i_{jk} \delta_i, \tag{1.26}$$

then G_e also becomes an r-dimensional Lie algebra. Thus G_e and $\mathcal{G}$ are isomorphic as Lie algebras. Usually, the Lie algebra G_e with respect to the multiplication defined in (1.26) is also called the **Lie algebra** of the Lie group G.

Remark. We can develop in exactly the same way the left fundamental differential form $\tilde{\omega}$ and the left-invariant vector fields on a Lie group G. Suppose

$$\tilde{\omega} = \sum_{i=1}^{r} \tilde{\omega}^i \delta_i, \tag{1.27}$$

and $\tilde{X}_i$ is the left-invariant vector field of δ_i generated by left translations. Then

$$\tilde{\omega}^i(\tilde{X}_j) = \left\langle \tilde{X}_j , \tilde{\omega}^i \right\rangle = \delta^i_j. \tag{1.28}$$

Suppose the structure equation is

$$d\tilde{\omega}^i = -\frac{1}{2} \sum_{j,k=1}^{r} \tilde{c}^i_{jk} \tilde{\omega}^j \wedge \tilde{\omega}^k, \tag{1.29}$$

or

$$\left[\tilde{X}_j , \tilde{X}_k\right] = \sum_{i=1}^{r} \tilde{c}^i_{jk} \tilde{X}_i. \tag{1.30}$$

Then through a simple calculation we see that there is only a difference in sign between $\tilde{c}^i_{jk}$ and c^i_{jk}, that is,

$$\tilde{c}^i_{jk} = -c^i_{jk} \tag{1.31}$$

(the proof is left to the reader).

It is worth mentioning that the bracket product of right-invariant vector fields and that of left-invariant vector fields define two operations of multiplication on G_e, denoted by $[\,,\,]_{\text{left}}$ and $[\,,\,]_{\text{right}}$ respectively, that is,

$$\begin{cases} [\delta_i, \delta_j]_{\text{right}} = [X_i, X_j]_e, \\ [\delta_i, \delta_j]_{\text{left}} = [\tilde{X}_i, \tilde{X}_j]_e. \end{cases} \tag{1.32}$$

From (1.31) we see that these two operations differ by just a sign:

$$[\delta_i, \delta_j]_{\text{right}} = -[\delta_i, \delta_j]_{\text{left}}. \tag{1.33}$$

When referring to G_e as the Lie algebra of G in this book, we use the multiplication operation given by (1.26), which is based on $[\,,\,]_{\text{right}}$

Example 6. Calculation of the structure constants for the general linear group GL $(n; \mathbb{R})$.

The elements of GL $(n; \mathbb{R})$ are $n \times n$ nondegenerate real matrices. Suppose $A = (A^j_i) \in$ GL $(n; \mathbb{R})$. Then (A^j_i), $(1 \leq i, j \leq n)$ is a coordinate system on the manifold GL $(n; \mathbb{R})$, and dA^j_i, $(1 \leq i, j \leq n)$ gives a coframe field on GL $(n; \mathbb{R})$, where $dA = (dA^j_i)$ is any tangent vector at A on GL $(n; \mathbb{R})$. Thus the right fundamental differential form of GL $(n; \mathbb{R})$, by (1.13), is

$$\omega = dA \cdot A^{-1}. \tag{1.34}$$

Exteriorly differentiating (1.34) we get

$$d\omega = -dA \wedge dA^{-1} = \omega \wedge \omega. \tag{1.35}$$

Let gl $(n; \mathbb{R})$ denote the tangent space at the identity element (identity matrix) I in the Lie group GL $(n; \mathbb{R})$. It is the n^2-dimensional vector space $\mathbb{R}^{n^2}$, and its elements are $n \times n$ real matrices. In this representation, gl $(n; \mathbb{R})$ has a basis E^j_i, $1 \leq i, j \leq n$, where E^j_i denote the $n \times n$ matrix with the value 1 for the element at the intersection of the j-th row and the i-th column, and 0 for other entries. Hence we may write

$$\omega = \sum_{i,j=1}^{n} \omega^j_i E^i_j = (\omega^j_i). \tag{1.36}$$

From (1.35) we have

$$dω_i^j = \sum_{k=1}^{n} ω_i^k \wedge ω_k^j$$

$$= \frac{1}{2} \sum_{p,q,r,s=1}^{n} (δ_i^p δ_q^j δ_s^r - δ_i^r δ_s^j δ_q^p) ω_p^s \wedge ω_r^q.$$

Hence the structure constants of the Lie group GL $(n; \mathbb{R})$ are

$$c_{(p,s)(r,q)}^{(i,j)} = -δ_i^p δ_q^j δ_s^r + δ_i^r δ_s^j δ_q^p. \tag{1.37}$$

The multiplication table for the Lie algebra gl $(n; \mathbb{R})$ is

$$\left[E_s^p, E_q^r \right] = δ_q^p E_s^r - δ_s^r E_q^p. \tag{1.38}$$

Suppose $A, B \in$ gl $(n; \mathbb{R})$, represented by components as

$$A = (A_p^s) = \sum_{p,s=1}^{n} A_p^s E_s^p,$$

$$B = (B_r^q) = \sum_{r,q=1}^{n} B_r^q E_q^r.$$

Then by (1.38) we have

$$[A, B] = B \cdot A - A \cdot B. \tag{1.39}$$

Definition 1.3. Suppose G, H are two Lie groups. If there is a smooth map $f : H \longrightarrow G$ which is also a homomorphism between the groups, then f is called a **homomorphism** (of Lie groups) from H to G. If f is also a diffeomorphism, then it is called an **isomorphism** (of Lie groups) from H to G.

Theorem 1.4. *Suppose $f : H \longrightarrow G$ is a Lie group homomorphism. Then f induces a homomorphism $f_* : \mathcal{H} \longrightarrow \mathcal{G}$ between the Lie algebras. If f is a Lie group isomorphism, then f_* is an isomorphism between the Lie algebras.*

Proof. Let f_* denote the tangent map of the smooth map f. First we show that f_* maps the right-invariant vector fields of the Lie group H to the right-invariant vector fields of the Lie group G. Choose any $X_e \in H_e$, and let

$$Y_{\bar{e}} = f_* X_e \in G_{\bar{e}},$$

where e is the identity element of H and $\bar{e} = f(e)$ is the identity element of G. Let X, Y be right-invariant vector fields generated by X_e, $Y_{\bar{e}}$ on their respective Lie groups. Then for any $a \in H$, we have

$$
\begin{aligned}
f_* X_a &= f_* \circ (R_a)_* X_e \\
&= (R_{\bar{a}})_* \circ f_* X_e \\
&= Y_{\bar{a}},
\end{aligned}
$$

where $\bar{a} = f(a)$. Thus the image of a right-invariant vector field on H under f_* can be extended to a right-invariant vector field on G. We still use the notation $f_* : \mathcal{H} \longrightarrow \mathcal{G}$ for such a correspondence.

Moreover the tangent map f_* commutes with the Poisson bracket product. Hence for $X_1, X_2 \in \mathcal{H}$ we have

$$
f_* [X_1, X_2]_a = [Y_1, Y_2]_{\bar{a}}, \qquad a \in H,
$$

where Y_i is the right-invariant vector field on G extended from $f_* X_i$. Thus $f_* : \mathcal{H} \longrightarrow \mathcal{G}$ is a homomorphism between Lie algebras.

When f is an isomorphism between Lie groups, f_* is also invertible. Hence $f_* : \mathcal{H} \longrightarrow \mathcal{G}$ is an isomorphism between Lie algebras. □

Definition 1.4. Suppose H, G are two Lie groups. If

1) H is a subgroup of G; and
2) the map $id : H \longrightarrow G$ is a manifold imbedding,

then H is called a **Lie subgroup** of G.

We have mentioned in Example 5 that if H is a regular submanifold of the Lie group G and H is a subgroup of G, then H must be a Lie group and hence a Lie subgroup of G. For instance SL $(n; \mathbb{R})$ and O $(n; \mathbb{R})$ are both Lie subgroups of GL $(n; \mathbb{R})$. But in general a Lie subgroup H of a Lie group G need not be a regular submanifold of G.

Example 7. Suppose $G = \mathbb{T}^2$. Choose an irrational number α, and let

$$
H = \{(t, \alpha t), t \in \mathbb{R}\}/L,
$$

where $L = \{(n_1, n_2), n_i \in \mathbb{Z}\}$. Then H is a Lie subgroup of G, but not a regular submanifold of G.

By Theorem 1.4, because the imbedding map $id : H \longrightarrow G$ is a homomorphism of Lie groups, it induces a homomorphism of Lie algebras from $\mathcal{H}$ to $\mathcal{G}$. Because H_e is a subspace of G_e, the Lie algebra multiplication (1.26) of G_e restricted to H_e is closed.

The general linear groups are typical Lie groups whose structures were already calculated in Example 6. Thus we usually study Lie groups via the general linear groups. Suppose G is an r-dimensional Lie group. A homomorphism from the Lie group G to GL $(n; \mathbb{R})$ is called a **representation** of order n of the Lie group G. Every r-dimensional Lie group has a natural representation of order r—the "adjoint" representation— defined below.

Suppose $x \in G$, and let

$$\alpha_x(g) = xgx^{-1} = L_x \circ R_{x^{-1}}(g). \tag{1.40}$$

Then α_x is an automorphism of the Lie group G, called an **inner automorphism** of G. By Theorem 1.4, the tangent map $(\alpha_x)_*$ of α_x determines an inner automorphism of the Lie algebra G_e. Let

$$\text{Ad}(x) = (\alpha_x)_* : G_e \longrightarrow G_e. \tag{1.41}$$

$\text{Ad}(x)$ is a nondegenerate linear transformation on the linear space G_e, and is therefore an element of GL $(r; \mathbb{R})$. Hence we obtain a map

$$\text{Ad} : G \longrightarrow \text{GL}(r; \mathbb{R}).$$

Obviously Ad is a homomorphism between groups, since for any $x, y \in G$ we have

$$\begin{aligned}
\text{Ad}(x \cdot y) &= (\alpha_{(x \cdot y)})_* \\
&= (\alpha_x \circ \alpha_y)_* \\
&= \text{Ad}(x) \circ \text{Ad}(y).
\end{aligned}$$

If we use local coordinates, Ad is given by smooth functions of the local coordinates. Hence Ad : $G \longrightarrow$ GL $(r; \mathbb{R})$ is a homomorphism between Lie groups.

Definition 1.5. The Lie group homomorphism Ad : $G \longrightarrow$ GL $(r; \mathbb{R})$ given by (1.41) is called the **adjoint representation** of the Lie group G.

By Theorem 1.4 the tangent map of the adjoint representation Ad : $G \longrightarrow$ GL $(r; \mathbb{R})$ induces a homomorphism, ad, from the Lie algebra G_e to gl $(r; \mathbb{R})$, called the **adjoint representation** of the Lie algebra G_e of the Lie group G. Thus gl $(r; \mathbb{R})$ can be viewed as a set of linear transformations from G_e to itself. For any $X \in G_e$, $\text{ad}(X)$ is a linear transformation on G_e. We shall prove in the next section that

$$\text{ad}(X) \cdot Y = -[X, Y]. \tag{1.42}$$

§6–2   Lie Transformation Groups

Transformation groups play an important role in geometry. According to Klein's viewpoint, the objects of study in geometry are the invariant properties of geometrical figures under the actions of specific transformation groups. The consideration of different transformation groups then leads to different kinds of geometry, such as Euclidean geometry, affine geometry and projective geometry. The action of a Lie group on a manifold, that is, of a Lie transformation group, is a generalization of these classical transformation groups. Its study has exerted great influence on modern differential geometry.

Definition 2.1. Suppose M is an m-dimensional smooth manifold. If there is a smooth map $\varphi : \mathbb{R} \times M \longrightarrow M$, denoted for any $(t, p) \in \mathbb{R} \times M$ by

$$\varphi_t(p) = \varphi(t, p),$$

such that the following conditions are satisfied:

1) $\varphi_0(p) = p$;
2) $\varphi_s \circ \varphi_t = \varphi_{s+t}$ for any real numbers s, t,

then we say $\mathbb{R}$ (left) operates on the manifold M **smoothly**, and call φ_t a **one-parameter group of diffeomorphisms** on M.

Obviously $\varphi_t : M \longrightarrow M$ is a smooth map. It follows immediately from the above conditions that $\varphi_t^{-1} = \varphi_{-t}$, that is, every φ_t is invertible. Thus φ_t is a diffeomorphism from M to itself. Choose $p \in M$ and let

$$\gamma_p(t) = \varphi_t(p). \tag{2.1}$$

Then γ_p is a parametrized curve through p on M, called the **orbit** of φ_t through p.

If we use X_p to denote a tangent vector of the orbit γ_p at p (i.e., $t = 0$), then we get a tangent vector field X on the manifold M called the **induced tangent vector field** of the one-parameter group of diffeomorphisms, φ_t, on M. It is obvious that X is smooth. Suppose f is a smooth function on M. Then

$$
\begin{aligned}
(Xf)(p) &= X_p f \\
&= \lim_{t \to 0} \frac{f(\gamma_p(t)) - f(p)}{t} \\
&= \lim_{t \to 0} \frac{f(\varphi(t, p)) - f(p)}{t}.
\end{aligned}
\tag{2.2}
$$

Thus Xf is a smooth function on M, which implies that X is smooth. The important point is that the orbit γ_p is the integral curve of the tangent vector field X, that is, for any point $q = \gamma_p(s)$ on the orbit γ_p, X_q is the tangent vector of γ_p at $t = s$. In fact, since $\gamma_q(t) = \gamma_p(t + s)$, we have

$$X_q = \lim_{t \to 0} \frac{\gamma_q(t) - q}{t} = \lim_{t \to 0} \frac{\gamma_p(t + s) - \gamma_p(s)}{t}. \tag{2.3}$$

From (2.3) we obtain

$$
\begin{aligned}
X_q f &= \lim_{t \to 0} \frac{f(\varphi(t + s, p)) - f(\varphi(s, p))}{t} \\
&= \lim_{t \to 0} \frac{f \circ \varphi_s(\gamma_p(t)) - f \circ \varphi_s(p)}{t} \\
&= X_p(f \circ \varphi_s) \\
&= ((\varphi_s)_* X_p) f,
\end{aligned}
$$

that is,

$$(\varphi_s)_* X_p = X_{\gamma_p(s)}. \tag{2.4}$$

The converse question is: given a tangent vector field X on M, does there exist a one-parameter group of diffeomorphisms such that X is the tangent vector field induced by φ_t? In other words, can a tangent vector field X determine a one-parameter group of diffeomorphisms? Theorem 2.1 below answers this question.

Definition 2.2. Suppose U is an open neighborhood in the smooth manifold M. If there is a smooth map $\varphi : (-\epsilon, \epsilon) \times U \longrightarrow M$, denoted by $\varphi_t(p) = \varphi(t, p)$ for any $p \in U$, $|t| < \epsilon$, which satisfies

1) for any $p \in U$, $\varphi_0(p) = p$;
2) if $|s| < \epsilon$, $|t| < \epsilon$, $|t+s| < \epsilon$ and $p, \varphi_t(p) \in U$, then $\varphi_{t+s}(p) = \varphi_s \circ \varphi_t(p)$,

then φ_t is called a **local one-parameter group of diffeomorphisms** acting on U.

A local one-parameter group of diffeomorphisms also induces a smooth tangent vector field on U. Suppose $p \in U$, and choose a local coordinate system $(V; x^i)$, $V \subset U$, at p. Due to the smoothness of φ, for sufficiently small positive $\epsilon_0 < \epsilon$, if $|t| < \epsilon_0$, then we always have $\varphi_t(p) \in V$. From (2.2) we have

$$X_p = \sum_{i=1}^{m} X_p^i \left(\frac{\partial}{\partial x^i} \right)_p, \tag{2.5}$$

where

$$X_p^i = \left. \frac{dx^i(\gamma_p(t))}{dt} \right|_{t=0}. \tag{2.6}$$

When p and $q = \gamma_p(s)$ are both in V, we also have

$$X_q = \sum_{i=1}^{m} X_q^i \left(\frac{\partial}{\partial x^i} \right)_q, \tag{2.7}$$

where

$$X_q^i = \frac{dx^i(\gamma_p(t))}{dt} \bigg|_{t=s}. \tag{2.8}$$

Theorem 2.1. *Suppose X is a smooth tangent vector field on M. Then for any point $p \in M$ there exists a neighborhood U and a local one-parameter group φ_t of diffeomorphisms on U, $|t| < \epsilon$, such that $X|_U$ is precisely the tangent vector field induced by φ_t on U.*

Proof. Choose a local coordinate system $(V; x^i)$ at p. Consider the system of ordinary differential equations

$$\frac{dx^i}{dt} = X^i, \qquad 1 \le i \le m, \tag{2.9}$$

where X^i are the components of the tangent vector field X with respect to the natural basis $\left\{ \frac{\partial}{\partial x^i}, 1 \le i \le m \right\}$, i.e.,

$$X = \sum_{i=1}^{m} X^i \frac{\partial}{\partial x^i}.$$

According to the theory of ordinary differential equations, there exist $\epsilon_1 > 0$ and a neighborhood $U_1 \subset V$ of p such that for any point $q \in U_1$, (2.9) has a unique integral curve $x_q(t)$ ($|t| < \epsilon_1$) passing through q, that is, it satisfies the following equations and initial conditions:

$$\begin{cases} \dfrac{dx_q^i(t)}{dt} &= X^i(x_q(t)), \qquad |t| < \epsilon_1, \quad 1 \le i \le m, \\[2mm] x_q(0) &= q. \end{cases} \tag{2.10}$$

Furthermore the solution $x_q(t)$ depends on (t, q) smoothly. Let

$$\varphi(t, q) = \varphi_t(q) = x_q(t), \qquad q \in U_1, \qquad |t| < \epsilon_1. \tag{2.11}$$

Then φ is a smooth map from $(-\epsilon_1, \epsilon_1) \times U_1$ to M. Now we show this is a local one-parameter group of diffeomorphisms.

Suppose $|t| < \epsilon_1, |s| < \epsilon_1, |t + s| < \epsilon_1$, and $q, \varphi_s(q) \in U_1$. Since $x_q(t + s)$ and $x_{\varphi_s(q)}(t)$ are both integral curves of (2.9) passing through $x_q(s) = \varphi_s(q)$, by the uniqueness property of the solution we have

$$\varphi_{t+s}(q) = x_q(t + s) = x_{\varphi_s(q)}(t) = \varphi_t \circ \varphi_s(q).$$

Thus φ_t is a local one-parameter group of diffeomorphisms which induces $X|_{U_1}$.
□

Suppose $X_p \neq 0$ at the point p. Then by Theorem 4.3 in Chapter 1, there exist local coordinates u^i near p such that $X = \partial/\partial u^1$. Then φ_t has the very simple expression:

$$\varphi_t(u^1, \dots, u^m) = (u^1 + t, u^2, \dots, u^m), \tag{2.12}$$

i.e., φ_t manifests itself as a displacement along the u^1-axis.

Remark. Obviously (2.9) is independent of the choice of local coordinates. If there are two coordinate neighborhoods V_1, V_2 such that $V_1 \cap V_2 \neq \varnothing$, and there are local one-parameter groups of diffeomorphisms $\varphi_t^{(1)}, \varphi_t^{(2)}$ acting on V_1, V_2, respectively, such that both are determined by the same tangent vector field X, then from the uniqueness of the solution of (2.9) we know that the actions of $\varphi_t^{(1)}$ and $\varphi_t^{(2)}$ are the same on $V_1 \cap V_2$.

Corollary. *Suppose X is a smooth tangent vector field on a smooth compact manifold M. Then X determines a one-parameter group of diffeomorphisms on M.*

Proof. By Theorem 2.1, there exist for every point p a neighborhood $U(p)$ and a positive number $\epsilon(p)$ such that there is a local one-parameter group of diffeomorphisms $\varphi_t(p)$ on $U(p)$. By the above remark, in the intersection of any two such $U(p)$, the actions of the corresponding one-parameter groups are the same. Due to the compactness of M, there is a finite subcovering, say $\{U_\alpha, 1 \leq \alpha \leq r\}$, of $\{U(p), p \in M\}$, with corresponding positive numbers ϵ_α. Let $\epsilon = \min\{\epsilon_\alpha | 1 \leq \alpha \leq r\}$. Now we can define a map $\varphi : (-\epsilon, \epsilon) \times M \longrightarrow M$ as follows: If $p \in U_\alpha$ then let

$$\varphi(t, p) = \varphi_t^{(\alpha)}(p), \qquad |t| < \epsilon. \tag{2.13}$$

It is easy to extend φ to a map from $\mathbb{R} \times M$ to M. Suppose t is an arbitrary real number. Then there is a positive integer N such that $|t|/N < \epsilon$. Hence

$$\varphi(t, p) = \left[\varphi_{t/N}\right]^N(p) \tag{2.14}$$

is independent of the choice of N, where on the right hand side we compose the local transformation $\varphi_{t/N}$ on M with itself N times. Obviously $\varphi : \mathbb{R} \times M \longrightarrow M$ is the one-parameter group of diffeomorphisms determined by the tangent vector field X.
□

Theorem 2.2. *Suppose φ_t is a one-parameter group of diffeomorphisms on a smooth manifold M, and X is the induced tangent vector field of φ_t on M. If $\psi : M \longrightarrow M$ is a diffeomorphism, then $\psi_* X$ is the tangent vector field induced by the one-parameter group of diffeomorphisms $\psi \circ \varphi_t \circ \psi^{-1}$ on M.*

Proof. Suppose f is any smooth function on M. Then by definition we have

$$
\begin{aligned}
(\psi_* X_p)f &= X_p(f \circ \psi) \\
&= \frac{d}{dt} f \circ \psi(\varphi_t(p))\Big|_{t=0} \\
&= \frac{d}{dt} f(\psi \circ \varphi_t \circ \psi^{-1}(\psi(p)))\Big|_{t=0},
\end{aligned}
$$

which means that $\psi_* X_p$ is the tangent vector at $\psi(p)$ of the orbit of the one-parameter transformation group $\psi \circ \varphi_t \circ \psi^{-1}$ through the point $\psi(p)$. Therefore $\psi_* X$ is the induced tangent vector field of $\psi \circ \varphi_t \circ \psi^{-1}$ on M.   □

Definition 2.3. Suppose X is a smooth tangent vector field on M, and $\psi : M \longrightarrow M$ is a diffeomorphism. If

$$\psi_* X = X, \tag{2.15}$$

then we say the vector field X is **invariant under** ψ.

From Theorem 2.2 we obtain

Corollary. *A necessary and sufficiently condition for a tangent vector field X to be invariant under a diffeomorphism $\psi : M \longrightarrow M$ is that the local one-parameter group of diffeomorphism φ_t determined by X commutes with ψ.*

Theorem 2.3. *Suppose X, Y are any two smooth tangent vector fields on a manifold M. If the local one-parameter group of diffeomorphisms generated by X is φ_t, then*

$$
\begin{aligned}
[X, Y] &= \lim_{t \to 0} \frac{Y - (\varphi_t)_* Y}{t} \\
&= \lim_{t \to 0} \frac{(\varphi_t^{-1})_* Y - Y}{t}.
\end{aligned}
\tag{2.16}
$$

Proof. We only need to show the first equality. Suppose $p \in M$, and f is a smooth function defined near p. Let

$$F(t) = f(\varphi_t(p)). \tag{2.17}$$

Since

$$F(t) - F(0) = \int_0^1 \frac{dF(st)}{ds} ds = t \int_0^1 F'(u)|_{u=st} \, ds,$$

it follows that

$$f(\varphi_t(p)) = f(p) + t g_t(p), \qquad (2.18)$$

where

$$g_t(p) = \int_0^1 F'(u)|_{u=st} \, ds = \int_0^1 \frac{df(\varphi_u(p))}{du} \bigg|_{u=st} ds, \qquad (2.19)$$

and

$$g_0(p) = \int_0^1 F'(u)|_{u=0} \, ds = \frac{df(\varphi_u(p))}{du} \bigg|_{u=0} = X_p f. \qquad (2.20)$$

Applying the middle operator in (2.16) to f we obtain

$$\begin{aligned}
\left(\lim_{t \to 0} \frac{Y - (\varphi_t)_* Y}{t} \right)_p f &= \lim_{t \to 0} \frac{Y_p f - Y_{\varphi_t^{-1}(p)} (f \circ \varphi_t)}{t} \\
&= \lim_{t \to 0} \frac{Y_p f - Y_{\varphi_t^{-1}(p)} f}{t} - \lim_{t \to 0} Y_{\varphi_t^{-1}(p)} (g_t) \\
&= X_p(Yf) - Y_p(Xf) \\
&= [X, Y]_p f.
\end{aligned}$$

Therefore

$$[X, Y] = \lim_{t \to 0} \frac{Y - (\varphi_t)_* Y}{t}.$$

□

Remark 1. Suppose γ_p is the orbit through p of the one-parameter group of diffeomorphisms φ_t. Because φ_t^{-1} maps the point $q = \gamma_p(t) = \varphi_t(p)$ in γ_p to the point p, $(\varphi_t^{-1})_*$ establishes a homomorphism from the tangent space $T_q(M)$ to the tangent space $T_p(M)$. If Y is a tangent vector field on M defined on the orbit γ_p, then $(\varphi_t^{-1})_* Y_{\varphi_t(p)}$ is a curve on the tangent space $T_p(M)$. Theorem 2.3 tells us that $[X, Y]_p$ is precisely the tangent vector of this curve at $t = 0$, hence it is the rate of change of the tangent vector field Y along the orbit of X. We usually call the operator on the right hand side of (2.16) the **Lie derivative** of the tangent vector field Y with respect to X, and denote it by $L_X Y$. Thus Theorem 2.3 becomes

$$L_X Y = [X, Y]. \tag{2.21}$$

The concept of the Lie derivative can be generalized to any tensor field on M. In fact, the map $(\varphi_t)^*$ establishes a homomorphism from the cotangent space $T_q^*(M)$ to the cotangent space $T_p^*(M)$. This map and $(\varphi_t^{-1})_*$ together then induce a homomorphism $\Phi_t : T_s^r(\varphi_t(p)) \longrightarrow T_s^r(p)$ between tensor spaces so that for any $v_1, \ldots, v_r \in T_{\varphi_t(p)}(M)$, and $v^{*1}, \ldots, v^{*s} \in T_{\varphi_t(p)}^*(M)$, we have

$$\begin{aligned} \Phi_t(v_1 \otimes \cdots \otimes v_r \otimes v^{*1} \otimes \cdots \otimes v^{*s}) \\ = (\varphi_t^{-1})_* v_1 \otimes \cdots \otimes (\varphi_t^{-1})_* v_r \otimes \varphi_t^* v^{*1} \otimes \cdots \otimes \varphi_t^* v^{*s}. \end{aligned} \tag{2.22}$$

Thus the Lie derivative $L_X \xi$ of the type-(r, s) tensor field ξ with respect to X is defined by

$$L_X \xi = \lim_{t \to 0} \frac{\Phi_t(\xi) - \xi}{t}. \tag{2.23}$$

Obviously $L_X \xi$ is also a type-(r, s) tensor field.

The Lie derivative $L_X f$ of the scalar field f with respect to the tangent vector field X is defined to be the directional derivative of f with respect to X.

Remark 2. The Lie derivative of an exterior differential form is a special case of the definition (2.23). Suppose ω is a differential r-form on M. Then $L_X \omega$ is still a differential r-form, and is defined by

$$L_X \omega = \lim_{t \to 0} \frac{\varphi_t^* \omega - \omega}{t}. \tag{2.24}$$

It is not difficult to verify that for any r smooth tangent vector fields $Y_1, \ldots, Y_r$ on M we have

$$\begin{aligned} \langle Y_1 \wedge \cdots \wedge Y_r, L_X \omega \rangle \\ = X \langle Y_1 \wedge \cdots \wedge Y_r, \omega \rangle - \sum_{i=1}^r \langle Y_1 \wedge \cdots \wedge L_X Y_i \wedge \cdots \wedge Y_r, \omega \rangle. \end{aligned} \tag{2.25}$$

For a smooth tangent vector field X on M, we can define a linear operator $i(X) : A^r(M) \longrightarrow A^{r-1}(M)$ as follows:

If $r = 0$, then $i(X)$ acts on $A^0(M)$ as the zero map.
If $r = 1$, $\omega \in A^r(M)$, then define

$$i(X)\omega = \langle X , \omega \rangle . \tag{2.26}$$

If $r > 1$, then for any $r - 1$ smooth tangent vector fields $Y_1, \ldots Y_{r-1}$, we have

$$\langle Y_1 \wedge \cdots \wedge Y_{r-1} , i(X)\omega \rangle = \langle X \wedge Y_1 \wedge \cdots \wedge Y_{r-1} , \omega \rangle . \tag{2.27}$$

It is then easy to verify the following formulas:

1) $L_X \circ i(Y) - i(Y) \circ L_X = i([X , Y])$;
2) $L_X \circ L_Y - L_Y \circ L_X = L_{[X , Y]}$;
3) $d \circ i(X) + i(X) \circ d = L_X$;
4) $d \circ L_X = L_X \circ d$.

This set of formulas is called the **H. Cartan formulas**, which play an important role in the theory of exterior differential forms. We leave the proofs to the reader.

Now we apply the discussion of one-parameter groups of diffeomorphisms to Lie groups. Suppose X is a right-invariant vector field on the r-dimensional Lie group G, and the local one-parameter group of diffeomorphisms determined by X is denoted by φ_t. Since a right translation R_a ($a \in G$) preserves the tangent vector field X, by the Corollary to Theorem 2.2, R_a commutes with φ_t, i.e.,

$$R_a \circ \varphi_t = \varphi_t \circ R_a. \tag{2.28}$$

Thus we see that if $\varphi_t(p)$ is defined in a neighborhood U of the identity element e and for $|t| < \epsilon$, then for any point $a \in G$, $\varphi_t(p)$ is defined in the neighborhood $U \cdot a$ of a and for $|t| < \epsilon$. This means that there exists a common $\epsilon > 0$ such that $\varphi_t(p)$ is defined for all $p \in M$ and $|t| < \epsilon$. Hence the right-invariant vector field X determines a one-parameter group of diffeomorphisms on the Lie group G (see the Corollary of Theorem 2.1). Let

$$a_t = \varphi_t(e). \tag{2.29}$$

Then

$$\begin{aligned}
a_{t+s} &= \varphi_{t+s}(e) = \varphi_t \circ \varphi_s(e) = \varphi_t \circ R_{a_s}(e) \\
&= R_{a_s} \circ \varphi_t(e) = R_{a_s}(a_t) = a_t \cdot a_s.
\end{aligned}$$

Hence a_t is a one-parameter subgroup of the Lie group G (that is, a 1-dimensional Lie subgroup).

From (2.28) we obtain

$$\varphi_t(x) = \varphi_t \circ R_x(e) = R_x \circ \varphi_t(e) = a_t \cdot x.$$

Thus the action of φ_t on G is the left translation on G determined by a_t, i.e.,

$$\varphi_t = L_{a_t}. \tag{2.30}$$

Precisely because of this, we usually also call a right-invariant vector field an **infinitesimal left translation**.

The above discussion shows that any right-invariant vector field X on a Lie group G determines a one-parameter subgroup a_t of G, and that the one-parameter group of diffeomorphisms φ_t determined by X on G is the left translation determined by a_t on G.

Theorem 2.4. *Suppose* $\mathrm{Ad} : G \longrightarrow GL(r; \mathbb{R})$ *is the adjoint representation of the r-dimensional Lie group G, and*

$$\mathrm{ad} = (\mathrm{Ad})_* : G_e \longrightarrow gl(r; \mathbb{R})$$

is the adjoint representation of the Lie algebra G_e of G. Then for any $X, Y \in G_e$ we have

$$\mathrm{ad}(X) \cdot Y = -[X, Y]. \tag{2.31}$$

Proof. Suppose the one-parameter subgroup determined by X is a_t. Then the one-parameter group of diffeomorphisms determined by the corresponding right-invariant vector field $\tilde{X}$ is $\varphi_t = L_{a_t}$. Suppose the corresponding right-invariant vector field for Y is $\tilde{Y}$. Since

$$\mathrm{ad}(X) = (\mathrm{Ad})_* X = \lim_{t \to 0} \frac{\mathrm{Ad}(a_t) - \mathrm{Ad}(e)}{t},$$

it follows from Theorem 2.3 that

$$
\begin{aligned}
\mathrm{ad}(X) \cdot Y &= \lim_{t \to 0} \frac{(L_{a_t})_* \circ (R_{a_t^{-1}})_* Y - Y}{t} \\
&= \lim_{t \to 0} \frac{(\varphi_t)_* \tilde{Y}_{a_t^{-1}} - \tilde{Y}_e}{t} \\
&= -[\tilde{X}, \tilde{Y}]_e \\
&= -[X, Y].
\end{aligned}
$$

$\square$

In the following we will focus on general Lie transformation groups.

Definition 2.4. Suppose M is an m-dimensional smooth manifold, and G is an r-dimensional Lie group. If there is a smooth map $\theta : G \times M \longrightarrow M$, denoted by

$$\theta(g, x) = g \cdot x, \qquad (g, x) \in G \times M,$$

that satisfies the following conditions:

1) if e is the identity element of G, then for any $x \in M$ we have

$$e \cdot x = x;$$

2) if $g_1, g_2 \in G$, then for any $x \in M$ we have

$$g_1 \cdot (g_2 \cdot x) = (g_1 \cdot g_2) \cdot x,$$

then we say that G is a **Lie transformation group** which acts on M (on the left).

Obviously, a one-parameter group of diffeomorphisms is a special example of a Lie transformation group, i.e., $G = \mathbb{R}$. The Lie group G acting on itself by left translations is also a Lie transformation group.

If for any element g in G that is not the identity element there exists a point $x \in M$ such that $g \cdot x \neq x$, then we say G acts on M **effectively**. If for any $g \neq e$ and any $x \in M$ we have $g \cdot x \neq x$, then we say that there is no fixed point for the action of G on M, or that G acts on M **freely**.

For a fixed $g \in G$, let

$$L_g(x) = g \cdot x, \qquad x \in M. \tag{2.32}$$

Then $L_g : M \longrightarrow M$ is a smooth map. Since $L_g^{-1} = L_{g^{-1}}$, L_g is a diffeomorphism from M to itself. Obviously $\{L_g | g \in G\}$ forms a subgroup of the group of diffeomorphisms on M. When G acts on M effectively, G is isomorphic to this subgroup.

A basic fact of Lie transformation groups is that there exists a finite-dimensional Lie algebra on M which is homomorphic to the Lie algebra of the Lie group G. First we construct a map from the Lie algebra G_e to the space of smooth tangent vector fields on M.

Suppose $X \in G_e$, and a_t is the one-parameter subgroup determined by X. Then L_{a_t} is a one-parameter group of diffeomorphisms acting on M. The tangent vector field $\tilde{X}$ induced by L_{a_t} on M is called the **fundamental tangent vector field** determined by X on M. By definition,

$$\tilde{X}_p = \lim_{t \to 0} \frac{L_{a_t}(p) - p}{t}. \tag{2.33}$$

Theorem 2.5. *Suppose G is a Lie transformation group acting on M. Then the set of all fundamental tangent vector fields on M forms a Lie algebra which is homomorphic to the Lie algebra G_e of G. If the action on M is effective, then the Lie algebra formed by the fundamental tangent vector fields is isomorphic to G_e.*

Proof. We know that the set $\Gamma(T(M))$ of smooth tangent vector fields on M forms an infinite-dimensional Lie algebra with respect to the Poisson bracket product. We need to show that the map

$$\sigma : G_e \longrightarrow \Gamma(T(M))$$

given in (2.33) is a homomorphism between Lie algebras.

The linear properties of σ are not obvious from (2.33). So we first introduce another representation for σ. For a fixed $p \in M$, let the map $\sigma_p : G \longrightarrow M$ be defined as follows:

$$\sigma_p(g) = L_g(p) = g \cdot p. \tag{2.34}$$

We will show that the tangent map $(\sigma_p)_* : G_e \longrightarrow T_p(M)$ is exactly the map given in (2.33), that is,

$$(\sigma_p)_* X = \tilde{X}_p, \qquad X \in G_e. \tag{2.35}$$

For this purpose, we only need to carry out a direct calculation. Suppose f is any smooth function on M. Then

$$
\begin{aligned}
((\sigma_p)_* X)f &= X(f \circ \sigma_p) \\
&= \frac{d}{dt} f \circ \sigma_p(a_t)\Big|_{t=0} \\
&= \frac{d}{dt} f(L_{a_t}(p))\Big|_{t=0} \\
&= \tilde{X}_p f,
\end{aligned}
$$

which implies (2.35). Therefore the map $\sigma : G_e \longrightarrow \Gamma(T(M))$ is given by

$$(\sigma(X))_p = (\sigma_p)_* X = \tilde{X}_p, \qquad X \in G_e. \tag{2.36}$$

Since the tangent map $(\sigma_p)_*$ is linear, σ is also linear.

σ can also be understood as a linear map from $\mathcal{G}$ to $\Gamma(T(M))$. Suppose X is a right-invariant vector field on G, and $\tilde{X} = \sigma(X_e)$. Then for any point $g \in G$ we have

$$
\begin{aligned}
(\sigma_p)_* X_g &= (\sigma_p)_* \circ (R_g)_* X_e \\
&= (\sigma_{(g \cdot p)})_* X_e \\
&= \tilde{X}_{g \cdot p}.
\end{aligned}
$$

Thus the fundamental tangent vector field $\tilde{X}$ can be viewed as an extension of the image of the right-invariant vector field X on G under the map $(\sigma_p)_*$. It follows that for any two right-invariant vector fields X, Y on G,

$$(\sigma_p)_* [X, Y]_g = [\tilde{X}, \tilde{Y}]_{\sigma_p(g)}.$$

Hence

$$\sigma([X_e, Y_e]) = [\tilde{X}, \tilde{Y}], \tag{2.37}$$

that is, $\sigma : G_e \longrightarrow \Gamma(T(M))$ is a homomorphism between Lie algebras whose image set is the Lie algebra formed by the fundamental tangent vector fields on M.

If $\tilde{X} = 0$, then the one-parameter group of diffeomorphisms L_{a_t} corresponding to $\tilde{X}$ is trivial, i.e., for any $X \in M$,

$$L_{a_t}(x) = a_t \cdot x = x.$$

If G acts on M effectively, then the above equation holds only if $a_t = e$; therefore $X = 0$, i.e., σ is a one-to-one map. This shows that the Lie algebra formed by the fundamental tangent vector fields on M is isomorphic to the Lie algebra of the Lie group G. □

It is easy to see that if G has no fixed point in M, then there exist exactly r fundamental tangent vector fields that are linearly independent everywhere. Any fundamental tangent vector field is then a linear combination of them with constant coefficients.

As an example, we consider the frame bundle P of a smooth manifold M. We mentioned in §4–3 that the structure group $GL(m; \mathbb{R})$ acts naturally on P as a Lie transformation group of left operators on P. Because locally P is a direct product $\pi^{-1}(U) \simeq U \times GL(m; \mathbb{R})$, under this representation, the structure group $GL(m; \mathbb{R})$ acts as left translations on fibers, that is,

$$A \cdot (p, B) = (p, A \cdot B), \tag{2.38}$$

where $p \in U$, and $A, B \in GL(m; \mathbb{R})$. Hence there are no fixed points for the action of $GL(m; \mathbb{R})$ on P, and a necessary and sufficient condition for any two elements of P to be equivalent under the action of $GL(m; \mathbb{R})$ is that these two elements (i.e. frames) have the same origin. The latter fact implies that the base manifold M is the quotient space of the frame bundle P with respect to the equivalence relation generated by the action of the group $GL(m; \mathbb{R})$.

The principal bundle is a generalization of the frame bundle. If we use the concept of a Lie transformation group, the principal bundle can be defined as follows. Suppose P and M are two smooth manifolds, and G an r-dimensional Lie transformation group acting on P from the left. If

1) there are no fixed points of the action of G on P;
2) M is the quotient space of the manifold P with respect to the equivalence relation defined by the group action of G, and the projection map $\pi : P \longrightarrow M$ is a smooth map; and
3) P is locally trivial, i.e., for every point $x \in M$ there exists a neighborhood U of x such that $\pi^{-1}(U)$ is isomorphic to $U \times G$, which means that there exists a diffeomorphism

$$p \in \pi^{-1}(U) \longrightarrow (\pi(p), \varphi(p)) \in U \times G$$

such that for any $a \in G$ we have

$$\varphi(a \cdot p) = a \cdot \varphi(p),$$

then we say P is the **principal bundle** on M with the Lie group G as its structure group.

The fiber $\pi^{-1}(x)$ above a point $x \in M$ is the orbit

$$G \cdot p = \{L_a(p)|a \in G\}$$

of the Lie transformation group G on P that passes through the point $p \in \pi^{-1}(x)$. The Lie algebra formed by the fundamental tangent vector fields on P is isomorphic to the Lie algebra of the Lie group G. Because G acts freely, the fundamental tangent vectors of P at each of its points span an r-dimensional tangent subspace which is precisely the tangent space of the fiber $\pi^{-1}(x)$ at that point, called the **vertical space**. This setup allows one to develop the theory of connections on principal bundles (see Kobayashi and Nomizu 1963 and 1969).

§6–3   The Method of Moving Frames

Suppose M is an m-dimensional connected smooth manifold, and G is an r-dimensional Lie group. Consider the right fundamental differential forms of G, $\omega^i (1 \le i \le r)$, which satisfy the structure equations

$$d\omega^i = -\frac{1}{2} \sum_{j,k=1}^{r} c_{jk}^i \, \omega^j \wedge \omega^k, \tag{3.1}$$

where the c_{jk}^i are the structure constants of the Lie group G. For a smooth map $f : M \longrightarrow G$, let

$$\psi^i = f^* \omega^i. \tag{3.2}$$

Then the ψ^i satisfy the same system of equations

$$d\psi^i = -\frac{1}{2} \sum_{j,k=1}^{r} c_{jk}^i \, \psi^j \wedge \psi^k. \tag{3.3}$$

This is also a sufficient condition for the map $f : M \longrightarrow G$ to exist locally.

Theorem 3.1. *Suppose there exist r differential 1-forms ψ^i $(1 \le i \le r)$ on M that satisfy Equations (3.3), where c_{jk}^i are the structure constants of a Lie group G. Then there exists for every point $p \in M$ a neighborhood U and a smooth map $f : U \longrightarrow G$ such that*

$$f^* \omega^i = \psi^i, \tag{3.4}$$

where ω^i is a right fundamental differential form of G. If f_1, f_2 are any two such maps, then there exists an element g in G such that

$$f_2 = R_g \circ f_1, \tag{3.5}$$

that is, the images of f_1 and f_2 differ by just a right translation of G.

Proof. Consider the system of Pfaffian equations of $m+r$ independent variables on $M \times G$:

$$\theta^i \equiv \psi^i - \omega^i = 0, \qquad 1 \le i \le r. \tag{3.6}$$

Since the ω^i are linearly independent everywhere, the θ^i are also linearly independent everywhere. Equation (3.6) gives an m-dimensional plane distribution on $M \times G$. Because

$$
\begin{aligned}
d\theta^i &= -\frac{1}{2} \sum_{j,k=1}^{r} c_{jk}^i \left(\psi^j \wedge \psi^k - \omega^j \wedge \omega^k \right) \\
&= -\frac{1}{2} \sum_{j,k=1}^{r} c_{jk}^i \left(\psi^j \wedge \theta^k + \theta^j \wedge \omega^k \right) \\
&\equiv 0 \bmod \left(\theta^1, \dots, \theta^r \right),
\end{aligned}
$$

by the Frobenius Theorem, the system of equations (3.6) is completely integrable. Hence, for any point $(x_0, a_0) \in M \times G$, there exist local coordinate systems $(U; x^\alpha)$ at x_0 and $(V; a^i)$ at a_0 such that (3.6) has a unique m-dimensional integral manifold

$$\varphi^i \left(x^1, \dots, x^m; a^1, \dots, a^r \right) = 0, \qquad 1 \le i \le r, \tag{3.7}$$

in $U \times V$ that passes through (x_0, a_0), where $x \in U$, $a \in V$.

Since the ω^i $(1 \leq i \leq r)$ are linearly independent, there must exist a neighborhood $U_1 \subset U$ of x_0 in which we can solve (3.7) to get

$$f^i = f^i \left(x^1, \dots, x^m \right), \qquad 1 \leq i \leq r, \tag{3.8}$$

such that

$$\varphi^i \left(x^1, \dots, x^m; f^1(x), \dots, f^r(x) \right) \equiv 0, \qquad x \in U_1, \tag{3.9}$$

$$f^i \left(x_0^1, \dots, x_0^m \right) = a_0^i, \qquad 1 \leq i \leq r.$$

Obviously, the map $f : U_1 \longrightarrow G$ given by (3.8) satisfies

$$\psi^i = f^* \omega^i, \qquad 1 \leq i \leq r.$$

Suppose $f_1, f_2 : U_1 \longrightarrow G$ are two such maps, with

$$f_1(x_0) = a_1, \qquad f_2(x_0) = a_2. \tag{3.10}$$

Let

$$g = a_1^{-1} \cdot a_2. \tag{3.11}$$

Then

$$(R_g \circ f_1)^* \omega^i = (f_1)^* \circ (R_g)^* \omega^i = (f_1)^* \omega^i = \psi^i, \tag{3.12}$$

and

$$(R_g \circ f_1) (x_0) = a_2. \tag{3.13}$$

Thus $R_g \circ f_1$ and f_2 are both solutions of (3.6) and satisfy the same initial condition. By the uniqueness property of solutions we have

$$f_2 = R_g \circ f_1.$$

$\square$

Now consider the rigid motion group $E(N)$ of the N-dimensional Euclidean space $\mathbb{R}^N$. Choose an orthogonal frame $(O; \delta_1, \dots, \delta_N)$ in $\mathbb{R}^N$. Then the action (denoted as a right action) by $\tilde{a} \in E(N)$ on $\mathbb{R}^N$ is

$$x \cdot \tilde{a} = x \cdot A + a, \tag{3.14}$$

where

$$
\begin{cases}
x = \left(x^1, \ldots, x^N\right) = \sum_{\alpha=1}^{N} x^\alpha \delta_\alpha, \\[2mm]
a = \left(a^1, \ldots, a^N\right) = \sum_{\alpha=1}^{N} a^\alpha \delta_\alpha, \\[2mm]
A = \begin{pmatrix} a_1^1 & \cdots & a_1^N \\ \vdots & \ddots & \vdots \\ a_N^1 & \cdots & a_N^N \end{pmatrix}, \\[2mm]
A \cdot {}^t\!A = I, \qquad \det A > 0,
\end{cases}
\tag{3.15}
$$

i.e., the matrix A is an orthogonal matrix with determinant $+1$. Hence an element $\tilde{a}$ in $E(N)$ can be expressed by a pair of matrices (A, a).

Suppose $\tilde{a} = (A, a)$, $\tilde{b} = (B, b) \in E(N)$, and the action $\tilde{a} \cdot \tilde{b}$ on $\mathbb{R}^N$ is defined to be the action by applying the actions $\tilde{a}$ and $\tilde{b}$ successively. Therefore the group multiplication in $E(N)$ is given by

$$
\tilde{a} \cdot \tilde{b} = (A \cdot B, a \cdot B + b), \tag{3.16}
$$

and the inverse element of $\tilde{a}$ is

$$
\tilde{a}^{-1} = (A^{-1}, -a \cdot A^{-1}). \tag{3.17}
$$

Obviously $E(N)$ is a $\frac{1}{2}N(N+1)$-dimensional Lie group.

Let $\mathcal{F}\left(\mathbb{R}^N\right)$ be the set of all orthogonal frames on $\mathbb{R}^N$, and $\mathcal{F}_+\left(\mathbb{R}^N\right)$ the set of orthogonal frames on $\mathbb{R}^N$ that have the same orientation as the fixed frame $(O; \delta_1, \ldots, \delta_N)$. They are both principal bundles on $\mathbb{R}^N$ with $O(N; \mathbb{R})$ and $SO(N; \mathbb{R})$ as the structure groups, respectively. [$SO(N; \mathbb{R})$ is called the **special orthogonal group** and is composed of orthogonal matrices with determinant $+1$.]

The manifold $\mathcal{F}_+(\mathbb{R}^N)$ can be identified with $E(N)$ because, for any orthogonal frame $(p; e_1, \ldots, e_N)$ that has the same orientation as $(O; \delta_1, \ldots, \delta_N)$ in $\mathbb{R}^N$, there exists a rigid motion $\tilde{a} \in E(N)$ that transforms $(O; \delta_1, \ldots, \delta_N)$ to $(p; e_1, \ldots, e_N)$. The correspondence relation is

$$
\tilde{a} = (A, a) \longleftrightarrow (p; e_1, \ldots, e_N), \tag{3.18}
$$

where

$$\begin{cases} \overrightarrow{Op} = \displaystyle\sum_{\alpha=1}^{N} a^{\alpha}\delta_{\alpha}, \\ e_{\alpha} = \displaystyle\sum_{\beta=1}^{N} a_{\alpha}^{\beta}\delta_{\beta}. \end{cases} \tag{3.19}$$

If we denote

$$\begin{cases} \delta = {}^{t}(\delta_{1},\dots,\delta_{N}), \\ e = {}^{t}(e_{1},\dots,e_{N}), \end{cases} \tag{3.20}$$

then (3.19) can be written as

$$\begin{cases} \overrightarrow{Op} = a \cdot \delta, \\ e = A \cdot \delta. \end{cases} \tag{3.21}$$

Apply an infinitesimal motion on the frame $(p; e_1, \dots, e_N)$ to get $(p + dp; e_\alpha + de_\alpha)$. The vectors dp and de_α can still be expressed with respect to the frame $(p; e_1, \dots, e_N)$. Let

$$\begin{cases} dp = \displaystyle\sum_{\alpha=1}^{N} \omega^{\alpha} e_{\alpha}, \\ de_{\alpha} = \displaystyle\sum_{\beta=1}^{N} \omega_{\alpha}^{\beta} e_{\beta}. \end{cases} \tag{3.22}$$

The one-forms ω^{α}, ω_{α}^{β} $(1 \leq \alpha, \beta \leq N)$ are called the **relative components** of the **moving frame**. If we denote $\theta = (\omega^1, \dots, \omega^N)$, $\omega = (\omega_{\alpha}^{\beta})$, then (3.22) can be written as

$$\begin{cases} dp = \theta \cdot e, \\ de = \omega \cdot e. \end{cases} \tag{3.23}$$

Differentiating (3.21) we immediately obtain

$$dp = da \cdot \delta = da \cdot A^{-1} \cdot e,$$
$$de = dA \cdot \delta = dA \cdot A^{-1} \cdot e.$$

Therefore

$$\begin{aligned} \theta &= da \cdot A^{-1}, \\ \omega &= dA \cdot A^{-1}. \end{aligned} \tag{3.24}$$

Since $A \cdot A^{-1} = I$, we have

$$dA \cdot A^{-1} + A \cdot dA^{-1} = dA \cdot A^{-1} + {}^t\left(dA \cdot A^{-1}\right) = 0,$$

that is,

$$\omega_\alpha^\beta + \omega_\beta^\alpha = 0. \tag{3.25}$$

From the viewpoint of the Lie group $E(N)$, the relative components ω^α and $\omega_\alpha^\beta = -\omega_\beta^\alpha$ are precisely the right fundamental differential forms on $E(N)$. In fact, for any element $\tilde{b} = (B, b) \in E(N)$ we have

$$R_{\tilde{b}}(\tilde{a}) = \tilde{a} \cdot \tilde{b} = (A \cdot B, a \cdot B + b).$$

Hence

$$\begin{aligned} \left(R_{\tilde{b}}\right)^* \theta = (da \cdot B) \cdot (A \cdot B)^{-1} &= da \cdot A^{-1} = \theta, \\ \left(R_{\tilde{b}}\right)^* \omega = (dA \cdot B) \cdot (A \cdot B)^{-1} &= dA \cdot A^{-1} = \omega. \end{aligned}$$

Since we are using orthogonal frames, the relative components can all be expressed with lower indices, i.e.,

$$\omega_\alpha = \omega^\alpha, \qquad \omega_{\alpha\beta} = \omega_\alpha^\beta,$$

which can also be viewed as a result of the **lowering** of indices by the metric tensor $g_{\alpha\beta} = e_\alpha \cdot e_\beta$ in $\mathbb{R}^N$. Exteriorly differentiating (3.23) we obtain

$$\begin{aligned} d\theta \cdot e - \theta \wedge de &= 0, \\ d\omega \cdot e - \omega \wedge de &= 0. \end{aligned}$$

Thus the structure equations of $E(N)$ are

$$\begin{cases} d\theta &= \theta \wedge \omega, \\ d\omega &= \omega \wedge \omega, \end{cases} \tag{3.26}$$

or

$$\begin{cases} d\omega_\alpha = \displaystyle\sum_{\gamma=1}^{N} \omega_\beta \wedge \omega_{\beta\alpha}, \\ d\omega_{\alpha\beta} = \displaystyle\sum_{\gamma=1}^{N} \omega_{\alpha\gamma} \wedge \omega_{\gamma\beta}. \end{cases} \qquad (3.27)$$

Applying Theorem 3.1 to $E(N)$, we immediately obtain the Fundamental Theorem for Moving Frames in $\mathbb{R}^N$:

Theorem 3.2. *Suppose ψ_α and $\psi_{\beta\gamma} = -\psi_{\gamma\beta}$ $(1 \leq \alpha, \beta, \gamma \leq N)$ are differential 1-forms depending on n variables. Then there exists a family of orthogonal frames depending on n parameters and having the given differential forms as their relative components if and only if the differential forms satisfy the following equations:*

$$\begin{cases} d\psi_\alpha = \displaystyle\sum_{\beta=1}^{N} \psi_\beta \wedge \psi_{\beta\alpha}, \\ d\psi_{\alpha\beta} = \displaystyle\sum_{\gamma=1}^{N} \psi_{\alpha\gamma} \wedge \psi_{\gamma\beta}. \end{cases} \qquad (3.28)$$

Moreover any two such families of orthogonal frames are related by a rigid motion in $\mathbb{R}^N$.

Proof. Let M denote the space of variables $x = (x^1, \ldots, x^n)$, and $G = E(N)$. By theorem 3.1 there exists a map $f : M \longrightarrow E(N)$ such that

$$f^*\omega_\alpha = \psi_\alpha, \qquad f^*\omega_{\alpha\beta} = \psi_{\alpha\beta}.$$

Let

$$\tilde{a} = f(x) = (A(x), a(x)),$$

and

$$\begin{cases} \overrightarrow{Op}(x) = a(x) \cdot \delta, \\ e(x) = A(x) \cdot \delta. \end{cases} \qquad (3.29)$$

Then $(p(x); e_1(x), \ldots, e_N(x))$ gives the required family of orthogonal frames.

$\square$

The concept of moving frames originates from mechanics. For example, in studying the motion of a rigid body, an orthogonal frame is attached rigidly to the moving object. As the rigid body moves, the attached frame is carried along. In this way, a family of orthogonal frames parametrized by the time t is obtained. This family completely describes the motion of the rigid body. The French mathematicians Cotton and Darboux generalized the concept of a one-parameter family of frames to the multi-parameter case; but it was E. Cartan who developed and pushed this theory to new heights, and successfully applied it to geometrical studies. At present, the combined use of the method of moving frames and exterior differentiation has become a powerful tool in differential geometry. In what follows we will use this approach to study submanifolds of Euclidean spaces.

Suppose $f : M \longrightarrow \mathbb{R}^N$ is an imbedding of an m-dimensional oriented smooth submanifold in $\mathbb{R}^N$. The range of values for the indices used below are:

$$1 \leq i, j, k, l \leq m,$$

$$m + 1 \leq A, B, C, D \leq N, \tag{3.30}$$

$$1 \leq \alpha, \beta, \gamma, \delta \leq N.$$

For convenience in writing, we will not distinguish between M and $f(M)$. Attach an orthogonal frame $(p; e_1, \ldots, e_N)$ to every point p in M such that e_i is a tangent vector of M at p, e_A is a normal vector of M at p, $(e_1, \ldots, e_m)$ has the same orientation as M, and $(e_1, \ldots, e_N)$ has the same orientation as a fixed frame $(O; \delta_1, \ldots, \delta_N)$ in $\mathbb{R}^N$. Suppose there is a frame field on an open neighborhood U of M which depends continuously and smoothly on the local coordinates of U. Then we usually call such a local orthogonal frame field a **Darboux frame** on the submanifold M. Obviously, there always exists a Darboux frame in a sufficiently small neighborhood of every point in M. Furthermore they are susceptible to the following transformations:

$$\begin{cases} e'_i &= \displaystyle\sum_{j=1}^{m} a_{ij} e_j, \\ e'_A &= \displaystyle\sum_{B=m+1}^{N} a_{AB} e_B, \end{cases} \tag{3.31}$$

where a_{ij}, a_{AB} are smooth functions on U, and $(a_{ij}) \in \mathrm{SO}\,(m; \mathbb{R})$, $(a_{AB}) \in \mathrm{SO}\,(N - m; \mathbb{R})$.

If we choose a Darboux frame in a neighborhood U in M, then it provides a smooth map f from U to $\mathcal{F}_+(\mathbb{R}^N)$. We will still use ω_α, $\omega_{\alpha\beta}$ to denote the differential 1-forms obtained by pulling the relative components of moving frames in $\mathbb{R}^N$ back to U by f^*. Obviously these 1-forms on U still satisfy the structure equations (3.27).

Since the origin p of a Darboux frame is in M, and e_i is a tangent vector of M at p, we have

$$dp = \sum_{i=1}^{m} \omega_i e_i, \qquad \omega_A = 0, \tag{3.32}$$

and the ω_i $(1 \le i \le m)$ are linearly independent everywhere. Suppose

$$I = dp \cdot dp = \sum_{i=1}^{m} (\omega_i)^2, \tag{3.33}$$

$$dA = \omega_1 \wedge \cdots \wedge \omega_m. \tag{3.34}$$

It is easy to verify that these quantities are independent of the transformation (3.31) of Darboux frames, that is, they are defined on the whole manifold M, and are called the **first fundamental form** and the **area element** of M, respectively. With I as the Riemannian metric, the manifold M becomes a Riemannian manifold. We say that the Riemannian manifold M has a Riemannian metric induced from $\mathbb{R}^N$.

The equations of motion for a Darboux frame can be written

$$\begin{cases} de_i = \displaystyle\sum_{j=1}^{m} \omega_{ij} e_j + \sum_{A=m+1}^{N} \omega_{iA} e_A, \\[2mm] de_B = \displaystyle\sum_{j=1}^{m} \omega_{Bj} e_j + \sum_{A=m+1}^{N} \omega_{BA} e_A, \end{cases} \tag{3.35}$$

where $\omega_\alpha, \omega_{\alpha\beta} = -\omega_{\beta\alpha}$ are the relative components mentioned previously, which satisfy the structure equations

$$\begin{cases} d\omega_i = \displaystyle\sum_{j=1}^{m} \omega_j \wedge \omega_{ji} \\[2mm] 0 = \displaystyle\sum_{j=1}^{m} \omega_j \wedge \omega_{jA}, \end{cases} \tag{3.36}$$

$$\begin{cases} d\omega_{ij} = \displaystyle\sum_{k=1}^{m} \omega_{ik} \wedge \omega_{kj} + \sum_{A=m+1}^{N} \omega_{iA} \wedge \omega_{Aj}, \\[2mm] d\omega_{iB} = \displaystyle\sum_{k=1}^{m} \omega_{ik} \wedge \omega_{kB} + \sum_{A=m+1}^{N} \omega_{iA} \wedge \omega_{AB}, \\[2mm] d\omega_{AB} = \displaystyle\sum_{k=1}^{m} \omega_{Ak} \wedge \omega_{kB} + \sum_{C=m+1}^{N} \omega_{AC} \wedge \omega_{CB}. \end{cases} \qquad (3.37)$$

By the Fundamental Theorem of Riemannian Geometry, the first formula of (3.36) and the skew-symmetry $\omega_{ij} + \omega_{ji} = 0$ together imply that ω_{ij} is the Levi-Civita connection on the Riemannian manifold M:

$$De_i = \sum_{j=1}^{m} \omega_{ij} e_j. \qquad (3.38)$$

By the first formula in (3.35) we know that De_i is the orthogonal projection of de_i on a tangent plane of M.

By Cartan's Lemma (see Theorem 3.4 of Chapter 2) we obtain from the second formula of (3.36) that

$$\omega_{jA} = \sum_{i=1}^{m} h_{Aji} \omega_i, \qquad h_{Aji} = h_{Aij}. \qquad (3.39)$$

Let

$$\begin{aligned} \mathrm{II} &= \sum_{i,A} \omega_i \omega_{iA} e_A \\ &= \sum_{A=m+1}^{N} \left(\sum_{i,j=1}^{m} h_{Aij} \omega_i \omega_j \right) e_A. \end{aligned} \qquad (3.40)$$

Then II is independent of the transformation (3.31) of Darboux frames. It is thus a differential 2-form defined on the whole manifold M, taking values on the space of normal vectors to M, and is called the **second fundamental form** of the submanifold M.

The curvature form of the Levi-Civita connection on M is

$$\begin{aligned} \Omega_{ij} &= d\omega_{ij} - \sum_{k=1}^{m} \omega_{ik} \wedge \omega_{kj} \\ &= \frac{1}{2} \sum_{k,l=1}^{m} R_{ijkl} \omega_k \wedge \omega_l, \end{aligned}$$

where R_{ijkl} is the curvature tensor. From the first formula in (3.37) we obtain

$$R_{ijkl} = \sum_{A=m+1}^{N} (h_{Ail}h_{Ajk} - h_{Aik}h_{Ajl}),\qquad (3.41)$$

which is the **Gauss equation** for the submanifold M. The last two formulas in (3.37) are the **Codazzi equations** from the theory of surfaces.

For hypersurfaces in a Euclidean space, the above formulas can be greatly simplified. Suppose M is an oriented hypersurface in $\mathbb{R}^{m+1}$. Then the Darboux frame of M has only one normal vector e_{m+1}. In this case the equations of motion for the Darboux frame are

$$dp = \sum_{i=1}^{m} \omega_i e_i, \qquad \omega_{m+1} = 0,$$

$$de_i = \sum_{j=1}^{m} \omega_{ij} e_j + \omega_{i\,m+1}\, e_{m+1},$$

$$de_{m+1} = \sum_{j=1}^{m} \omega_{m+1\,j}\, e_j.$$

The structure equations are

$$d\omega_i = \sum_{j=1}^{m} \omega_j \wedge \omega_{ji},\qquad (3.42)$$

$$\sum_{j=1}^{m} \omega_j \wedge \omega_{j\,m+1} = 0,\qquad (3.43)$$

$$d\omega_{ij} = \sum_{k=1}^{m} \omega_{ik} \wedge \omega_{kj} + \omega_{i\,m+1} \wedge \omega_{m+1\,j},\qquad (3.44)$$

$$d\omega_{i\,m+1} = \sum_{k=1}^{m} \omega_{ik} \wedge \omega_{k\,m+1}.\qquad (3.45)$$

Equation (3.39) then becomes

$$\omega_{j\,m+1} = \sum_{i=1}^{m} h_{ji}\omega_i, \qquad h_{ji} = h_{ij}.\qquad (3.46)$$

Hence we obtain the following expression for the second fundamental form of the hypersurface M:

$$\text{II} = \sum_{i=1}^{m} \omega_i \omega_{i\ m+1} = \sum_{i,j=1}^{m} h_{ij} \omega_i \omega_j. \tag{3.47}$$

Equations (3.44) and (3.45) are the Gauss equation and the Codazzi equation, respectively. Plugging (3.46) into the Gauss equation (3.44) we obtain

$$R_{ijkl} = h_{il}h_{jk} - h_{ik}h_{jl}, \tag{3.48}$$

which is the usual form for the Gauss equation.

Plugging (3.46) into the Codazzi equation we obtain

$$\sum_{j=1}^{m} \left(dh_{ij} - \sum_{k=1}^{m} h_{ik}\omega_{jk} - \sum_{k=1}^{m} h_{kj}\omega_{ik} \right) \wedge \omega_j = 0.$$

Therefore, it follows from Cartan's Lemma that

$$\begin{cases} dh_{ij} - \displaystyle\sum_{k=1}^{m} h_{ik}\omega_{jk} - \sum_{k=1}^{m} h_{kj}\omega_{ik} = \sum_{k=1}^{m} h_{ijk}\omega_k, \\ h_{ijk} = h_{ikj}. \end{cases} \tag{3.49}$$

If we let

$$\begin{cases} dh_{ij} = \displaystyle\sum_{k=1}^{m} h_{ij,k}\omega_k, \\ \omega_{ij} = \displaystyle\sum_{k=1}^{m} \Gamma_{ijk}\omega_k, \end{cases} \tag{3.50}$$

then

$$h_{ijk} = h_{ij,k} - \sum_{l=1}^{m} h_{il}\Gamma_{jlk} - \sum_{l=1}^{m} h_{lj}\Gamma_{ilk}. \tag{3.51}$$

Thus the Codazzi equation becomes

$$h_{ij,k} - h_{ik,j} = \sum_{l=1}^{m} \left(h_{il}\Gamma_{jlk} + h_{lj}\Gamma_{ilk} - h_{il}\Gamma_{klj} - h_{lk}\Gamma_{ilj} \right). \tag{3.52}$$

As a direct corollary of Theorem 3.2, we have the following fundamental theorem for hypersurfaces in $\mathbb{R}^{m+1}$:

Theorem 3.3. *Suppose there exist two differential 2-forms*

$$I = \sum_{i=1}^{m}(\omega_i)^2, \qquad II = \sum_{i,j=1}^{m} h_{ij}\omega_i\omega_j, \qquad (3.53)$$

where the ω_i ($1 \leq i \leq m$) are linearly independent differential 1-forms depending on m variables; and $h_{ij} = h_{ji}$ are functions of these m variables. Then a necessary and sufficient condition for a hypersurface to exist in $\mathbb{R}^{m+1}$ with I and II as its first and second fundamental forms is: I and II satisfy the Gauss–Codazzi equations (3.48) and (3.52), where Γ_{ijk} is the Levi-Civita connection determined by I, and R_{ijkl} is the corresponding curvature tensor. Moreover any two such hypersurfaces in $\mathbb{R}^{m+1}$ are related by a rigid motion.

§6–4   Theory of Surfaces

In the previous section, we used the method of moving frames to show that the first fundamental form I and second fundamental form II constitute a complete invariant system on hypersurfaces in $\mathbb{R}^{m+1}$, and that the Gauss–Codazzi equations are the integrability conditions satisfied by I and II. Hypersurfaces have very rich geometrical contents. Here we will use surfaces in $\mathbb{R}^3$ as an example to discuss their geometry.

Suppose $x : M \longrightarrow \mathbb{R}^3$ is a smooth surface in $\mathbb{R}^3$. If we choose local coordinates u^1, u^2 in a coordinate neighborhood U in M, then the surface x can be expressed by the parametrized equations

$$x^i = x^i(u^1, u^2), \qquad 1 \leq i \leq 3. \qquad (4.1)$$

Since x is an imbedding, the rank of the matrix

$$\begin{pmatrix} \dfrac{\partial x^1}{\partial u^1} & \dfrac{\partial x^2}{\partial u^1} & \dfrac{\partial x^3}{\partial u^1} \\[2ex] \dfrac{\partial x^1}{\partial u^2} & \dfrac{\partial x^2}{\partial u^2} & \dfrac{\partial x^3}{\partial u^2} \end{pmatrix}$$

is 2, and x is a one to one map.

Choose a Darboux frame $(x; e_1, e_2, e_3)$ on M such that e_1, e_2 are tangent to M, e_3 is a normal vector to M, and the orientation of (e_1, e_2, e_3) is the same as a chosen orientation of $\mathbb{R}^3$. (e_1, e_2) determines the orientation of M. Suppose the corresponding relative components for the frame field are ω_i, ω_{ij}, i.e.,

$$dx = \omega_1 e_1 + \omega_2 e_2, \qquad \omega_3 = 0, \qquad (4.2)$$

$$\begin{cases} de_1 = & \omega_{12}e_2 & +\omega_{13}e_3, \\ de_2 = \omega_{21}e_1 & & +\omega_{23}e_3, \\ de_3 = \omega_{31}e_1 & +\omega_{32}e_2, \end{cases} \tag{4.3}$$

$$\omega_{ij} + \omega_{ji} = 0,$$

where ω_i, ω_{ij} are differential 1-forms of the parameters u^1, u^2. The structure equations are

$$\begin{cases} d\omega_1 = & \omega_2 \wedge \omega_{21}, \\ d\omega_2 = & \omega_1 \wedge \omega_{12}. \end{cases} \tag{4.4}$$

$$0 = \omega_1 \wedge \omega_{13} + \omega_2 \wedge \omega_{23}, \tag{4.5}$$

$$d\omega_{12} = \omega_{13} \wedge \omega_{32}, \tag{4.6}$$

$$\begin{cases} d\omega_{13} = & \omega_{12} \wedge \omega_{23}, \\ d\omega_{23} = & \omega_{21} \wedge \omega_{13}. \end{cases} \tag{4.7}$$

As described in §6–3, (4.6) and (4.7) are the Gauss equation and the Codazzi equation of the surface, respectively. Each of these equations contains a wealth of information. The study of the local geometry of a surface depends on a proper understanding of them.

The first fundamental form of M is

$$I = dx \cdot dx = (\omega_1)^2 + (\omega_2)^2, \tag{4.8}$$

and the area element is

$$dA = \omega_1 \wedge \omega_2. \tag{4.9}$$

They are both invariant under admissible transformations of Darboux frames.

By Cartan's Lemma, it follows from (4.5) that

$$\begin{cases} \omega_{13} = & h_{11}\omega_1 + h_{12}\omega_2, \\ \omega_{23} = & h_{21}\omega_1 + h_{22}\omega_2, \qquad h_{12} = h_{21}. \end{cases} \tag{4.10}$$

By using the inner product we can obtain the **second** and **third fundamental forms** :

$$II = d^2x \cdot e_3 = -dx \cdot de_3 = \omega_1\omega_{13} + \omega_2\omega_{23}, \tag{4.11}$$

or

$$II = h_{11}(\omega_1)^2 + 2h_{12}\omega_1\omega_2 + h_{22}(\omega_2)^2; \tag{4.12}$$

and

$$III = de_3 \cdot de_3 = (\omega_{13})^2 + (\omega_{23})^3. \tag{4.13}$$

At a fixed point $x \in M$, the ratio $\omega_1 : \omega_2$ determines a tangent direction v at x on M. Obviously

$$\kappa_n = \frac{II}{I} \tag{4.14}$$

is a function of (x, v), called the **normal curvature** of M at x along the direction v. It is easy to show that the normal curvature has the following geometric meaning. The plane determined by the tangent direction v and the normal vector e_3 intersects M at a planar curve called the **normal sectional curve** of the surface M along v. Then the normal curvature κ_n at the point x is exactly the curvature of this normal sectional curve at x.

The second fundamental form II can also be interpreted as a linear transformation field on the tangent spaces of the surface M. The third formula of equation (4.3) can be viewed as a differential 1-form defined on M with tangent vector values, so it defines at every point $x \in M$ a linear transformation W from the tangent space $T_x(M)$ to itself as follows: Suppose $X \in T_x(M)$, then

$$W(X) = -\langle X, de_3 \rangle = \omega_{13}(X)e_1 + \omega_{23}(X)e_2. \tag{4.15}$$

The above definition is independent of the choice of Darboux frames. The transformation W is usually called the **Weingarten transformation** of the tangent plane of the surface at x. It is obvious that, from (4.10), we have

$$\begin{cases} W(e_1) &= h_{11}e_1 + h_{12}e_2, \\ W(e_2) &= h_{21}e_1 + h_{22}e_2. \end{cases} \tag{4.16}$$

Thus the matrix of the transformation W under the basis (e_1, e_2) is exactly the coefficient matrix (h_{ij}) of the second fundamental form II. Because $h_{ij} = h_{ji}$,

W is a self-adjoint transformation. In fact, for any $X, Y \in T_x(M)$, we have

$$
\begin{aligned}
W(X) \cdot Y &= \omega_{13}(X)\omega_1(Y) + \omega_{23}(X)\omega_2(Y) \\
&= \omega_1(X)\omega_{13}(Y) + \omega_2(X)\omega_{23}(Y) \qquad (4.17) \\
&= X \cdot W(Y).
\end{aligned}
$$

According to a result of linear algebra, the linear transformation W has two real eigenvalues κ_1, κ_2 and two corresponding eigenvectors that are orthogonal to each other. The eigenvalues κ_1, κ_2 satisfy the quadratic equation

$$
\begin{vmatrix} h_{11} - \kappa & h_{12} \\ h_{21} & h_{22} - \kappa \end{vmatrix} = \kappa^2 - 2H\kappa + K = 0, \qquad (4.18)
$$

where

$$
H = \frac{1}{2}(h_{11} + h_{22}), \qquad K = h_{11}h_{22} - h_{12}{}^2. \qquad (4.19)
$$

H and K are both independent of the choice of Darboux frames, and are called the **mean curvature** and the **total curvature** of the surface M, respectively.

Solving equation (4.18) we get

$$
\kappa_1, \kappa_2 = H \pm \sqrt{H^2 - K}. \qquad (4.20)
$$

Thus we see that κ_1, κ_2 are both continuous functions on M. Therefore

$$
H = \frac{1}{2}(\kappa_1 + \kappa_2), \qquad K = \kappa_1 \cdot \kappa_2. \qquad (4.21)
$$

Choose a special Darboux frame $(x; e_1, e_2, e_3)$ on M such that e_1, e_2 are eigenvectors of the transformation W at the point x on the surface M. Then

$$
h_{11} = \kappa_1, \quad h_{12} = h_{21} = 0, \quad h_{22} = \kappa_2.
$$

Hence equation (4.10) can be simplified as

$$
\omega_{13} = \kappa_1\omega_1, \qquad \omega_{23} = \kappa_2\omega_2, \qquad (4.22)
$$

and the fundamental form II becomes

$$
\mathrm{II} = \kappa_1(\omega_1)^2 + \kappa_2(\omega_2)^2. \qquad (4.23)
$$

Thus the normal curvature κ_n in the tangent direction v can be expressed as

$$\kappa_n = \kappa_1 \cos^2 \theta_1 + \kappa_2 \sin^2 \theta_1, \qquad (4.24)$$

where θ_1 is the angle between v and e_1. This implies that κ_1, κ_2 are exactly the normal curvatures of the surface in the directions e_1, e_2. We call the eigenvectors e_1, e_2 at x of the transformation W the **principal directions** of the surface at x, and the corresponding eigenvalues κ_1, κ_2 the **principal curvatures** of the surface at x. Formula (4.24) is none other than the famous **Euler formula**.

If we assume that $\kappa_1 \geq \kappa_2$, (4.24) can be rewritten as

$$\begin{aligned} \kappa_n &= \kappa_1 + (\kappa_2 - \kappa_1) \sin^2 \theta_1 \\ &= (\kappa_1 - \kappa_2) \cos^2 \theta_1 + \kappa_2. \end{aligned}$$

Therefore

$$\kappa_1 \geq \kappa_n \geq \kappa_2. \qquad (4.25)$$

Hence the principal directions are exactly the directions along which the normal curvature assumes extreme values, and the principal curvatures are exactly the maximum and minimum values of the normal curvature. If $\kappa_1 = \kappa_2$ at x, then the normal curvature at that point is the same along any direction, and we call such a point an **umbilical point** of the surface M. At an umbilical point, there is no uniquely determined direction along which the normal curvature assumes extremal values.

If $x \in M$ is not an umbilical point, then there exists a neighborhood of x in which $\kappa_1 \neq \kappa_2$, and hence $H^2 - K^2 \neq 0$. Equation (4.20) implies that κ_1, κ_2 are smooth functions in that neighborhood, and thus we have the following theorem.

Theorem 4.1. *Suppose M is a smooth surface in $\mathbb{R}^3$. Then the principal curvatures κ_1, κ_2 are continuous functions on M. The set of non-umbilical points of M forms an open set in M. Suppose $\kappa_1 > \kappa_2$. Then κ_1 and κ_2 are smooth functions on this open set.*

By the fundamental theorem of Riemannian geometry, the differential 1-form ω_{12} is uniquely determined by equation (4.4) and the skew-symmetry property $\omega_{12} + \omega_{21} = 0$. Here we can give an expression for ω_{12} directly. Suppose

$$\omega_{12} = -\omega_{21} = p\omega_1 + q\omega_2. \qquad (4.26)$$

Substituting in (4.4) we obtain

$$\begin{cases} d\omega_1 &= p\omega_1 \wedge \omega_2, \\[2mm] d\omega_2 &= q\omega_1 \wedge \omega_2. \end{cases} \qquad (4.27)$$

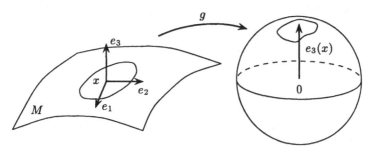

FIGURE 12.

Hence p and q are coefficients of $d\omega_1$ and $d\omega_2$, respectively, when the latter are expressed in terms of $\omega_1 \wedge \omega_2$. The differential form ω_{12} gives the Levi-Civita connection on the surface M:

$$
\begin{cases}
De_1 = \omega_{12}e_2, \\
De_2 = -\omega_{12}e_1.
\end{cases}
\tag{4.28}
$$

From equation (4.6) we obtain

$$
\begin{aligned}
d\omega_{12} &= -\omega_{13} \wedge \omega_{23} \\
&= -(h_{11}h_{22} - h_{12}{}^2)\omega_1 \wedge \omega_2, \\
&= -K\omega_1 \wedge \omega_2.
\end{aligned}
\tag{4.29}
$$

The total curvature $K - h_{11}h_{22} - h_{12}{}^2$ is defined by the second fundamental form II of the surface M, but the above formula shows that K is also determined by $d\omega_{12}$, ω_1, ω_2, i.e., it is determined by the first fundamental form of M, and hence is an intrinsic invariant quantity of the surface M. Thus the total curvature K of a surface M is independent of its shape changes in $\mathbb{R}^3$ which preserve the first fundamental form. This result was discovered by Gauss, and he called it an "astonishing" theorem (Theorema egregium).

While studying surfaces, Gauss often used the map $g : M \longrightarrow S^2 \subset \mathbb{R}^3$, such that for any $x \in M$,

$$
g(x) = e_3(x).
\tag{4.30}
$$

At present we usually call this the **Gauss map** (Figure 12). Obviously

(e_1, e_2, e_3) can still be viewed as an orthogonal frame at the point $g(x)$ on S^2. Since

$$de_3 = \omega_{31}e_1 + \omega_{32}e_2,$$

the third fundamental form

$$\text{III} = de_3 \cdot de_3 = (\omega_{31})^2 + (\omega_{32})^2$$

is the differential 2-form obtained by pulling the first fundamental form on S^2 by the Gauss map back to the surface. Similarly,

$$g^*d\sigma = \omega_{31} \wedge \omega_{32}, \qquad (4.31)$$

where $d\sigma$ is the area element on S^2. From (4.29) we obtain

$$K = \frac{g^*d\sigma}{dA}. \qquad (4.32)$$

This gives a geometric interpretation of the total curvature K: Suppose the image of a domain D on the surface M under the Gauss map is denoted by D', and the area of D is A while that of D' is A'. Then

$$|K| = \lim_{D \to P} \frac{A'}{A}, \qquad P \in M.$$

Thus the total curvature K is a measure of the bending of a surface at some point.

In §5–4 we have used the intrinsic method to prove the Gauss–Bonnet formula: If M is a 2-dimensional oriented compact Riemannian manifold, then

$$\frac{1}{2\pi} \int_M K dA = \chi(M). \qquad (4.33)$$

The study and generalization of this formula has stimulated the development of several branches of mathematics. The global properties of surfaces are active topics of study in recent years. The reader may consult, for example, Chern 1967 (a), which contains some of the most basic theorems of global differential geometry. Here we introduce the Liebmann theorem, which describes the characteristics of spherical surfaces.

Lemma 1. *Suppose M is a 2-dimensional compact surface with constant total curvature K. Then K must be greater than zero.*

Proof. Consider the function $r(x) = x \cdot x$. Since M is compact, there must exist a point $x_0 \in M$ such that $r(x)$ attains its maximum value at $x = x_0$. Therefore

$$dr\big|_{x_0} = 0, \qquad d^2r\big|_{x_0} \le 0. \qquad (4.34)$$

Since

$$\frac{1}{2}dr = x \cdot dx = (x \cdot e_1)\omega_1 + (x \cdot e_2)\omega_2,$$

we have

$$x \cdot e_1|_{x_0} = x \cdot e_2|_{x_0} = 0, \tag{4.35}$$

that is, the radius vector x_0 is exactly the normal vector to the surface M at the point x_0.

Differentiating $x \cdot e_i$ $(i = 1, 2)$ we get

$$\begin{aligned} d(x \cdot e_i) &= dx \cdot e_i + x \cdot de_i \\ &= \omega_i + \sum_{k=1}^{3} \omega_{ik}(x \cdot e_k). \end{aligned}$$

Using (4.35), we have, on the other hand,

$$d(x \cdot e_i)|_{x_0} = [\omega_i + \omega_{i3}(x \cdot e_3)]_{x_0}. \tag{4.36}$$

Therefore

$$\begin{aligned} 0 \geq \frac{1}{2} d^2 r\Big|_{x_0} &= [d(x \cdot e_1)\omega_1 + d(x \cdot e_2)\omega_2]_{x_0} \\ &= \left[(\omega_1)^2 + (\omega_2)^2\right]_{x_0} + (\omega_1\omega_{13} + \omega_2\omega_{23})_{x_0} \cdot (x \cdot e_3)_{x_0}. \end{aligned}$$
$$\tag{4.37}$$

Since $[(\omega_1)^2 + (\omega_2)^2]_{x_0} > 0$, and x_0 is the farthest point on M from the origin O, we have

$$x_0 \neq 0, \qquad x_0 \| e_3, \qquad x_0 \cdot e_3 \neq 0.$$

Thus the second fundamental form of M at x_0

$$\mathrm{II} = (\omega_1\omega_{13} + \omega_2\omega_{23})x_0$$

is a fixed 2-form, and hence the determinant of its coefficients satisfies

$$K(x_0) = (h_{11}h_{22} - h_{12}{}^2)x_0 > 0.$$

Since we have assumed that the total curvature K of M is constant, it must be a positive constant. $\qquad\square$

Lemma 2. *If M is a connected surface on which every point is an umbilical point, then M must be a sphere or a plane.*

Proof. The condition for x to be an umbilical point on M is that the characteristic directions of the Weingarten transformation at x are not well-defined. Thus

$$de_3 + \kappa dx = 0, \qquad (4.38)$$

where e_3 is the normal vector to the surface and κ is the principal curvature. Since every point of M is an umbilical point, the above formula holds everywhere on M, and $\kappa = H$ is a smooth function on M. Exteriorly differentiating (4.38) we have

$$\begin{aligned} d\kappa \wedge dx &= (d\kappa \wedge \omega_1)e_1 + (d\kappa \wedge \omega_2)e_2 = 0, \\ d\kappa \wedge \omega_1 &= 0, \qquad d\kappa \wedge \omega_2 = 0. \end{aligned}$$

Since $\omega_1 \wedge \omega_2 \neq 0$,

$$d\kappa = 0, \qquad \kappa = \text{const}. \qquad (4.39)$$

We distinguish between two cases as follows: 1) Suppose $\kappa = 0$. Then by (4.38)

$$de_3 = 0, \qquad e_3 = e_3^0 \text{ (a constant vector)}.$$

Hence

$$\begin{aligned} d(x \cdot e_3^0) &= dx \cdot e_3^0 = 0, \\ x \cdot e_3^0 &= \text{const}. \end{aligned} \qquad (4.40)$$

Thus M is a plane in $\mathbb{R}^3$.

2) Suppose $\kappa \neq 0$. Then by (4.38)

$$d\left(\frac{e_3}{\kappa} + x\right) = 0.$$

Hence

$$\begin{aligned} \frac{e_3}{\kappa} + x &= x_0 \text{ (a constant vector)}, \\ (x - x_0)^2 &= \frac{1}{\kappa^2}. \end{aligned} \qquad (4.41)$$

Thus, M is a sphere centered at x_0 with radius $1/|\kappa|$. $\qquad \square$

Theorem 4.2 (Liebmann Theorem). *Suppose M is a compact connected surface in $\mathbb{R}^3$ with constant total curvature K. Then M is a sphere.*

Proof. By Lemma 1, K must be a positive constant. Suppose κ_1, κ_2 are the principal curvatures of M, $\kappa_1 \geq \kappa_2$. Then $K = \kappa_1 \kappa_2 > 0$. By Theorem 4.1, κ_1 is a continuous function on M. Since M is compact, we may assume that κ_1 attains its maximum value at $x_0 \in M$. Hence κ_2 attains its minimum value at x_0.

We consider the following two cases:

1) Suppose $\kappa_1(x_0) = \kappa_2(x_0)$. Since $\kappa_1(x_0) \geq \kappa_1 \geq \kappa_2 \geq \kappa_2(x_0)$, it follows that at every point on M, $\kappa_1 = \kappa_2 = \pm\sqrt{K}$. By Lemma 2, M is a sphere.

2) Suppose $\kappa_1(x_0) > \kappa_2(x_0)$. Then x_0 is not an umbilical point. Thus there exists a neighborhood U of x_0 such that no point of U is an umbilical point. Hence we can choose a Darboux frame $(x; e_1, e_2, e_3)$ in U such that e_1, e_2 are the mutually orthogonal principal directions corresponding to the principal curvatures κ_1, κ_2, respectively. Therefore

$$\omega_{13} = \kappa_1 \omega_1, \qquad \omega_{23} = \kappa_2 \omega_2. \tag{4.42}$$

Suppose $\omega_{12} = p\omega_1 + q\omega_2$. Let

$$dp = p_1\omega_1 + p_2\omega_2, \qquad dq = q_1\omega_1 + q_2\omega_2. \tag{4.43}$$

Then

$$d\omega_{12} = (q_1 - p_2)\omega_1 \wedge \omega_2. \tag{4.44}$$

Comparing with the Gauss equation (4.29) we obtain

$$K = -(q_1 - p_2). \tag{4.45}$$

Exteriorly differentiating (4.42) and using the Codazzi equation (4.7), we have

$$
\begin{aligned}
d\kappa_1 \wedge \omega_1 + p\kappa_1\omega_1 \wedge \omega_2 &= p\kappa_2\omega_1 \wedge \omega_2, \\
d\kappa_2 \wedge \omega_2 + q\kappa_2\omega_1 \wedge \omega_2 &= q\kappa_1\omega_1 \wedge \omega_2.
\end{aligned}
$$

Thus

$$
\begin{cases}
(d\kappa_1 - p(\kappa_1 - \kappa_2)\omega_2) \wedge \omega_1 &= 0, \\
(d\kappa_2 - q(\kappa_1 - \kappa_2)\omega_1) \wedge \omega_2 &= 0.
\end{cases}
\tag{4.46}
$$

Since $\kappa_1 \cdot \kappa_2 = K = \text{const} > 0$, it follows that $\kappa_2 = K/\kappa_1$. Hence

$$d\kappa_2 = -\frac{K}{\kappa_1^2}d\kappa_1. \tag{4.47}$$

Plugging this into the second formula of (4.46) we get

$$\left(d\kappa_1 + \frac{q\kappa_1^2}{K}(\kappa_1 - \kappa_2)\omega_1\right) \wedge \omega_2 = 0. \qquad (4.48)$$

Combining (4.48) and the first formula in (4.46), we have

$$d\kappa_1 = -\frac{q\kappa_1^2}{K}(\kappa_1 - \kappa_2)\omega_1 + p(\kappa_1 - \kappa_2)\omega_2. \qquad (4.49)$$

From the assumption of the theorem, κ_1 attains its maximum value at x_0. Thus

$$d\kappa_1|_{x_0} = 0, \qquad d^2\kappa_1|_{x_0} \leq 0. \qquad (4.50)$$

Moreover, at x_0, $\kappa_1 \neq 0$, $\kappa_1 - \kappa_2 \neq 0$. Thus

$$p(x_0) = q(x_0) = 0. \qquad (4.51)$$

Differentiating (4.49) again, and evaluating the result at the point x_0, we have

$$0 \geq d^2\kappa_1|_{x_0} = \left[-\frac{\kappa_1^2}{K}(\kappa_1 - \kappa_2)q_1(\omega_1)^2 + (\kappa_1 - \kappa_2)p_2(\omega_2)^2 \right.$$
$$\left. + \left(-\frac{\kappa_1^2}{K}q_2(\kappa_1 - \kappa_2) + (\kappa_1 - \kappa_2)p_1\right)\omega_1\omega_2\right]_{x_0}, \qquad (4.52)$$

where ω_1, ω_2 can assume any real values. At the point x_0 we have

$$K > 0, \qquad \kappa_1 - \kappa_2 > 0.$$

If we let the 2-forms on the right hand side of (4.52) assume values along the directions e_1, e_2, we have

$$-q_1(x_0) \leq 0, \qquad p_2(x_0) \leq 0. \qquad (4.53)$$

Plugging this into (4.45) we get

$$K(x_0) \leq 0,$$

which contradicts $K > 0$. Therefore the case $\kappa_1(x_0) > \kappa_2(x_0)$ cannot occur.

□

Chapter 7

Complex Manifolds

§7–1   Complex Manifolds

The definition of a complex manifold is formally the same as that for a real manifold; but a complex structure imposes much stronger restrictions on a manifold and hence gives it richer contents.

Suppose $\mathbb{C}$ represents the field of complex numbers, and $\mathbb{C}_m$ is the complex m-dimensional vector space of m-tuples $(z^1, \dots, z^m)$ $(z^i \in \mathbb{C})$.

Definition 1.1. Let M be a Hausdorff space with a countable basis. If a family of coordinate charts $\{(U_\alpha, \varphi_\alpha)\}$ on M is given such that $\{U_\alpha\}$ forms an open covering of M, and every φ_α is a homeomorphism from U_α to an open set in $\mathbb{C}_m$ [a] satisfying the condition that, for any U_α, U_β with $U_\alpha \cap U_\beta \neq \varnothing$,

$$\varphi_\beta \circ \varphi_\alpha^{-1} : \varphi_\alpha(U_\alpha \cap U_\beta) \longrightarrow \varphi_\beta(U_\alpha \cap U_\beta)$$

is a holomorphic map between two open sets in $\mathbb{C}_m$, then M is called an **m-dimensional complex manifold**.

Suppose $(z^1, \dots, z^m)$ $(z^i \in \mathbb{C})$ is a local coordinate system on U_α, and $(w^1, \dots, w^m)$ $(w^i \in \mathbb{C})$ is a local coordinate system on U_β. When $U_\alpha \cap U_\beta \neq \varnothing$, the map $\varphi_\beta \circ \varphi_\alpha^{-1}$ can be represented by local coordinates as

$$w^k = w^k \left(z^1, \dots, z^m\right), \qquad 1 \leq k \leq m. \tag{1.1}$$

---

[a]Let $z^i = x^i + i \cdot y^i$. Then $\mathbb{C}_m$ can be viewed as the $2m$-dimensional vector space $\mathbb{R}^{2m}$ of real $2m$-tuples $(x^1, \dots, x^m, y^1, \dots, y^m)$. The topological structure of $\mathbb{C}_m$ is identical to that of $\mathbb{R}^{2m}$. Obviously the family of subsets of $\mathbb{C}_m$ of the form

$$\left\{ (z^1, \dots, z^m) \,\middle|\, \sum_{j=1}^{m} (z^j - z_0^j)(\bar{z}^j - \bar{z}_0^j) < r, \; r \in \mathbb{R}, \; r > 0 \right\}$$

constitutes a topological basis of $\mathbb{C}_m$.

Thus the statement that $\varphi_\beta \circ \varphi_\alpha^{-1}$ is a holomorphic map means that every function $w^k\left(z^1, \ldots, z^m\right)$ is holomorphic on the open set $\varphi_\alpha\left(U_\alpha \cap U_\beta\right)$ of $\mathbb{C}_m$.

By a holomorphic function we mean the following. Suppose U is an open set in $\mathbb{C}_m$ where the coordinates z^k are expressed by $z^k = x^k + iy^k$. Let f be a smooth complex-valued function on U given by the expression

$$f(z^1, \ldots, z^m) = g(x^1, \ldots, x^m, y^1, \ldots, y^m) + ih(x^1, \ldots, x^m, y^1, \ldots, y^m).$$
(1.2)

If the **Cauchy–Riemann conditions**

$$\frac{\partial g}{\partial x^k} = \frac{\partial h}{\partial y^k}, \qquad \frac{\partial g}{\partial y^k} = -\frac{\partial h}{\partial x^k}, \qquad 1 \leq k \leq m$$
(1.3)

are satisfied, then we call f a **holomorphic function** on U. The following three conditions are equivalent:

1) f is a holomorphic function on U;
2) for every point $a \in U$ there exists a neighborhood $V \subset U$ such that f can be expressed in V as a convergent series

$$f(z) = \sum_{k_1, \ldots, k_m = 0}^{\infty} c_{k_1 \cdots k_m} \left(z^1 - a^1\right)^{k_1} \cdots \left(z^m - a^m\right)^{k_m};$$
(1.4)

3) the complex derivatives $\dfrac{\partial f}{\partial z^k}$ $(1 \leq k \leq m)$ exist in U.

Now the map $\varphi_\beta \circ \varphi_\alpha^{-1}$ is holomorphic, and is also a homeomorphism from $\varphi_\alpha\left(U_\alpha \cap U_\beta\right)$ to $\varphi_\beta\left(U_\alpha \cap U_\beta\right)$. Therefore

$$\frac{\partial\left(w^1, \ldots w^m\right)}{\partial\left(z^1, \ldots, z^m\right)} \neq 0.$$
(1.5)

Definition 1.2. Suppose M and N are m-dimensional and n-dimensional complex manifolds, respectively, and $f : M \longrightarrow N$ is a continuous map. If, for every point $p \in M$, there exists a neighborhood U such that f can be expressed in U by local coordinates as

$$w^k = w^k\left(z^1, \ldots, z^m\right), \qquad 1 \leq k \leq n,$$
(1.6)

where w^k are all holomorphic functions, then f is called a **holomorphic map**.

Suppose $f : M \longrightarrow \mathbb{C}$ is a holomorphic function on the complex manifold M. By the maximum modulus theorem, if the modulus of f assumes its maximum value at p_0 in a neighborhood U of $p_0 \in M$, i.e., $|f(p)| \le |f(p_0)|$ $(p \in U)$, then we have, in U,

$$f(p) = f(p_0).$$

Suppose M is a compact, connected complex manifold. If $|f(p)|$ $(p \in M)$ is a continuous function on M, then it must assume a maximum value on M. By the previous conclusion, a holomorphic function f on M must be a constant. Thus we know that a holomorphic map $f : M \longrightarrow \mathbb{C}_n$ from a compact, connected complex manifold M to $\mathbb{C}_n$ must map M to a point in $\mathbb{C}_n$.

Example 1. $\mathbb{C}_m$ is an m-dimensional complex manifold. $\mathbb{C}_1$ is called the **Gauss complex plane**.

Example 2 (The m-dimensional complex projective space $\mathbb{C}P_m$). Define a relation $\sim$ among the elements in $\mathbb{C}_{m+1} - \{0\}$ as follows:

$$\left(z^0, z^1, \dots, z^m\right) \sim \left(w^0, w^1, \dots, w^m\right)$$

if and only if there exists a nonzero complex number λ such that

$$\left(z^0, z^1, \dots, z^m\right) = \lambda \left(w^0, w^1, \dots, w^m\right). \tag{1.7}$$

It is easy to verify that this is an equivalence relation. The m-dimensional complex projective space $\mathbb{C}P_m$ is the quotient space $(\mathbb{C}_{m+1} - \{0\})/\sim$. An element in this space is denoted by $\left[z^0, z^1, \dots, z^m\right]$. The $(m+1)$-tuple $(z^0, z^1, \dots, z^m)$ of numbers are called the **homogeneous coordinates** of the point $\left[z^0, z^1, \dots, z^m\right]$, and are determined by a point in $\mathbb{C}P_m$ up to a nonzero complex factor. As in real projective spaces, $\mathbb{C}P_m$ can be covered by $m+1$ open sets U_j $(0 \le j \le m)$, where

$$U_j = \left\{ \left[z^0, z^1, \dots, z^m\right] \in \mathbb{C}P_m, |z^j \ne 0\right\}, \tag{1.8}$$

and the coordinates on U_j are

$$_j\zeta^k = z^k/z^j, \qquad 0 \le k \le m, \quad k \ne j. \tag{1.9}$$

Since $_j\zeta^k$ can assume any complex value, every U_j is homeomorphic to $\mathbb{C}_m$. The formula for the change of coordinates in $U_j \cap U_k$ is

$$\begin{cases} _j\zeta^h &= _k\zeta^h/_k\zeta^j, \quad h \ne j, k, \\ _j\zeta^k &= 1/_k\zeta^j. \end{cases} \tag{1.10}$$

These are all holomorphic functions, so $\mathbb{C}P_m$ is an m-dimensional complex manifold.

When the 1-dimensional complex projective space $\mathbb{C}P_1$ is viewed as a 2-dimensional real manifold, it is usually called the **Riemann sphere**. Because $\mathbb{C}P_1$ can be covered by two coordinate neighborhoods U_0, U_1, and U_0 is just $\mathbb{C}P_1$ without the point $p = [0,1]$, U_0 is homeomorphic to the Gauss complex plane, and the Riemann sphere $\mathbb{C}P_1$ is hence the 2-dimensional sphere S^2.

Consider the natural projection $\pi : \mathbb{C}_{m+1} - \{0\} \longrightarrow \mathbb{C}P_m$ such that

$$\pi\left(z^0, z^1, \ldots, z^m\right) = \left[z^0, z^1, \ldots, z^m\right]. \tag{1.11}$$

For $p \in \mathbb{C}P_m$, we can identify $\pi^{-1}(p)$ with $\mathbb{C}^* = \mathbb{C}_1 - \{0\}$. If we substitute for the coordinates $\left(z^0, z^1, \ldots, z^m\right)$ of $\pi^{-1}(U_j)$ the quantities $_j\zeta^k = z^k/z^j$ $(0 \le k \le m, k \ne j)$, and z^j, then

$$\pi^{-1}(U_j) \cong U_j \times \mathbb{C}^*.$$

This shows that locally $\mathbb{C}_{m+1} - \{0\}$ has a product structure, and z^j gives the coordinates of the fiber $\pi^{-1}(p)$.

If $p \in U_j \cap U_k$, then U_j and U_k give coordinate systems z^j and z^k, respectively, on $\pi^{-1}(p)$. At the same point $x \in \pi^{-1}(p)$, there is a relation between the two coordinates z^j and z^k:

$$z^j = z^k \cdot {}_k\zeta^j = z^k/{}_j\zeta^k, \tag{1.12}$$

where $_k\zeta^j : U_j \cap U_k \longrightarrow \mathbb{C}^*$ is a nonzero holomorphic function on $U_j \cap U_k$. Hence $\mathbb{C}_{m+1} - \{0\}$ is a holomorphic fiber bundle on the m-dimensional complex projective space $\mathbb{C}P_m$ whose typical fiber and structure group are both $\mathbb{C}^*$.

Consider the following equation in $\mathbb{C}_{m+1}$:

$$\sum_{k=0}^{m} z^k \bar{z}^k = 1. \tag{1.13}$$

If we view $\mathbb{C}_{m+1}$ as the real vector space $\mathbb{R}^{2(m+1)}$, then equation (1.13) defines a $(2m+1)$-dimensional unit sphere S^{2m+1} in $\mathbb{R}^{2(m+1)}$. (To distinguish these structures, we denote the real dimension on the upper right and the complex dimension on the lower right.)

Restricting the natural projection (1.11) on S^{2m+1} we get

$$\pi : S^{2m+1} \longrightarrow \mathbb{C}P_m. \tag{1.14}$$

For any $p \in \mathbb{C}P_m$, the complete preimage $\pi^{-1}(p)$ is a circle. The map π is called the **Hopf fibering** of S^{2m+1}.

When $m = 1$, (1.14) can be written

$$\pi : S^3 \longrightarrow \mathbb{C}P_1 \approx S^2, \tag{1.15}$$

where $\mathbb{C}P_1$ is topologically homeomorphic to S^2. This is an example of an **essential map** [b] from a higher to lower dimension, which plays an important historical role in the development of homotopy theory in topology.

Example 3. The orbit determined by the system of equations

$$P_l\left(z^0, z^1, \ldots, z^m\right) = 0, \qquad 1 \le l \le q,$$

where each P_l is a homogeneous polynomial on $\mathbb{C}P_m$, is called an **algebraic variety**. For instance, the complex manifold given by the equation

$$(z^0)^2 + \cdots + (z^m)^2 = 0 \tag{1.16}$$

is called a **hyperquadric**. A theorem of W.–L. Chow (**Chow's theorem**) states that every compact submanifold imbedded in $\mathbb{C}P_m$ is an algebraic variety.

Example 4 (Complex torus). $\mathbb{C}_m$ can be viewed as a $2m$-dimensional real vector space $\mathbb{R}^{2m}$. Choose $2m$ real-linearly independent vectors $\{v_\alpha\}$ in $\mathbb{R}^{2m}$. They form the lattice

$$L = \left\{ \sum_{\alpha=1}^{2m} n_\alpha v_\alpha, \quad n_\alpha \in \mathbb{Z} \right\}. \tag{1.17}$$

$\mathbb{C}_m$ and L are both groups under addition. The quotient group $\mathbb{C}_m/L$ is an m-dimensional complex manifold, called the m-**dimensional complex torus**.

Topologically, the m-dimensional complex torus and the $2m$-dimensional real torus are homeomorphic. Yet the former has a complex manifold structure, and thus has richer contents. For instance, when $m = 1$, a holomorphic map from a complex torus to itself is conformal (angle preserving). Hence the angle between two vectors v_1, v_2 and the ratio of their lengths are invariant under holomorphic maps.

If a complex torus can be imbedded in a complex projective space as a non-singular submanifold, then for a sufficiently large N there exists a nondegenerate holomorphic map

$$f : \mathbb{C}_m/L \longrightarrow \mathbb{C}P_N. \tag{1.18}$$

Such a complex torus is called an **Abelian variety**. The study of Abelian varieties forms an important branch of Algebraic Geometry and Number Theory.

---

[b]A continuous map $f : X \longrightarrow Y$ is called **essential** if it is not homotopic to a constant map $X \longrightarrow y_0 \in Y$, that is, not homotopic to zero.

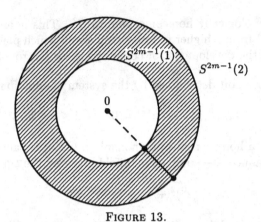

FIGURE 13.

Example 5 (Hopf Manifolds). Consider the transformation $\alpha : \mathbb{C}_m - \{0\} \longrightarrow \mathbb{C}_m - \{0\}$ such that

$$\alpha\left(z^1, \ldots, z^m\right) = 2\left(z^1, \ldots, z^m\right). \tag{1.19}$$

Let the discrete group generated by α be denoted by Δ. Then the quotient space $(\mathbb{C}_m - \{0\})/\Delta$ is an m-dimensional complex manifold, called a **Hopf manifold**.

Topologically a Hopf manifold is homeomorphic to $S^{2m-1} \times S^1$. To see this, we need only consider the annulus between the two concentric spheres $S^{2m-1}(1)$, $S^{2m-1}(2)$ with radii 1 and 2, respectively, in $\mathbb{C}_m = \mathbb{R}^{2m}$. (see Figure 13). Identifying the two end points of the line segment resulting from the intersection of the spheres with every radius, we obtain a space that is obviously homeomorphic to a Hopf manifold.

A Hopf manifold is the simplest example of a non-algebraic variety.

Example 6. Suppose M is a 2-dimensional oriented surface with the Riemannian metric

$$ds^2 = (\omega_1)^2 + (\omega_2)^2. \tag{1.20}$$

If we assume that ds^2 is analytic, then

$$ds^2 = (\omega_1 + i\omega_2)(\omega_1 - i\omega_2).$$

The differential equation $\omega_1 + i\omega_2 = 0$ has an integrating factor λ such that

$$\lambda(\omega_1 + i\omega_2) = dz. \tag{1.21}$$

Thus

$$ds^2 = \frac{1}{|\lambda|^2} dz d\bar{z}. \tag{1.22}$$

If we let $z = x + iy$, then

$$ds^2 = \frac{1}{|\lambda|^2} (dx^2 + dy^2). \tag{1.23}$$

This shows that locally an analytic Riemannian metric on a surface always corresponds conformally to the Euclidean metric. The parameters x, y that allow the Riemannian metric ds^2 to be written in the form (1.23) are called **isothermal parameters**. When ds^2 is smooth, according to the Korn–Lichtenstein Theorem (cf. S. S. Chern 1955 concerning this difficult theorem), the orientation of the surface is determined by

$$dx \wedge dy = \frac{1}{2} dz \wedge d\bar{z}. \tag{1.24}$$

If ds^2 can also be written

$$ds^2 = \frac{1}{|\mu|^2} dw d\bar{w}, \tag{1.25}$$

then dw is either a multiple of dz or a multiple of $d\bar{z}$. If the complex coordinates z and w give the same orientation of the surface, then dw must be a multiple of dz. In this case w is a holomorphic function of z. Thus we see that every 2-dimensional oriented surface must have a complex manifold structure such that it becomes a 1-dimensional complex manifold.

A 1-dimensional complex manifold is also called a **Riemann surface**. It is the fundamental object of study in the theory of functions of one complex variable.

§7–2   The Complex Structure on a Vector Space

To further study the structure of complex manifolds, we must first be acquainted with complex structures on vector spaces.

Definition 2.1. Suppose V is an m-dimensional real vector space. A **complex structure** J on V is a linear transformation $J : V \longrightarrow V$ from V to itself such that

$$J^2 = -\,\mathrm{id} \; : V \longrightarrow V. \tag{2.1}$$

Remark. In fact, J is tantamount to multiplying vectors by i. If we let $i \cdot X = JX$, then the vector space V becomes a vector space over the complex number field. Conversely, if V is a complex vector space and we let $JX = i \cdot X$, then J is a complex structure on V when V is viewed as a real vector space.

Suppose V^* is the dual space of V. Then a complex structure J on V also induces a complex structure on V^*, also denoted by J, with the following definition. Suppose $\alpha \in V^*$, $x \in V$. Then

$$\langle x, J\alpha \rangle = \langle Jx, \alpha \rangle. \tag{2.2}$$

Clearly $J^2\alpha = -\alpha$, hence J is indeed a complex structure on V^*.

Choose a basis e_r, $1 \leq r \leq m$, for V, and suppose that the matrix for the complex structure J with respect to the basis $\{e_r\}$ is $A = \left(a_i^j \right)$, that is,

$$J \begin{pmatrix} e_1 \\ \vdots \\ e_m \end{pmatrix} = A \cdot \begin{pmatrix} e_1 \\ \vdots \\ e_m \end{pmatrix}, \tag{2.3}$$

where the a_i^j are real numbers. Since $J^2 = - \text{id}$, we have

$$A^2 = -I, \tag{2.4}$$

where I is the $m \times m$ unit matrix. It is obvious that the eigenvalues of A are $\pm i$, and they must appear in pairs. Therefore the dimension of V must be even. Suppose $m = 2n$.

Denote the basis for V^* dual to $\{e_r, 1 \leq r \leq 2n\}$ by $\{e^{*r}, 1 \leq r \leq 2n\}$. By (2.2) we obtain

$$J \begin{pmatrix} e^{*r} \\ \vdots \\ e^{*2n} \end{pmatrix} = {}^t A \cdot \begin{pmatrix} e^{*1} \\ \vdots \\ e^{*2n} \end{pmatrix}, \tag{2.5}$$

i.e., the matrix of the complex structure J on V^* with respect to the basis $\{e^{*r}, 1 \leq r \leq 2n\}$ is ${}^t A$, and has the same eigenvalues as A.

Because the eigenvalues of A are purely imaginary, it is more convenient to consider the complexified space $V^* \otimes \mathbb{C}$ of V^*. $V^* \otimes \mathbb{C}$ is the set of complex-valued linear functions on V, and is an m-dimensional complex vector space. Suppose λ is any element in $V^* \otimes \mathbb{C}$. Then λ can be expressed as

$$\lambda = \alpha + i\beta, \tag{2.6}$$

where $\alpha, \beta \in V^*$. Obviously a basis for V^* can also be a basis for $V^* \otimes \mathbb{C}$, and a complex structure J on V^* can be extended to a complex structure J on the complexified space $V^* \otimes \mathbb{C}$ simply by defining

$$J\lambda = J\alpha + iJ\beta. \tag{2.7}$$

Since $V^* \otimes \mathbb{C}$ is a complex vector space and the eigenvalues of the complex structure J are $\pm i$, there must exist eigenvectors corresponding to the eigenvalues $\pm i$. The eigenvectors corresponding to i of the complex structure J on $V^* \otimes \mathbb{C}$ are called **elements of type (1,0)**, and those corresponding to $-i$ are called **elements of type (0,1)**. It is obvious that all elements of type $(1,0)$ in $V^* \otimes \mathbb{C}$ form a complex subspace of $V^* \otimes \mathbb{C}$, denoted by $V_{\mathbb{C}}$. The set of all elements of type $(0, 1)$ in $V^* \otimes \mathbb{C}$ also form a complex subspace of $V^* \otimes \mathbb{C}$, denoted by $\overline{V}_{\mathbb{C}}$. There is a one to one correspondence between the spaces $V_{\mathbb{C}}$ and $\overline{V}_{\mathbb{C}}$ under complex conjugation. In fact, if $\lambda = \alpha + i\beta \in V_{\mathbb{C}}$, then it follows from (2.7) that

$$J\lambda = J\alpha + iJ\beta = i(\alpha + i\beta) = -\beta + i\alpha.$$

Hence

$$J\alpha = -\beta, \qquad J\beta = \alpha. \tag{2.8}$$

Thus we see that

$$J\overline{\lambda} = J(\alpha - i\beta) = -i(\alpha - i\beta) = -i\overline{\lambda},$$

i.e., $\overline{\lambda} \in \overline{V}_{\mathbb{C}}$. It is easy to verify that $V_{\mathbb{C}} \cap \overline{V}_{\mathbb{C}} = \{0\}$. Moreover, any element of $V^* \otimes \mathbb{C}$ can be expressed as a sum of elements of type $(1,0)$ and type $(0,1)$. Suppose $f \in V^* \otimes \mathbb{C}$. Let

$$f_1 = \frac{1}{2}(f - i \cdot Jf), \qquad f_2 = \frac{1}{2}(f + i \cdot Jf). \tag{2.9}$$

Then $f = f_1 + f_2$, and

$$Jf_1 = i \cdot f_1, \qquad Jf_2 = -i \cdot f_2, \tag{2.10}$$

that is, $f_1 \in V_{\mathbb{C}}$, $f_2 \in \overline{V}_{\mathbb{C}}$. Therefore $V^* \otimes \mathbb{C}$ can be expressed as the direct sum of $V_{\mathbb{C}}$ and $\overline{V}_{\mathbb{C}}$. This also shows that both $V_{\mathbb{C}}$ and $\overline{V}_{\mathbb{C}}$ are complex vector spaces of dimension $n = m/2$.

Choose any basis λ^i, $1 \le j \le n$, for $V_{\mathbb{C}}$. Then $\left\{ \lambda^j, \overline{\lambda}^j, 1 \le j \le n \right\}$ forms a basis for $V^* \otimes \mathbb{C}$. With respect to this basis, the matrix of the complex structure J has the normal form

$$
\begin{pmatrix}
i & & & & & \\
& \ddots & & & & \\
& & i & & & \\
& & & -i & & \\
& & & & \ddots & \\
& & & & & -i
\end{pmatrix}. \tag{2.11}
$$

Now we decompose the complex-valued linear function λ^j on V into real and imaginary parts. Let

$$
\lambda^j = e^{*j} + i \cdot e^{*n+j}, \tag{2.12}
$$

where e^{*j} and e^{*n+j} are elements in V^*. Because $J\lambda^j = i \cdot \lambda^j$, it follows from (2.7) that

$$
Je^{*j} = -e^{*n+j}, \qquad Je^{*n+j} = e^{*j}. \tag{2.13}
$$

Furthermore we obtain from (2.12) that

$$
\begin{cases}
e^{*j} & = \; \dfrac{1}{2}\left(\lambda^j + \overline{\lambda}^j\right), \\[2mm]
e^{*n+j} & = \; -\dfrac{i}{2}\left(\lambda^j - \overline{\lambda}^j\right).
\end{cases} \tag{2.14}
$$

Thus we see that the $2n$ real-valued linear functions e^{*j}, e^{*n+j}, $1 \le j \le n$, on V can be expressed as complex linear combinations of $\left\{\lambda^j, \overline{\lambda}^j, 1 \le j \le n\right\}$, and vice versa. Therefore they form a basis for $V^* \otimes \mathbb{C}$, as well as a basis for V^*.

Suppose $\{e_j, e_{n+j}, 1 \le j \le n\}$ is the basis for V dual to $\{e^{*j}, e^{*n+j}, 1 \le j \le n\}$. Then it is easy to verify that

$$
Je_j = e_{n+j}, \qquad Je_{n+j} = -e_j. \tag{2.15}
$$

Theorem 2.1. *Suppose J is a complex structure on a real vector space V. Then the dimension m of V must be even, say $m = 2n$. Moreover there exists a basis $\{e_j, Je_j, 1 \le j \le n\}$ for V; and any two such bases give the same orientation to V.*

Proof. The existence of bases of the form $\{e_j, Je_j, 1 \leq j \leq n\}$ was proved in the previous discussion. We only need to show that they determine the same orientation for V.

As described above, there is a dual basis $\{e^{*j}, -Je^{*j}, 1 \leq j \leq n\}$ of V^*. Suppose λ^j is defined as in (2.12). Then

$$
\begin{aligned}
e^{*j} \wedge Je^{*j} &= -e^{*j} \wedge e^{*n+j} \\
&= -\frac{i}{2}\lambda^j \wedge \overline{\lambda}^j.
\end{aligned}
$$

Hence

$$
\bigwedge_{1 \leq j \leq n} \left(e^{*j} \wedge Je^{*j}\right) = \left(-\frac{i}{2}\right)^n \bigwedge_{1 \leq j \leq n} \left(\lambda^j \wedge \overline{\lambda}^j\right). \tag{2.16}
$$

If we choose another basis $\{\mu^j, \overline{\mu}^j, 1 \leq j \leq n\}$ for $V^* \otimes \mathbb{C}$, where μ^j is a type $(1,0)$ element in $V^* \otimes \mathbb{C}$, then there is an $n \times n$ nondegenerate complex matrix G such that

$$
\left(\mu^1, \ldots, \mu^n\right) = \left(\lambda^1, \ldots, \lambda^n\right) \cdot G. \tag{2.17}
$$

Hence

$$
\left\{
\begin{aligned}
\mu^1 \wedge \cdots \wedge \mu^n &= (\det G)\, \lambda^1 \wedge \cdots \wedge \lambda^n, \\
\bigwedge_{1 \leq j \leq n} \left(\mu^j \wedge \overline{\mu}^j\right) &= |\det G|^2 \bigwedge_{1 \leq j \leq n} \left(\lambda^j \wedge \overline{\lambda}^j\right).
\end{aligned}
\right. \tag{2.18}
$$

The above equation shows that the difference between the two $2n$-exterior forms on the two sides of the equation is just a positive factor $|\det G|^2$, which proves that if $\{a_j, Ja_j, 1 \leq j \leq n\}$ is another basis of V determined by $\{\mu^j, \overline{\mu}^j, 1 \leq j \leq n\}$, then it gives V the same orientation as given by $\{e_j, Je_j, 1 \leq j \leq n\}$. $\qquad\square$

We have shown above that if V is given a complex structure J, then $V^* \otimes \mathbb{C}$ has a unique direct sum decomposition $V_{\mathbb{C}} \oplus \overline{V}_{\mathbb{C}}$, and there is a one to one correspondence between these two subspaces under complex conjugation. Conversely, any such direct sum decomposition for $V^* \otimes \mathbb{C}$ also determines a complex structure on V.

Theorem 2.2. *Suppose V is a real $2n$-dimensional vector space. If there is a direct sum decomposition $V_{\mathbb{C}} \oplus \overline{V}_{\mathbb{C}}$ of $V^* \otimes \mathbb{C}$ such that there is a one to one correspondence between $V_{\mathbb{C}}$ and $\overline{V}_{\mathbb{C}}$ under complex conjugation, then there is a unique complex structure J on V such that the type $(1,0)$ elements with respect to J are the elements of $V_{\mathbb{C}}$, and the type $(0,1)$ elements with respect to J are those of $\overline{V}_{\mathbb{C}}$.*

Proof. Define a linear transformation $J : V^* \otimes \mathbb{C} \longrightarrow V^* \otimes \mathbb{C}$ as follows:

$$\begin{cases} Jf = i \cdot f, & f \in V_\mathbb{C}, \\ Jf = -i \cdot f, & f \in \overline{V}_\mathbb{C}. \end{cases} \tag{2.19}$$

Since $V^* \otimes \mathbb{C} = V_\mathbb{C} \oplus \overline{V}_\mathbb{C}$, the map J is completely determined by (2.19). Choose a basis λ^j, $1 \le j \le n$ for $V_\mathbb{C}$. Let

$$e^{*j} = \frac{1}{2} \left(\lambda^j + \overline{\lambda}^j \right), \qquad e^{*n+j} = -\frac{i}{2} \left(\lambda^j - \overline{\lambda}^j \right). \tag{2.20}$$

Then e^{*j} and e^{*n+j} are real-valued linear functions on V. Hence $\{e^{*j}, e^{*n+j}, 1 \le j \le n\}$ is not only a basis for $V^* \otimes \mathbb{C}$, but also a basis for V^*. Because

$$\begin{cases} Je^{*j} = \frac{1}{2} \left(i \cdot \lambda^j - i \cdot \overline{\lambda}^j \right) = -e^{*n+j}, \\ Je^{*n+j} = -\frac{i}{2} \left(i \cdot \lambda^j + i \cdot \overline{\lambda}^j \right) = e^{*j}, \end{cases} \tag{2.21}$$

J is a complex structure on V^*, and hence defines a complex structure J on V. By (2.19) we see that $V_\mathbb{C}$ and $\overline{V}_\mathbb{C}$ are, with respect to J, complex subspaces of type $(1,0)$ elements and type $(0,1)$ elements, respectively. Uniqueness is obvious: if J is a complex structure that satisfies the conditions of the theorem, then (2.19) holds. $\qquad\qquad\square$

We now consider the space $\Lambda^r (V^* \otimes \mathbb{C})$ whose elements are complex-valued r-linear alternating functions on V. Obviously, it can be expressed as the direct sum:

$$\Lambda^r (V^* \otimes \mathbb{C}) = \sum_{p+q=r} (\Lambda^p V_\mathbb{C}) \wedge (\Lambda^q \overline{V}_\mathbb{C}), \tag{2.22}$$

where elements in $(\Lambda^p V_\mathbb{C}) \wedge (\Lambda^q \overline{V}_\mathbb{C})$ can be expressed as

$$\sum C_{i_1 \cdots i_p, \overline{j}_1 \cdots \overline{j}_q} \lambda^{i_1} \wedge \cdots \wedge \lambda^{i_p} \wedge \overline{\lambda}^{j_1} \wedge \cdots \wedge \overline{\lambda}^{j_q}. \tag{2.23}$$

These elements are called exterior (p,q)-forms.

If we use $\Pi_{p,q}$ to denote the natural projection from $\Lambda^r (V^* \otimes \mathbb{C})$ to the space $(\Lambda^p V_\mathbb{C}) \wedge (\Lambda^q \overline{V}_\mathbb{C})$ $(p+q=r)$ of exterior (p,q)-forms, and let

$$\alpha_{p,q} = \prod_{p,q} \alpha, \tag{2.24}$$

then

$$\alpha = \sum_{p+q=r} \alpha_{p,q}. \tag{2.25}$$

The following properties are obvious:

1) If α is an exterior (p, q)-form, then $\bar{\alpha}$ is an exterior (q, p)-form.
2) If α is an exterior (p, q)-form and β is an exterior (r, s)-form, then $\alpha \wedge \beta$ is an exterior $(p + r, q + s)$-form.
3) If either p or $q > n$, then all exterior (p, q)-forms are zero.

Remark. In the above description we emphasized the complexification of the dual space V^* of V because elements in V^* are real-valued linear functions on V, and hence the meaning of multiplying by i (the imaginary unit) is clear. Conversely, we can also consider the complexification space $V \otimes \mathbb{C}$ of V if we view V as the dual space of V^*. The eigenvectors of the complex structure J in $V \otimes \mathbb{C}$ corresponding to the eigenvalues $+i$ and $-i$ are called **type** $(1, 0)$ **vectors** and **type** $(0, 1)$ **vectors**, respectively. By Theorem 2.1, there exists a basis $\{e_j, Je_j, 1 \leq j \leq n\}$ for V. If we let

$$\begin{cases} \xi_j &= \dfrac{1}{2}(e_j - iJe_j), \\ \bar{\xi}_j &= \dfrac{1}{2}(e_j + iJe_j), \end{cases} \quad 1 \leq j \leq n, \tag{2.26}$$

then the ξ_j, $\bar{\xi}_j$ are type $(1, 0)$ vectors and type $(0, 1)$ vectors, respectively, in $V \otimes \mathbb{C}$, and they form a basis for $V \otimes \mathbb{C}$. Introducing the complex linear extension of the pairing determined by the dual relation between V and V^*, we see that $\left\{\lambda^j, \bar{\lambda}^j, 1 \leq j \leq n\right\}$ in Theorem 2.1 and $\{\xi_j, \bar{\xi}_j, 1 \leq j \leq n\}$ are precisely dual bases of each other. In fact,

$$\begin{aligned} \langle \xi_j, \lambda^k \rangle &= \frac{1}{2} \langle e_j - iJe_j, e^{*k} - iJe^{*k} \rangle \\ &= \frac{1}{2} \{ \langle e_j, e^{*k} \rangle - \langle Je_j, Je^{*k} \rangle - i \langle Je_j, e^{*k} \rangle - i \langle e_j, Je^{*k} \rangle \} \\ &= \delta_j^k. \end{aligned}$$

Similarly,

$$\langle \bar{\xi}_j, \lambda^k \rangle = \langle \xi_j, \bar{\lambda}^k \rangle = 0, \qquad \langle \bar{\xi}_j, \bar{\lambda}^k \rangle = \delta_j^k.$$

Definition 2.2. Suppose V is a real vector space with complex structure J. If $H : V \times V \longrightarrow \mathbb{C}$ is a complex-valued function in two variables which satisfies the following conditions:

1) for any $x_1, x_2, y \in V$, $a_1, a_2 \in \mathbb{R}$ we have

$$H(a_1 x_1 + a_2 x_2, y) = a_1 H(x_1, y) + a_2 H(x_2, y);$$

2) for any $x, y \in V$ we have

$$H(y, x) = \overline{H(x, y)};$$

3) $H(Jx, y) = iH(x, y),$

then H is called a **Hermitian structure** on the real vector space V.

If we separate the real and imaginary parts of $H(x, y)$ and write

$$H(x, y) = F(x, y) + iG(x, y), \tag{2.27}$$

then F and G are both real-valued bilinear functions on V. From condition 2) above we have

$$F(y, x) + iG(y, x) = F(x, y) - iG(x, y).$$

Hence

$$F(y, x) = F(x, y), \qquad G(y, x) = -G(x, y), \tag{2.28}$$

that is, F is a symmetric bilinear function and G is an anti-symmetric bilinear function. From condition 3) above we have

$$F(Jx, y) = -G(x, y), \qquad G(Jx, y) = F(x, y). \tag{2.29}$$

It follows that

$$F(Jx, Jy) = F(x, y), \qquad G(Jx, Jy) = G(x, y), \tag{2.30}$$

that is, F and G are both invariant under J. Thus a given Hermitian structure on V determines two real-valued bilinear functions on V which are invariant under J, one symmetric and the other anti-symmetric. Each of these functions can be expressed in terms of the other through J.

Conversely, if there is a given real-valued symmetric bilinear function F (or an anti-symmetric bilinear function G) on a real vector space V with complex structure J which is invariant under J, then by (2.29) and (2.27), a Hermitian structure H is determined on V (The reader should verify this).

If the real-valued symmetric bilinear function $F(x, y)$ corresponding to a Hermitian structure H is positive definite, then we call H a **positive definite** Hermitian structure. Obviously, a positive definite Hermitian structure H defines an inner product on V that is invariant under J: for any $x, y \in V$, let

$$x \cdot y = F(x, y) = \frac{1}{2}\left(H(x, y) + \overline{H(x, y)}\right). \tag{2.31}$$

Now we express the Hermitian structure H using a basis λ^k for $V_{\mathbb{C}}$. Suppose $x, y \in V$. Then they can be expressed as

$$
\begin{cases}
x = \displaystyle\sum_{j=1}^{n} \left(x^j e_j + x^{n+j} J e_j\right), \\
y = \displaystyle\sum_{j=1}^{n} \left(y^j e_j + y^{n+j} J e_j\right),
\end{cases}
\qquad x^j, y^j \in \mathbb{R}. \tag{2.32}
$$

Therefore

$$
H(x, y) = \sum_{j,k=1}^{n} \left[\left(x^j + i x^{n+j}\right)\left(y^k - i y^{n+k}\right) H(e_j, e_k)\right]. \tag{2.33}
$$

On the other hand, because of the duality between $\{e_j, J e_j, 1 \leq j \leq n\}$ and $\{e^{*j}, -J e^{*j}, 1 \leq j \leq n\}$, we have

$$
\begin{aligned}
\lambda^j(x) &= e^{*j}(x) - i \cdot J e^{*j}(x) \\
&= x^j + i \cdot x^{n+j}.
\end{aligned}
$$

Similarly,

$$
\overline{\lambda}^k(y) = y^k - i \cdot y^{n+k}.
$$

Therefore

$$
H(x, y) = \sum h_{j\overline{k}} \lambda^j(x) \overline{\lambda}^k(y), \tag{2.34}
$$

where

$$
h_{j\overline{k}} = H(e_j, e_k), \qquad \overline{h}_{j\overline{k}} = h_{k\overline{j}}, \tag{2.35}
$$

that is,

$$
H = \sum h_{j\overline{k}} \lambda^j \otimes \overline{\lambda}^k. \tag{2.36}
$$

Since $G(x, y)$ is a real-valued anti-symmetric bilinear function on V it corresponds to an exterior 2-form $\hat{H}$ such that

$$
\left\langle x \wedge y, \hat{H} \right\rangle = -G(x, y). \tag{2.37}
$$

$\hat{H}$ is called the **Kählerian form** of the Hermitian structure H. Since

$$
\begin{aligned}
-G(x, y) &= \frac{i}{2}\left(H(x, y) - \overline{H(x, y)}\right) \\
&= \frac{i}{2} \sum_{j,k} h_{j\overline{k}}\left(\lambda^j(x)\overline{\lambda}^k(y) - \lambda^j(y)\overline{\lambda}^k(x)\right) \\
&= \left\langle x \wedge y, \frac{i}{2} \sum_{j,k} h_{j\overline{k}} \lambda^j \wedge \overline{\lambda}^k \right\rangle,
\end{aligned}
$$

we have

$$\hat{H} = \frac{i}{2} \sum_{j,\overline{k}} h_{j\overline{k}} \lambda^j \wedge \overline{\lambda}^k. \tag{2.38}$$

§7–3   Almost Complex Manifolds

Definition 3.1. Suppose M is an m-dimensional smooth manifold, and J is a smooth tensor field of $(1,1)$-type on M, that is, for every $x \in M$, J_x is a linear transformation from the tangent space $T_x(M)$ to itself. If every J_x ($x \in M$) is a complex structure on the tangent space $T_x(M)$, then the tensor field J is called an **almost complex structure** on M. A smooth manifold with a given almost complex structure is called an **almost complex manifold**.

By a smooth tensor field J we mean that, if X is a smooth tangent vector field on M, then JX is also a smooth tangent vector field on M. Obviously not every manifold has an almost complex structure. For instance, by Theorem 2.1 we have the following.

Theorem 3.1. *An almost complex manifold must be an orientable manifold of even dimension.*

Remark. Orientability with even dimensionality is not a sufficient condition for a manifold to have an almost complex structure. Ehresmann and Hopf proved that the 4-dimensional sphere S^4 cannot have an almost complex structure (Steenrod 1951, p.217).

Suppose M is an almost complex manifold of dimension $m = 2n$. Let A denote the space of smooth complex-valued differential $(1,0)$-forms, and $\overline{A}$ the space of forms which are complex conjugates of elements of A. Then at every point $x \in M$ there is a direct sum decomposition

$$T_x^*(M) \otimes \mathbb{C} = A_x \oplus \overline{A}_x. \tag{3.1}$$

Suppose x^α ($1 \leq \alpha \leq 2n$) is a local coordinate system on M. Under the natural basis $\partial/\partial x^\alpha$ for the tangent space, an almost complex structure J can be expressed as

$$J_x \left(\frac{\partial}{\partial x^\alpha} \right) = \sum_\beta a_\alpha^\beta(x) \frac{\partial}{\partial x^\beta}, \tag{3.2}$$

where a_α^β are smooth functions on a neighborhood in M, and

$$\sum_{\gamma=1}^{2n} a_\alpha^\gamma a_\gamma^\beta = -\delta_\alpha^\beta. \tag{3.3}$$

Obviously, at every point $x \in M$, the forms

$$\sum_{\beta} \left(a_{\beta}^{\alpha} + i \delta_{\beta}^{\alpha} \right) dx^{\beta}, \qquad 1 \leq \alpha \leq 2n \tag{3.4}$$

are exterior $(1,0)$-forms. By the discussion in §7–2, among these $2n$ exterior $(1,0)$-forms there are n which are complex-linearly independent.

Theorem 3.2. *A complex manifold is naturally an almost complex manifold.*

Proof. An n-dimensional complex manifold M can be viewed as a $2n$-dimensional real manifold. Suppose $\{z^k, 1 \leq k \leq n\}$ is a local coordinate system in the complex manifold M. Denote $z^k = x^k + iy^k$. Then $\{x^k, y^k, 1 \leq k \leq n\}$ is a local coordinate system in the real manifold M, and $\left\{ \frac{\partial}{\partial x^k}, \frac{\partial}{\partial y^k}, 1 \leq k \leq n \right\}$ gives the natural basis in a coordinate neighborhood of M.

Suppose at every point $x \in M$, a linear transformation $J_x : T_x(M) \longrightarrow T_x(M)$ is defined by

$$J_x \left(\frac{\partial}{\partial x^k} \right) = \frac{\partial}{\partial y^k}, \qquad J_x \left(\frac{\partial}{\partial y^k} \right) = -\frac{\partial}{\partial x^k}. \tag{3.5}$$

Obviously $J_x^2 = -\operatorname{id} : T_x(M) \longrightarrow T_x(M)$. We will prove that the definition of J_x is independent of the choice of complex coordinates z^k. Hence the linear transformation field defined above is an almost complex structure defined globally on M.

To show this, suppose w^k is another local coordinate system near x. Then the z^j are holomorphic functions of w^k. Suppose $w^k = u^k + iv^k$. Then we have the Cauchy–Riemann equations

$$\frac{\partial x^j}{\partial u^k} = \frac{\partial y^j}{\partial v^k}, \qquad \frac{\partial x^j}{\partial v^k} = -\frac{\partial y^j}{\partial u^k}. \tag{3.6}$$

Thus the action of J_x as defined by (3.5) on $\frac{\partial}{\partial u^k}, \frac{\partial}{\partial v^k}$ yields

$$\begin{cases} J_x \left(\dfrac{\partial}{\partial u^k} \right) = J_x \left(\displaystyle\sum_j \dfrac{\partial x^j}{\partial u^k} \dfrac{\partial}{\partial x^j} + \sum_j \dfrac{\partial y^j}{\partial u^k} \dfrac{\partial}{\partial y^j} \right) = \dfrac{\partial}{\partial v^k}, \\[4mm] J_x \left(\dfrac{\partial}{\partial v^k} \right) = J_x \left(\displaystyle\sum_j \dfrac{\partial x^j}{\partial v^k} \dfrac{\partial}{\partial x^j} + \sum_j \dfrac{\partial y^j}{\partial v^k} \dfrac{\partial}{\partial y^j} \right) = -\dfrac{\partial}{\partial u^k}. \end{cases} \tag{3.7}$$

Hence, the action of J_x on $\{ \frac{\partial}{\partial u^k}, \frac{\partial}{\partial v^k}, 1 \leq k \leq n \}$ has the same form as defined in (3.5). $\qquad\square$

The almost complex structure defined in (3.5) is called the **canonical almost complex structure** of the complex manifold M. Thus

$$J_x(dx^k) = -dy^k, \qquad J_x(dy^k) = dx^k. \tag{3.8}$$

Hence $dz^k = dx^k + idy^k$ is a differential $(1,0)$-form, and $d\bar{z}^k = dx^k - idy^k$ is a differential $(0,1)$-form. In the complexified space $T_x(M) \otimes \mathbb{C}$ of a tangent space, let

$$\frac{\partial}{\partial z^k} = \frac{1}{2}\left(\frac{\partial}{\partial x^k} - i \cdot \frac{\partial}{\partial y^k}\right), \tag{3.9}$$

$$\frac{\partial}{\partial \bar{z}^k} = \frac{1}{2}\left(\frac{\partial}{\partial x^k} + i \cdot \frac{\partial}{\partial y^k}\right), \qquad 1 \le k \le n. \tag{3.10}$$

These are tangent vectors of type $(1,0)$ and $(0,1)$, respectively. Together they form a basis for $T_m(M) \otimes \mathbb{C}$. [see (2.26)].

The very natural question arises: Does every almost complex manifold of dimension $2n$ have a complex structure? When $n = 1$, the answer is yes; but for the general case it is negative.

Suppose locally an almost complex structure is determined by n complex-linearly independent differential 1-forms θ^k ($1 \le k \le n$), such that θ^k are the corresponding differential $(1,0)$-forms. $d\theta^k$ is then an exterior differential 2-form that can be expressed as

$$d\theta^k = \frac{1}{2}\sum_{j,l} A^k_{jl}\theta^j \wedge \theta^l + \sum_{j,l} B^k_{jl}\theta^j \wedge \bar{\theta}^l + \frac{1}{2}\sum_{j,l} C^k_{jl}\bar{\theta}^j \wedge \bar{\theta}^l, \tag{3.11}$$

where A^k_{jl} and C^k_{jl} are anti-symmetric with respect to the lower indices. The condition

$$d\theta^k \equiv 0 \bmod \theta^j \tag{3.12}$$

is independent of the choice of $\{\theta^k\}$. Indeed, if there are another n differential $(1,0)$-forms λ^k that are complex-linearly independent, then they can be written as

$$\lambda^k = \sum_j \mu^k_j \theta^j, \tag{3.13}$$

where the μ^k_j are smooth complex-valued functions, and $\det\left(\mu^k_j\right) \ne 0$. Suppose

$$d\lambda^k = \frac{1}{2}\sum_{j,l} A'^k_{jl}\lambda^j \wedge \lambda^l + \sum_{j,l} B'^k_{jl}\lambda^j \wedge \bar{\lambda}^l + \frac{1}{2}\sum_{j,l} C'^k_{jl}\bar{\lambda}^j \wedge \bar{\lambda}^l. \tag{3.14}$$

Then

$$C'^{k}_{pq}\mu^{p}_{j}\mu^{q}_{l} = C^{r}_{jl}\mu^{k}_{r}. \tag{3.15}$$

Therefore $C^{k}_{jl} = 0$ if and only if $C'^{k}_{jl} = 0$, that is, (3.12) is equivalent to

$$d\lambda^{k} \equiv 0 \mod \lambda^{j}.$$

Thus (3.12) is a meaningful condition on the entire almost complex manifold.

Definition 3.2. The condition (3.12) is called the **integrability condition** for the almost complex manifold M. If the integrability condition holds on an almost complex manifold , then the almost complex manifold is said to be **integrable**.

A 2-dimensional almost complex manifold is always integrable. When the dimension ≥ 4, there always exists a non-integrable almost complex structure on any almost complex manifold. Moreover, in this case, an integrable almost complex structure will become non-integrable if the structure is perturbed slightly. None of these results are easy to prove.

Suppose M is a complex manifold. Then with respect to a canonical almost complex structure on M, dz^{k} is a differential $(1,0)$-form. Choose $\theta^{k} = dz^{k}$. Then obviously the integrability condition holds. Therefore a canonical almost complex structure on a complex manifold is always integrable. The important point is that the converse statement is also true.

Theorem 3.3. *If there is an integrable almost complex structure on a manifold M, then it must be a canonical almost complex structure induced from a complex manifold structure.*

Newlander and Nirenberg gave a proof of the above theorem under the assumption of smoothness of the almost complex structure (Newlander and Nirenberg, 1957). Nijenhuis and Woolf, Kohn, and Hörmander further proved the theorem under weaker differentiability conditions. These theorems are all very difficult, and will not be discussed here.

When an almost complex structure is real-analytic, Theorem 3.3 is easy to prove. Because (3.12) holds, by the Frobenius theorem, there exists a local complex coordinate system z^{k} such that the differential $(1,0)$-forms are all linear combinations of dz^{k}. If there are two such coordinate systems z^{k} and w^{j} in the same neighborhood, then the dw^{j} are linear combinations of the dz^{k}. This implies that the w^{j} are holomorphic functions of the z^{k}. These coordinate systems then define a complex manifold structure on M.

In the following we express the integrability condition in terms of tensors of an almost complex structure. By (3.2), the matrix for an almost complex structure J under the local coordinate system x^α $(1 \le \alpha \le 2n)$ is (a_α^β), and it satisfies condition (3.3). Therefore all differential $(1,0)$-forms can be written locally as linear combinations of

$$\sum_\beta \left(a_\beta^\alpha + i\delta_\beta^\alpha\right) dx^\beta, \qquad 1 \le \alpha \le 2n.$$

Hence the integrability condition (3.12) becomes

$$d\left(\sum_\beta \left(a_\beta^\alpha + i\delta_\beta^\alpha\right) dx^\beta\right) = \frac{1}{2} \sum_{\beta,\gamma} a_{\beta\gamma}^\alpha dx^\gamma \wedge dx^\beta$$

$$\equiv 0 \bmod \left(\sum_\lambda \left(a_\lambda^\theta + i\delta_\lambda^\theta\right) dx^\lambda\right), \quad (3.16)$$

$$(1 \le \theta \le 2n)$$

where

$$a_{\beta\gamma}^\alpha = \frac{\partial a_\beta^\alpha}{\partial x^\gamma} - \frac{\partial a_\gamma^\alpha}{\partial x^\beta}. \tag{3.17}$$

Because the system of linear equations for dx^β

$$\sum_\beta \left(a_\beta^\alpha + i\delta_\beta^\alpha\right) dx^\beta = 0, \qquad 1 \le \alpha \le 2n \tag{3.18}$$

determines the complex subspace of tangent vectors of type $(0,1)$ in $T_x\left(M\right)\otimes\mathbb{C}$, and the latter is spanned by the tangent vectors of type $(0,1)$:

$$\sum_\beta \left(a_\beta^\alpha - i\delta_\beta^\alpha\right) \frac{\partial}{\partial x^\alpha}, \qquad 1 \le \beta \le 2n,$$

the maximal linearly independent set in the solution set for (3.18),

$$y_{(\beta)} = \left(a_\beta^1 - i\delta_\beta^1, \ldots, a_\beta^{2n} - i\delta_\beta^{2n}\right), \qquad 1 \le \beta \le 2n, \tag{3.19}$$

provides a fundamental solution set for (3.18). Therefore it follows from (3.16) that

$$\sum_{\beta,\gamma} a_{\beta\gamma}^\alpha \left(a_\lambda^\beta - i\delta_\lambda^\beta\right)\left(a_\mu^\gamma - i\delta_\mu^\gamma\right) = 0,$$

that is,

$$\sum_\beta \left(a_{\beta\mu}^\alpha a_\lambda^\beta - a_{\beta\lambda}^\alpha a_\mu^\beta\right) = 0. \tag{3.20}$$

Let

$$t^\alpha_{\beta\gamma} = a^\alpha_{\beta\rho} a^\rho_\gamma - a^\alpha_{\gamma\rho} a^\rho_\beta. \tag{3.21}$$

It is easy to verify that this is a tensor field of type $(1, 2)$ on M, called the **torsion tensor** of the almost complex structure J. Thus the above result can be stated as:

Theorem 3.4. *Suppose J is an almost complex structure on a manifold M. Then a necessary and sufficient condition for J to be integrable is that its torsion tensor is zero.*

Definition 3.3. Suppose ω is a smooth complex-valued exterior differential form on an almost complex manifold M. If at every point $x \in M$, $\omega(x)$ is an exterior (p, q)-form on $T_x(M)$, then we say ω is an **exterior differential form of type** (p, q). The set of all such forms is denoted by $A_{p,q}$.

Obviously, $A_{p,q}$ is a module over the ring of smooth complex-valued functions. It has the following simple properties:

1) If $\alpha \in A_{p,q}$, then $\bar\alpha \in A_{q,p}$;
2) if $\alpha \in A_{p,q}$, $\beta \in A_{r,s}$, then $\alpha \wedge \beta \in A_{p+r,q+s}$;
3) $dA_{p,q} \subset A_{p+2,q-1} + A_{p+1,q} + A_{p,q+1} + A_{p-1,q+2}$;
4) If p or $q > n \left(= \frac{1}{2} \dim M\right)$, then $A_{p,q} = 0$.

Property 3) requires some clarification. Locally $A_{p,q}$ is generated by smooth complex-valued functions and exterior differential $(1, 0)$- and $(0, 1)$-forms. On the other hand,

$$
\begin{aligned}
dA_{0,0} &\subset A_{1,0} + A_{0,1}, \\
dA_{1,0} &\subset A_{2,0} + A_{1,1} + A_{0,2}, \\
dA_{0,1} &\subset A_{2,0} + A_{1,1} + A_{0,2}.
\end{aligned}
$$

Hence, by the definition of exterior differentials, property 3) follows from induction.

Suppose $\omega \in A_{p,q}$. Let

$$\partial\omega = \prod_{p+1,q} d\omega, \qquad \bar\partial\omega = \prod_{p,q+1} d\omega. \tag{3.22}$$

Then $\partial : A_{p,q} \longrightarrow A_{p+1,q}$ and $\bar\partial : A_{p,q} \longrightarrow A_{p,q+1}$ are both linear maps.

Theorem 3.5. *Suppose J is an almost complex structure on a manifold M. Then a necessary and sufficient condition for J to be integrable is*

$$d = \partial + \bar\partial. \tag{3.23}$$

Proof. (Sufficiency) Suppose θ^k $(1 \leq k \leq n)$ are locally linearly independent differential $(1,0)$-forms with respect to J. Because $d = \partial + \bar{\partial}$, we have

$$\prod_{0,2} d\theta^k = 0,$$

that is, the integrability condition

$$d\theta^k \equiv 0 \bmod \theta^j, \qquad 1 \leq k \leq n, \tag{3.24}$$

holds.

(Necessity) If (3.24) holds, then

$$dA_{1,0} \subset A_{2,0} + A_{1,1}, \qquad dA_{0,1} \subset A_{1,1} + A_{0,2}. \tag{3.25}$$

By induction, it is not difficult to show that

$$dA_{p,q} \subset A_{p+1,q} + A_{p,q+1}.$$

Therefore

$$d = \partial + \bar{\partial}.$$

$\square$

Theorem 3.6. *An almost complex structure J on a manifold M is integrable if and only if*

$$\bar{\partial}^2 = 0. \tag{3.26}$$

Proof. (Necessity) Suppose J is integrable. Then $d = \partial + \bar{\partial}$. Hence

$$0 = d^2 = \partial^2 + \left(\partial \circ \bar{\partial} + \bar{\partial} \circ \partial\right) + \bar{\partial}^2.$$

Suppose $\omega \in A_{p,q}$. Then

$$\partial^2 \omega \in A_{p+2,q}, \qquad \left(\partial \circ \bar{\partial} + \bar{\partial} \circ \partial\right) \omega \in A_{p+1,q+1}, \qquad \bar{\partial}^2 \omega \in A_{p,q+2}.$$

Hence

$$\partial^2 \omega = 0, \qquad \left(\partial \circ \bar{\partial} + \bar{\partial} \circ \partial\right) \omega = 0, \qquad \bar{\partial}^2 \omega = 0. \tag{3.27}$$

Therefore (3.26) holds.

(Sufficiency) Suppose $\bar{\partial}^2 = 0$. If F is a smooth complex-valued function on M, then we can write

$$dF = \sum_k F_k \theta^k + \sum_k G_k \bar{\theta}^k. \tag{3.28}$$

Thus we obtain, by using (3.11),

$$\partial F = \sum_k F_k \theta^k, \qquad \overline{\partial} F = \sum_k G_k \overline{\theta}^k, \tag{3.29}$$

$$\overline{\partial}^2 F = \prod_{0,2} d\left(\overline{\partial} F\right) = \prod_{0,2} d\left(\overline{\partial} - d\right) F$$
$$= -\prod_{0,2} d\left(\partial F\right) = -\frac{1}{2} \sum_{k,j,l} F_k C_{jl}^k \overline{\theta}^j \wedge \overline{\theta}^l. \tag{3.30}$$

Since $\overline{\partial}^2 F = 0$ for any F, $C_{jl}^k = 0$. Hence the integrability condition holds. □

Now assume that M is an n-dimensional complex manifold with local coordinate system $z^k = x^k + iy^k$, $1 \le k \le n$. Then the dz^k are differential $(1,0)$-forms on M with respect to the canonical almost complex structure on M. Hence a smooth exterior differential (p,q)-form can be expressed locally as

$$a = \sum a_{k_1 \cdots k_p, \bar{l}_1 \cdots \bar{l}_q} dz^{k_1} \wedge \cdots \wedge dz^{k_p} \wedge d\bar{z}^{l_1} \wedge \cdots \wedge d\bar{z}^{l_q}, \tag{3.31}$$

where the $a_{k_1 \cdots k_p, \bar{l}_1 \cdots \bar{l}_q}$ are smooth, complex-valued functions.

If f is a smooth, complex-valued function on M, then

$$\begin{aligned} df &= \sum_{k=1}^n \left(\frac{\partial f}{\partial x^k} dx^k + \frac{\partial f}{\partial y^k} dy^k \right) \\ &= \sum_{k=1}^n \left(\frac{\partial f}{\partial z^k} dz^k + \frac{\partial f}{\partial \bar{z}^k} d\bar{z}^k \right), \end{aligned}$$

where $\dfrac{\partial}{\partial z^k}$ and $\dfrac{\partial}{\partial \bar{z}^k}$ are operators defined in (3.9) and (3.10), respectively. Therefore

$$\partial f = \sum_{k=1}^n \frac{\partial f}{\partial z^k} dz^k, \qquad \overline{\partial} f = \sum_{k=1}^n \frac{\partial f}{\partial \bar{z}^k} d\bar{z}^k. \tag{3.32}$$

Thus it follows from (3.31) that

$$\begin{aligned} \partial a &= \sum \partial a_{k_1 \cdots k_p, \bar{l}_1 \cdots \bar{l}_q} \wedge dz^{k_1} \wedge \cdots \wedge dz^{k_p} \wedge d\bar{z}^{l_1} \wedge \cdots \wedge d\bar{z}^{l_q} \tag{3.33} \\ &= \sum \frac{\partial a_{k_1 \cdots k_p, \bar{l}_1 \cdots \bar{l}_q}}{\partial z^k} dz^k \wedge dz^{k_1} \wedge \cdots \wedge dz^{k_p} \wedge d\bar{z}^{l_1} \wedge \cdots \wedge d\bar{z}^{l_q} \end{aligned}$$

Similarly,

$$\overline{\partial} a = (-1)^p \sum \frac{\partial a_{k_1 \cdots k_p, \bar{l}_1 \cdots \bar{l}_q}}{\partial \bar{z}^l} dz^{k_1} \wedge \cdots \wedge dz^{k_p} \wedge d\bar{z}^l \wedge d\bar{z}^{l_1} \wedge \cdots \wedge d\bar{z}^{l_q}. \tag{3.34}$$

If we denote

$$f\left(z^1, \ldots, z^n\right) = g\left(z^1, \ldots, z^n\right) + ih\left(z^1, \ldots, z^n\right), \qquad (3.35)$$

then

$$
\begin{aligned}
\frac{\partial f}{\partial \bar{z}^k} &= \frac{1}{2}\left(\frac{\partial}{\partial x^k} + i\frac{\partial}{\partial y^k}\right)(g + ih) \\
&= \frac{1}{2}\left(\frac{\partial g}{\partial x^k} - \frac{\partial h}{\partial y^k}\right) + \frac{i}{2}\left(\frac{\partial h}{\partial x^k} + \frac{\partial g}{\partial y^k}\right).
\end{aligned} \qquad (3.36)
$$

Thus we have:

Theorem 3.7. *Suppose f is a smooth, complex-valued function on a complex manifold M. Then a necessary and sufficient condition for f to be a holomorphic function is that $\bar{\partial} f = 0$.*

Proof. The Cauchy–Riemann condition for f is

$$\frac{\partial g}{\partial x^k} = \frac{\partial h}{\partial y^k}, \qquad \frac{\partial g}{\partial y^k} = -\frac{\partial h}{\partial x^k}, \qquad 1 \le k \le n.$$

By (3.32) and (3.36), the above condition is equivalent to $\bar{\partial} f = 0$, that is, the condition for f to be holomorphic is that $\bar{\partial} f = 0$.

If a is a differential $(p, 0)$-form with local expression

$$a = \sum a_{k_1 \cdots k_p} dz^{k_1} \wedge \cdots \wedge dz^{k_p},$$

then, when the $a_{k_1 \cdots k_p}$ are all holomorphic functions, we obtain from Theorem 3.6 that

$$da = \partial a = \sum \frac{\partial a_{k_1 \cdots k_p}}{\partial z^k} dz^k \wedge dz^{k_1} \wedge \cdots \wedge dz^{k_p}.$$

Thus the operator ∂ maps holomorphic differential $(p, 0)$-forms complex-linearly to holomorphic $(p + 1, 0)$-forms.   □

§7–4 Connections on Complex Vector Bundles

We have discussed vector bundles (E, M, π) on a manifold M in §3–1. When the typical fiber is a q-dimensional complex vector space V, the vector bundle is a q-dimensional complex vector bundle on M. In this case, the structure group is

$$GL\left(V\right) \simeq GL\left(q; \mathbb{C}\right).$$

Suppose M is an m-dimensional smooth manifold, and (E, M, π) a q-dimensional complex vector bundle on M. Then the section space $\Gamma(E)$ has a complex-linear structure, and is a module over the ring of all smooth complex-valued functions on M. The discussion on connections on real vector bundles given in §4–1 can be adapted directly to the case of complex vector bundles (E, M, π), provided we replace the real number field in the earlier discussion, wherever it occurs, by the complex number field. We will not repeat this discussion here.

Suppose $\{a_\alpha, 1 \le \alpha \le q\}$ is a local frame field of the complex vector bundle E in a neighborhood $U \subset M$. Then the action of a connection D on E can be expressed in U by

$$Ds_\alpha = \sum_\beta \omega_\alpha^\beta s_\beta, \tag{4.1}$$

where ω_α^β is a complex-valued differential 1-form. If we use the matrix notation, then (4.1) can be written as

$$DS = \omega \cdot S, \tag{4.2}$$

where

$$S = {}^t (s_1, \ldots, s_q), \tag{4.3}$$

$$\omega = \begin{pmatrix} \omega_1^1 & \cdots & \omega_1^q \\ \vdots & \ddots & \vdots \\ \omega_q^1 & \cdots & \omega_q^q \end{pmatrix}. \tag{4.4}$$

Hence the curvature matrix for the connection is

$$\Omega = (\Omega_\alpha^\beta) = d\omega - \omega \wedge \omega. \tag{4.5}$$

Exteriorly differentiating (4.5), we obtain the Bianchi identity:

$$d\Omega = \omega \wedge \Omega - \Omega \wedge \omega. \tag{4.6}$$

If we choose another local frame field S', and assume that

$$S' = A \cdot S, \tag{4.7}$$

where $\det A \ne 0$, then we have [see (1.29) in Chapter 4]

$$\Omega' = A \cdot \Omega \cdot A^{-1}. \tag{4.8}$$

The above transformation formula motivates the following definition.

Definition 4.1. If for every local frame field S of a vector bundle (E, M, π) there is a given $q \times q$ matrix Φ_S of exterior differential k-forms which satisfies the following transformation rule under a change (4.7) of the frame field S:

$$\Phi_{S'} = A \cdot \Phi_S \cdot A^{-1}, \tag{4.9}$$

then we call $\{\Phi_S\}$ a **tensorial matrix of adjoint type**.

Because the transformation formula for the connection matrix ω under the change (4.7) of the frame field S is given by

$$\omega' \cdot A = dA + A \cdot \omega, \tag{4.10}$$

it follows by exterior differentiation of (4.9) that

$$d\Phi_{S'} = dA \wedge \Phi_S \cdot A^{-1} + A \cdot d\Phi_S \cdot A^{-1} + (-1)^k A \cdot \Phi_S \wedge dA^{-1}.$$

By plugging (4.9) into the above equation and rearranging terms, we have

$$D\Phi_{S'} = A \cdot D\Phi_S \cdot A^{-1}, \tag{4.11}$$

where

$$D\Phi_S = d\Phi_s - \omega \wedge \Phi_S + (-1)^k \Phi_S \wedge \omega. \tag{4.12}$$

Thus we see that $\{D\Phi_S\}$ is still an adjoint type tensorial matrix whose elements are exterior differential $(k+1)$-forms. We call $D\Phi_S$ the **covariant derivative** of Φ_S.

By the above definition, the Bianchi identity (4.6) implies that the covariant derivative of the curvature matrix Ω is zero, that is,

$$D\Omega = 0. \tag{4.13}$$

For notational simplicity in discussions on tensorial matrices of adjoint type, we usually omit the lower index S specifying a given frame field when computations are carried out only with respect to that frame field.

Covariantly differentiating (4.12) again, we get

$$D^2\Phi = \Phi \wedge \Omega - \Omega \wedge \Phi \tag{4.14}$$

(The reader should verify this). On denoting the right hand side by

$$[\Phi, \Omega] = \Phi \wedge \Omega - \Omega \wedge \Phi, \tag{4.15}$$

(4.14) becomes

$$D^2\Phi = [\Phi, \Omega]. \tag{4.16}$$

Now consider a complex r-linear function $P(A_1, \ldots, A_r)$ of $q \times q$ matrices A_i $(1 \le i \le r)$. If we assume that

$$A_i = \left(a^i_{\alpha\beta} \right), \qquad 1 \le \alpha, \beta \le q, \quad 1 \le i \le r, \tag{4.17}$$

then the function P can be expressed as

$$P(A_1, \ldots, A_r) = \sum_{1 \le \alpha_i, \beta_i \le q} \lambda_{\alpha_1 \cdots \alpha_r, \beta_1 \cdots \beta_r} a^1_{\alpha_1 \beta_1} \cdots a^r_{\alpha_r \beta_r}, \tag{4.18}$$

where the $\lambda_{\alpha_1 \cdots \alpha_r, \beta_1 \cdots \beta_r}$ are complex numbers. If for any permutation σ of $\{1, \ldots, r\}$ we have

$$P\left(A_{\sigma(1)}, \ldots, A_{\sigma(r)} \right) = P(A_1, \ldots, A_r), \tag{4.19}$$

then we say P is **symmetric**. If for any $B \in \mathrm{GL}(q; \mathbb{C})$ we have

$$P\left(BA_1 B^{-1}, \ldots, BA_r B^{-1} \right) = P(A_1, \ldots, A_r), \tag{4.20}$$

then we say P is an **invariant polynomial**.

We can use the following method to obtain a sequence of symmetric invariant polynomials. Suppose I is the $q \times q$ identity matrix. Let

$$\det \left(I + \frac{i}{2\pi} A \right) = \sum_{0 \le j \le q} \binom{q}{j} P_j(A), \tag{4.21}$$

here $P_j(A)$ is a homogeneous polynomial of elements of A of order j. For any nondegenerate $q \times q$ matrix B, since

$$I + \frac{i}{2\pi} BAB^{-1} = B \left(I + \frac{i}{2\pi} A \right) B^{-1},$$

we have

$$\det \left(I + \frac{i}{2\pi} BAB^{-1} \right) = \det \left(I + \frac{i}{2\pi} A \right). \tag{4.22}$$

Therefore

$$P_j \left(BAB^{-1} \right) = P_j(A), \tag{4.23}$$

that is, $P_j(A)$ is an invariant polynomial.

Suppose $P_j(A_1, \ldots, A_j)$ is the **completely polarized polynomial** of $P_j(A)$, that is, $P_j(A_1, \ldots, A_j)$ is a symmetric j-linear function of $A_1, \ldots, A_j$, satisfying

$$P_j(A, \ldots, A) = P_j(A). \tag{4.24}$$

It is easy to show that $P_j(A_1, \ldots, A_j)$ can be expressed in terms of $P_j(A)$. For instance,

$$P_2(A_1, A_2) = \frac{1}{2}\{P_2(A_1 + A_2) - P_2(A_1) - P_2(A_2)\},$$

$$P_3(A_1, A_2, A_3) = \frac{1}{6}\{P_3(A_1 + A_2 + A_3) - P_3(A_1 + A_2) \quad\quad (4.25)$$
$$- P_3(A_1 + A_3) - P_3(A_2 + A_3) + P_3(A_1)$$
$$+ P_3(A_2) + P_3(A_3)\}.$$

Therefore $P_j(A_1, \ldots, A_j)$ is an invariant symmetric j-linear function.

Suppose $P(A_1, \ldots, A_r)$ is an invariant polynomial, and we express a non-degenerate matrix B as

$$B = I + B'. \quad\quad (4.26)$$

Then

$$B^{-1} = I - B' + \cdots, \qu\quad (4.27)$$

where the missing part contains higher order terms of the elements of the matrix B'. Plugging this into (4.20) and considering the linear part of B' we obtain

$$\sum_{1 \le i \le r} P(A_1, \ldots, B'A_i - A_iB', \ldots, A_r) = 0. \qu\quad (4.28)$$

This equation still holds if A_i is a matrix of exterior differential forms.[c]

Suppose A_i is a matrix of exterior differential d_i-forms. Then for any $q \times q$ matrix θ of differential 1-forms we have

$$\sum_{1 \le i \le r} (-1)^{d_1 + \cdots + d_{i-1}} P(A_1, \ldots, \theta \wedge A_i, \ldots, A_r)$$

$$+ \sum_{1 \le i \le r} (-1)^{d_1 + \cdots + d_i + 1} P(A_1, \ldots, A_i \wedge \theta, \ldots, A_r) = 0. \qu\quad (4.29)$$

To show this, we need only note that θ is a sum of matrices of the form $B' \cdot a$, where B' is a $q \times q$ matrix of numbers and a is a differential 1-form. Using the multilinear property of P, we need only show (4.29) for $\theta = B' \cdot a$. Plugging $\theta = B' \cdot a$ into the left hand side of (4.29), we obtain

$$a \wedge \left\{ \sum_{1 \le i \le r} P(A_1, \ldots, B' \cdot A_i, \ldots, A_r) \right.$$
$$\left. - \sum_{1 \le i \le r} P(A_1, \ldots, A_i \cdot B', \ldots, A_r) \right\}.$$

---

[c]Suppose $P(A_1, \ldots, A_r)$ can be expressed as in (4.18). Then we define $P(A_1, \ldots, A_r) = \sum_{1 \le \alpha_i, \beta_i \le q} \lambda_{\alpha_1 \cdots \alpha_r, \beta_1 \cdots \beta_r} a^1_{\alpha_1 \beta_1} \wedge \cdots \wedge a^r_{\alpha_r \beta_r}$ when $A_i = \left(a^i_{\alpha\beta}\right)$ is a matrix composed of exterior differential forms.

By (4.28), the quantity inside the brackets is zero, hence (4.29) holds.

Invariant polynomials establish a relation between the global and local properties of a connection. Suppose $P(A_1, \ldots, A_r)$ is an invariant polynomial, and choose A_i to be a tensorial matrix of adjoint type composed of differential d_i-forms. Obviously $P(A_1, \ldots, A_r)$ is an exterior differential $(d_1 + d_2 + \cdots + d_r)$-form independent of the choice of the local frame field. Hence it is an exterior differential form defined globally on M. By (4.12) and (4.29), we have

$$
\begin{aligned}
dP(A_1, \ldots, A_r) &= \sum_{1 \leq i \leq r} (-1)^{d_1 + \cdots + d_{i-1}} P(A_1, \ldots, dA_i, \ldots, A_r) \\
&= \sum_{1 \leq i \leq r} (-1)^{d_1 + \cdots + d_{i-1}} \{ P(A_1, \ldots, DA_i, \ldots, A_r) \quad (4.30) \\
&\qquad + P\left(A_1, \ldots, \omega \wedge A_i + (-1)^{d_i+1} A_i \wedge \omega, \ldots, A_r\right) \} \\
&= \sum_{1 \leq i \leq r} (-1)^{d_1 + \cdots + d_{i-1}} P(A_1, \ldots, DA_i, \ldots, A_r).
\end{aligned}
$$

In particular, for an invariant polynomial $P_j(A)$, if we choose A to be the curvature matrix Ω of a connection, then $D\Omega = 0$. Hence

$$dP_j(\Omega) = 0. \qquad (4.31)$$

Thus $P_j(\Omega)$ is a closed exterior differential $2j$-form defined globally on M.

Theorem 4.1. *Suppose (E, M, π) is a q-dimensional complex vector bundle on a smooth m-dimensional manifold M, and Ω and $\tilde{\Omega}$ are curvature forms corresponding to the connections ω and $\tilde{\omega}$, respectively. If $P(A_1, \ldots, A_r)$ is a symmetric invariant polynomial, then there exists an exterior differential $(2r - 1)$-form Q on M such that*

$$P(\tilde{\Omega}) - P(\Omega) = dQ. \qquad (4.32)$$

Proof. Let

$$\eta = \tilde{\omega} - \omega. \qquad (4.33)$$

If we choose another local frame field $S' = B \cdot S$, then

$$
\begin{aligned}
\tilde{\omega}' \cdot B &= dB + B \cdot \tilde{\omega}, \\
\omega' \cdot B &= dB + B \cdot \omega,
\end{aligned}
$$

where $\tilde{\omega}'$, ω' represent the corresponding connection matrices with respect to the local frame field S'. Therefore $\eta' = \tilde{\omega}' - \omega'$ and η satisfy the following relation

$$\eta' \cdot B = B \cdot \eta, \qquad (4.34)$$

that is, η is a tensorial matrix of adjoint type whose elements are differential 1-forms. Let

$$\omega_t = \omega + t\eta, \qquad 0 \le t \le 1. \tag{4.35}$$

Then ω_t generates a family of connections depending on the parameter t, which gives ω and $\tilde{\omega}$ at $t = 0$ and $t = 1$, respectively. The curvature matrix of the connection ω_t is

$$
\begin{aligned}
\Omega_t &= d\omega_t - \omega_t \wedge \omega_t \\
&= \Omega + tD\eta - t^2\eta \wedge \eta.
\end{aligned}
\tag{4.36}
$$

Hence

$$\frac{d}{dt}\Omega_t = D\eta - 2t\eta \wedge \eta, \tag{4.37}$$

where the covariant derivative is computed with respect to the connection ω. Suppose $P(A_1, \ldots, A_r)$ is a symmetric invariant polynomial, and let

$$
\begin{cases}
P(A) = P(A, \ldots A), \\
Q(B, A) = rP(B, A, \ldots, A).
\end{cases}
\tag{4.38}
$$

Then

$$
\begin{aligned}
\frac{d}{dt}P(\Omega_t) &= rP\left(\frac{d\Omega}{dt}, \Omega_t, \ldots, \Omega_t\right) \\
&= Q(D\eta, \Omega_t) - 2tQ(\eta \wedge \eta, \Omega_t).
\end{aligned}
\tag{4.39}
$$

By the definition of the covariant derivative, it follows from (4.36) that

$$
\begin{aligned}
D\Omega_t &= tD^2\eta - t^2 D(\eta \wedge \eta) \\
&= t(\eta \wedge \Omega - \Omega \wedge \eta) + t^2(\eta \wedge D\eta - D\eta \wedge \eta) \\
&= t[\eta, \Omega] + t^2[\eta, D\eta] \\
&= t[\eta, \Omega_t].
\end{aligned}
\tag{4.40}
$$

Thus

$$
\begin{aligned}
dQ\left(\eta, \Omega_t\right) &= r\, dP\left(\eta, \Omega_t, \ldots, \Omega_t\right) \\
&= r\, P\left(D\eta, \Omega_t, \ldots, \Omega_t\right) - r(r-1)P\left(\eta, D\Omega_t, \Omega_t, \ldots, \Omega_t\right) \\
&= Q\left(D\eta, \Omega_t\right) - r(r-1)tP\left(\eta, [\eta, \Omega_t], \Omega_t, \ldots, \Omega_t\right).
\end{aligned}
$$

$$(4.41)$$

Letting $\theta = A_1 = \eta$, $A_2 = \cdots = A_r = \Omega_t$ in (4.29), we obtain

$$
2P\left(\eta \wedge \eta, \Omega_t, \ldots, \Omega_t\right) - (r-1)P\left(\eta, [\eta, \Omega_t], \Omega_t, \ldots, \Omega_t\right) = 0,
$$

that is,

$$
2Q\left(\eta \wedge \eta, \Omega_t\right) - r(r-1)P\left(\eta, [\eta, \Omega_t], \Omega_t, \ldots, \Omega_t\right) = 0. \qquad (4.42)
$$

Comparing (4.41) with (4.42) we have

$$
\begin{aligned}
dQ\left(\eta, \Omega_t\right) &= Q\left(D\eta, \Omega_t\right) - 2tQ\left(\eta \wedge \eta, \Omega_t\right) \\
&= \frac{d}{dt}P\left(\Omega_t\right).
\end{aligned}
$$

$$(4.43)$$

Integrating both sides with respect to t, we get

$$
P(\tilde{\Omega}) - P(\Omega) = d\left(\int_0^1 Q\left(\eta, \Omega_t\right) dt\right).
$$

Finally, let

$$
Q = \int_0^1 Q\left(\eta, \Omega_t\right) dt. \qquad (4.44)
$$

Then Q is the desired exterior differential $(2r-1)$-form on M. $\qquad\square$

    As in (4.31), $P(\Omega)$ is a closed exterior differential form. If it is a real-valued exterior differential form, then it determines an element in the cohomology group $H^{2r}(M; \mathbb{R})$ on M. The essence of Theorem 4.1 is the following: even though the exterior differential form $P(\Omega)$ is defined by connections, the de Rham cohomology class determined by it is independent of the choice of connections on the complex vector bundle E. Now we introduce a Hermitian structure on E. For a compatible connection (see Def. 4.4 below) of a Hermitian structure, $P(\Omega)$ is in fact a real-valued exterior differential form. Thus we see that every symmetric invariant polynomial P corresponds to a de Rham cohomology class.

Definition 4.2. Suppose V is a complex vector space. If a complex-valued function $H(\xi, \eta)$, $\xi, \eta \in V$ defined on $V \times V$ satisfies the following conditions:

1) for any $\lambda_1, \lambda_2 \in \mathbb{C}$, $\xi_1, \xi_2 \in V$ we have

$$H(\lambda_1 \xi_1 + \lambda_2 \xi_2, \eta) = \lambda_1 H(\xi_1, \eta) + \lambda_2 H(\xi_2, \eta);$$

2) $\overline{H(\xi, \eta)} = H(\eta, \xi),$

then $H(\xi, \eta)$ is called a **Hermitian structure** on V. If for any $\xi \in V$, $\xi \neq 0$, we have

$$H(\xi, \xi) > 0,$$

then the Hermitian structure H is said to be **positive definite**.

Remark. Suppose J is a complex structure on a $2n$-dimensional real vector space V. Define

$$i \cdot X = JX, \qquad X \in V. \tag{4.45}$$

Then V becomes an n-dimensional complex vector space. In this way, the Hermitian structure on a real vector space V with a complex structure defined in Definition 2.2 corresponds to the Hermitian structure on a complex vector space V defined in Definition 4.2.

Definition 4.3. Suppose (E, M, π) is a q-dimensional complex vector bundle on an m-dimensional smooth manifold M. If at every point $x \in M$ there is a smooth assignment of a positive definite Hermitian structure on the fiber $\pi^{-1}(x)$, then we say that E is endowed with a Hermitian structure. A complex vector bundle with a given Hermitian structure is called a **Hermitian vector bundle**.

By "a smooth assignment" we mean that, if ξ, η are any two smooth sections of the bundle, then as a complex-valued function on M, $H(\xi, \eta)$ is smooth.

By the Partition of Unity Theorem, it is not difficult to show (the proof is similar to the existence proof for a Riemannian metric on M) that every complex vector bundle has a Hermitian structure. For any local frame field

$$S = {}^t(s_1, \ldots, s_q),$$

the Hermitian structure H corresponds to a Hermitian matrix

$$H_S = \left(h_{\alpha\bar\beta} \right) = {}^t \overline{H}_S, \tag{4.46}$$

where

$$h_{\alpha\bar{\beta}} = H\left(s_\alpha, s_\beta\right). \tag{4.47}$$

Suppose $\xi = \sum_\alpha \xi^\alpha s_\alpha$, $\eta = \sum_\beta \eta^\beta s_\beta$. Then

$$H\left(\xi, \eta\right) = \sum_{\alpha,\beta} h_{\alpha\bar{\beta}} \xi^\alpha \bar{\eta}^\beta. \tag{4.48}$$

Definition 4.4. Suppose D is a connection on a Hermitian vector bundle. If, for any two vector fields ξ, η parallel along an arbitrary curve, $H\left(\xi, \eta\right)$ is a constant, then we call D a **compatible connection** on the vector bundle.

Since ξ and η are parallel along a curve C, we have, along C,

$$
\begin{aligned}
D\xi^\alpha &= d\xi^\alpha + \sum_\beta \xi^\beta \omega_\beta^\alpha = 0, \\
D\eta^\alpha &= d\eta^\alpha + \sum_\beta \eta^\beta \omega_\beta^\alpha = 0.
\end{aligned}
$$

Therefore, from (4.48), we obtain

$$dH\left(\xi, \eta\right) = \sum_{\alpha,\beta} \left(dh_{\alpha\bar{\beta}} - \sum_\gamma h_{\gamma\bar{\beta}}\omega_\alpha^\gamma - \sum_\gamma h_{\alpha\bar{\gamma}}\bar{\omega}_\beta^\gamma\right)\xi^\alpha\bar{\eta}^\beta.$$

Thus, the condition for ω to be a compatible connection is

$$dh_{\alpha\bar{\beta}} - \sum_\gamma h_{\gamma\bar{\beta}}\omega_\alpha^\gamma - \sum_\gamma h_{\alpha\bar{\gamma}}\bar{\omega}_\beta^\gamma = 0, \tag{4.49}$$

or, in matrix notation,

$$dH = \omega \cdot H + H \cdot {}^t\bar{\omega}. \tag{4.50}$$

Remark. On a Hermitian bundle, a compatible connection necessarily exists. The reader should prove this.

Exteriorly differentiating (4.50) we obtain

$$\Omega \cdot H + H \cdot {}^t\bar{\Omega} = 0. \tag{4.51}$$

Thus the matrix $\Omega \cdot H$ is anti-Hermitian. It follows that, with respect to a compatible connection on a Hermitian vector bundle E,

$$
\begin{aligned}
\overline{\det\left(I + \frac{i}{2\pi}\Omega\right)} &= \det\left(I - \frac{i}{2\pi}\bar{\Omega}\right) \\
&= \det\left(I + \frac{i}{2\pi}H^{-1} \cdot \Omega \cdot H\right) \\
&= \det\left(I + \frac{i}{2\pi}\Omega\right).
\end{aligned}
$$

Hence

$$\overline{P_j(\Omega)} = P_j(\Omega).$$

Therefore $P_j(\Omega)$ is a real-valued closed exterior differential $2j$-form on M. The de Rham cohomology class $c_j(E)$ determined by $P_j(\Omega)$ is independent of the choice of the Hermitian structure and the compatible connection on M. It is called the j-th **Chern class** with real coefficients of the complex vector bundle E. (For more details consult Chern 1946.)

Expanding the determinant in (4.21), it is not difficult to obtain an expression for $P_j(\Omega)$:

$$P_j(\Omega) = \frac{1}{j!}\left(\frac{i}{2\pi}\right)^j \sum_{1 \le \alpha_r, \beta_r \le q} \delta^{\alpha_1 \cdots \alpha_j}_{\beta_1 \cdots \beta_j} \Omega^{\beta_1}_{\alpha_1} \wedge \cdots \wedge \Omega^{\beta_j}_{\alpha_j}, \qquad (4.52)$$

where $\Omega = \left(\Omega^\beta_\alpha\right)$ is the curvature matrix for a compatible connection on the Hermitian vector bundle E.

Remark. Suppose E is a q-dimensional real vector bundle on M with typical fiber V. Let $E \otimes \mathbb{C}$ be the complexification of E. It is then a complex vector bundle with the q-dimensional complex vector space $V \otimes \mathbb{C}$ as its typical fiber. Let c_{2j} be the $2j$-th Chern class of the complex vector bundle $E \times \mathbb{C}$. Then we call

$$p_j(E) = (-1)^j c_{2j} \in H^{4j}(M; \mathbb{R}) \qquad (4.53)$$

the j-th **Pontrjagin class** (c.f. Kobayashi and Nomizu 1963 and 1969). If the curvature matrix of a connection ω on a real vector bundle E is

$$\Omega = \left(\Omega^\beta_\alpha\right),$$

then the de Rham cohomology class $p_j(E)$ is determined by the real-valued closed exterior differential $4j$-form:

$$\frac{1}{(2j)!\,(2\pi)^{2j}} \sum_{1 \le \alpha_r, \beta_r \le q} \delta^{\alpha_1 \cdots \alpha_{2j}}_{\beta_1 \cdots \beta_{2j}} \Omega^{\beta_1}_{\alpha_1} \wedge \cdots \wedge \Omega^{\beta_{2j}}_{\alpha_{2j}}. \qquad (4.54)$$

Definition 4.5. Suppose M is an m-dimensional complex manifold, and $\pi : E \longrightarrow M$ is a q-dimensional complex vector bundle. If, for any two local coordinate neighborhoods U and W with nonempty intersection, the transition function

$$g_{UW} : U \cap W \longrightarrow \mathrm{GL}\,(q; \mathbb{C})$$

is holomorphic, then we call E a **holomorphic vector bundle** on M. If the typical fiber of a holomorphic vector bundle E is a 1-dimensional complex vector space ($q = 1$), then E is called a **holomorphic line bundle** on the complex manifold M.

Obviously the bundle space of a holomorphic vector bundle is a complex manifold.

If an m-dimensional complex manifold is viewed as a $2m$-dimensional real manifold, then the latter has a canonical almost complex structure. The set of all type $(1,0)$ tangent vectors with respect to this canonical almost complex structure is a holomorphic vector bundle on the complex manifold M, called the **tangent bundle** of M. This is so because for any local complex coordinate system $(z^1, \ldots, z^m)$, $\left(\frac{\partial}{\partial z^1}, \ldots, \frac{\partial}{\partial z^m}\right)$ constitutes a local frame field of the tangent bundle. Obviously, any two such frame fields are holomorphically related. Hence the tangent bundle is a holomorphic vector bundle.

Suppose $\gamma : U \longrightarrow E$ is a section of a holomorphic vector bundle E in a neighborhood $U \subset M$. If γ is a holomorphic map, then we call γ a **holomorphic section**. Suppose S and S' are two holomorphic local frame fields. Then on the intersection of their defining neighborhoods we have the following expression:

$$S' = A \cdot S, \tag{4.55}$$

where A is a nondegenerate matrix of holomorphic functions.

Definition 4.6. Suppose D is a connection on a holomorphic vector bundle (E, M, π). If for any holomorphic local frame field S, the connection matrix ω is composed of differential $(1,0)$-forms on M with respect to the canonical almost complex structure on M, then D is called a **type $(1,0)$ connection**.

The above definition is meaningful, since the fact that ω is a type $(1,0)$ matrix is independent of the choice of the local frame field S. Indeed, under a holomorphic transformation (4.55) of local frame fields, A is a matrix of holomorphic functions. Hence $\bar{\partial}A = 0$ by Theorem 3.6. Thus

$$\begin{aligned} \omega' &= dA \cdot A^{-1} + A \cdot \omega \cdot A^{-1} \\ &= \partial A \cdot A^{-1} + A \cdot \omega \cdot A^{-1}. \end{aligned}$$

Thus, if ω is a type $(1,0)$ matrix, then so is ω'. The converse is also true.

If E is a Hermitian holomorphic vector bundle, then there is a unique type $(1,0)$ compatible connection on E. In fact, the condition for ω to be a compatible connection is

$$dH = \omega \cdot H + H \cdot {}^t\bar{\omega}.$$

If ω is a type $(1,0)$ matrix, then $\bar{\omega}$ is a type $(0,1)$ matrix. Thus

$$\partial H = \omega \cdot H. \tag{4.56}$$

From this we see that the type $(1,0)$ compatible connection ω must be given by

$$\omega = \partial H \cdot H^{-1}. \tag{4.57}$$

It is easy to verify that the above equation actually gives a connection on E.

The curvature matrix for a type $(1,0)$ compatible connection on a Hermitian holomorphic vector bundle E is

$$
\begin{aligned}
\Omega &= d\left(\partial H \cdot H^{-1}\right) - \left(\partial H \cdot H^{-1}\right) \wedge \left(\partial H \cdot H^{-1}\right) \\
&= -\partial\bar{\partial}H \cdot H^{-1} + \partial H \cdot H^{-1} \wedge \bar{\partial}H \cdot H^{-1}.
\end{aligned} \tag{4.58}
$$

Hence Ω is a matrix of differential $(1,1)$-forms.

§7–5 Hermitian Manifolds and Kählerian Manifolds

Definition 5.1. Suppose M is an m-dimensional complex manifold. If a Hermitian structure H is given on the tangent bundle of M, then M is called a **Hermitian manifold**.

For a local complex coordinate system $(U; z^1, \ldots, z^m)$, the corresponding local frame field of the tangent bundle is

$$s_i = \frac{\partial}{\partial z^i}. \tag{5.1}$$

In this section we restrict the values assumed by indices as follows:

$$1 \le i, j, k, l \le m,$$

and adopt the Einstein summation convention notation, omitting the summation symbol. Let

$$h_{i\bar{k}} = h_{\bar{k}i} = H\left(\frac{\partial}{\partial z^i}, \frac{\partial}{\partial z^k}\right). \tag{5.2}$$

Then

$$\bar{h}_{k\bar{i}} = h_{\bar{k}i}, \tag{5.3}$$

and the matrix $H = \left(h_{i\bar{k}}\right)$ is positive definite. If ξ, η are two type $(1,0)$ tangent vector fields on U, then they can be expressed as

$$\xi = \xi^i \frac{\partial}{\partial z^i}, \qquad \eta = \eta^k \frac{\partial}{\partial z^k}, \tag{5.4}$$

where ξ^i, η^k are smooth complex-valued functions on U. Thus

$$H\left(\xi,\eta\right) = h_{i\overline{k}}\xi^i\overline{\eta}^k. \tag{5.5}$$

The **Kählerian form** on M,

$$\hat{H} = \frac{i}{2}h_{i\overline{k}}dz^i \wedge d\overline{z}^k, \tag{5.6}$$

is a real-valued differential $(1,1)$-form.

A connection on the tangent bundle has a torsion matrix. Suppose the coframe field dual to the local frame field $S = {}^t\left(s_1, \ldots, s_m\right)$ is $\sigma = \left(\sigma^1, \ldots, \sigma^m\right)$. If there is another local frame field

$$S' = A \cdot S, \tag{5.7}$$

then its dual coframe field is

$$\sigma' = \sigma \cdot A^{-1},$$

or

$$\sigma = \sigma' \cdot A. \tag{5.8}$$

Exteriorly differentiating (5.8) we obtain

$$\begin{aligned} d\sigma &= d\sigma' \cdot A - \sigma' \wedge dA \\ &= \left(d\sigma' - \sigma' \wedge \omega'\right)A + \sigma \wedge \omega, \end{aligned}$$

that is

$$\tau = \tau' \cdot A, \tag{5.9}$$

where

$$\tau = d\sigma - \sigma \wedge \omega, \qquad \tau' = d\sigma' - \sigma' \wedge \omega'. \tag{5.10}$$

τ is a $(1 \times m)$ matrix of complex-valued exterior differential 2-forms, called the **torsion matrix** of a connection on the tangent bundle of a complex manifold.

Theorem 5.1. *Suppose M is a Hermitian manifold. A necessary and sufficient condition for D to be a type $(1,0)$ connection on the tangent bundle of M is that its torsion matrix is composed of differential $(2,0)$-forms.*

Proof. Suppose $\sigma = (\sigma^1, \ldots, \sigma^m)$ is the coframe field dual to the holomorphic frame field S. Then every σ^i is a holomorphic differential $(1,0)$-form, that is, for the complex local coordinate system z^i, σ^i can be expressed as

$$\sigma^i = a^i_j dz^j,$$

where the a^i_j are holomorphic functions. Thus

$$\bar{\partial}\sigma = 0. \tag{5.11}$$

The connection matrix ω can be decomposed uniquely as

$$\omega = \omega_1 + \omega_2, \tag{5.12}$$

where ω_1, ω_2 are matrices of differential $(1,0)$-forms and $(0,1)$-forms, respectively. Thus the torsion matrix τ can be expressed as

$$\begin{aligned} \tau &= d\sigma - \sigma \wedge \omega \\ &= (\partial\sigma - \sigma \wedge \omega_1) - \sigma \wedge \omega_2. \end{aligned} \tag{5.13}$$

The right hand side is already decomposed into a sum of a type $(2,0)$ matrix and a type $(1,1)$ matrix. Thus we see that a necessary and sufficient condition for the torsion matrix to be composed of differential $(2,0)$-forms is

$$\sigma \wedge \omega_2 = 0. \tag{5.14}$$

Assume that $\omega_2 = \left(\theta^k_j\right)$, then (5.14) becomes

$$\sigma^j \wedge \theta^k_j = 0. \tag{5.15}$$

By Cartan's Lemma we have

$$\theta^k_j = a^k_{ij}\sigma^i, \tag{5.16}$$

where the a^k_{ij} are smooth complex-valued functions. Since σ^j is a differential $(1,0)$-form and θ^k_j is a differential $(0,1)$-form, condition (5.14) is equivalent to

$$a^k_{ij} = 0, \quad \text{that is,} \quad \omega_2 = 0, \tag{5.17}$$

which means that ω is a $(1,0)$-form.   □

In our definition of type $(1,0)$ connections we used holomorphic frame fields; yet the criterion given in Theorem 5.1 requires only the smoothness of local frames fields. This is because (5.9) implies that a change of local frame fields as in (5.7) does not change the even order of the torsion form τ^i, i.e., the type of the torsion matrix. This is a convenient fact in the study of Hermitian manifolds. In fact, a regular frame field $\{s_i; H(s_i, s_j) = \delta_{ij}\}$ is smooth, but not necessarily holomorphic.

According to our discussion in the last paragraph of the previous section, there exists a unique type $(1,0)$ compatible connection on the tangent bundle of a Hermitian manifold. Its torsion matrix must be of $(2,0)$-type, and its curvature matrix $(1,1)$-type. We usually call this connection the **Hermitian connection**.

Definition 5.2. If the Kählerian form $\hat{H}$ of a Hermitian manifold M is a closed exterior differential form, that is,

$$d\hat{H} = 0, \tag{5.18}$$

then we call M a **Kählerian manifold**.

Theorem 5.2. *A necessary and sufficient condition for a Hermitian manifold M to be a Kählerian manifold is that the torsion matrix of the Hermitian connection on M is zero.*

Proof. Obviously, the two conditions mentioned in this theorem are both independent of the choice of frame fields. Hence we only need to prove the theorem with respect to the natural frame field (5.1). Suppose the coframe field dual to (5.1) is

$$\sigma = \left(dz^1, \ldots, dz^m\right).$$

Then $d\sigma = 0$. Since the Hermitian connection is $\omega = \partial H \cdot H^{-1}$, where H is given by (5.2), a necessary and sufficient condition for the torsion matrix τ to be zero is:

$$\sigma \wedge \partial H = 0, \tag{5.19}$$

or

$$\sum_{i,j} \frac{\partial h_{i\bar{k}}}{\partial z^j} dz^j \wedge dz^i = 0,$$

$$\frac{\partial h_{i\bar{k}}}{\partial z^j} - \frac{\partial h_{j\bar{k}}}{\partial z^i} = 0, \qquad 1 \le i, j, k \le m. \tag{5.20}$$

But the exterior derivative of the Kählerian form $\hat{H}$ is

$$\begin{aligned}
d\hat{H} &= \frac{i}{2} \left(\frac{\partial h_{i\bar{k}}}{\partial z^j} dz^j + \frac{\partial h_{i\bar{k}}}{\partial \bar{z}^j} d\bar{z}^j \right) \wedge dz^i \wedge d\bar{z}^k \\
&= \frac{i}{2} \left\{ \frac{\partial h_{i\bar{k}}}{\partial z^j} dz^j \wedge dz^i \wedge d\bar{z}^k - \overline{\frac{\partial h_{i\bar{k}}}{\partial z^j}} dz^j \wedge dz^k \wedge d\bar{z}^i \right\}.
\end{aligned} \tag{5.21}$$

Therefore $d\hat{H} = 0$ is equivalent to (5.20).   □

Under a change of local frame fields, we have the following:

$$
\begin{aligned}
S' &= A \cdot S, \\
\sigma &= \sigma' \cdot A, \\
\Omega' \cdot A &= A \cdot \Omega, \\
H' &= A \cdot H \cdot {}^t\overline{A}.
\end{aligned}
\tag{5.22}
$$

Hence

$$
\Omega' \cdot H' = A \cdot (\Omega \cdot H) \cdot {}^t\overline{A}.
\tag{5.23}
$$

Moreover, the curvature matrix $\Omega \cdot H$ is Hermitian anti-symmetric [(4.51)], that is,

$$
\Omega \cdot H = -{}^t\overline{(\Omega \cdot H)}.
\tag{5.24}
$$

Since $\Omega \cdot H$ is a matrix of differential $(1,1)$-forms, we may assume that

$$
\begin{aligned}
\Omega \cdot H &= (\Omega_{i\overline{k}}), \\
\Omega_{i\overline{k}} &= \sum_{j,l} R_{i\overline{k}j\overline{l}}\,\sigma^j \wedge \overline{\sigma}^l.
\end{aligned}
\tag{5.25}
$$

From (5.24) we obtain

$$
\Omega_{i\overline{k}} = -\overline{\Omega}_{k\overline{i}}.
\tag{5.26}
$$

Thus

$$
\begin{aligned}
R_{i\overline{k}j\overline{l}}\,\sigma^j \wedge \overline{\sigma}^l &= -\overline{R}_{k\overline{i}j\overline{l}}\,\overline{\sigma}^j \wedge \sigma^l \\
&= \overline{R}_{k\overline{i}l\overline{j}}\,\sigma^j \wedge \overline{\sigma}^l,
\end{aligned}
$$

that is,

$$
R_{i\overline{k}j\overline{l}} = \overline{R}_{k\overline{i}l\overline{j}}.
\tag{5.27}
$$

Let $A = \left(A_i^j\right)$. Then (5.23) becomes

$$
\Omega'_{i\overline{k}} = \sum_{p,q} A_i^p \overline{A}_k^q \Omega_{p\overline{q}},
$$

where $\Omega' \cdot H' = \left(\Omega'_{i\bar{k}} \right)$. If we still denote

$$\Omega'_{i\bar{k}} = \sum_{j,l} R'_{i\bar{k}j\bar{l}} \sigma'^j \wedge \bar{\sigma}'^l,$$

then

$$R'_{i\bar{k}j\bar{l}} = \sum_{p,q,u,v} A_i^p \overline{A}_k^q A_j^u \overline{A}_l^v R_{p\bar{q}u\bar{v}}. \tag{5.28}$$

Choose a type $(1,0)$ tangent vector ξ at x in M with components ξ^i and ξ'^i under the frame fields S and S', respectively. Then

$$\xi^i = \sum_j A_j^i \xi'^j. \tag{5.29}$$

Thus

$$\sum_{i,j,k,l} R'_{i\bar{k}j\bar{l}} \xi'^i \overline{\xi}'^k \xi'^j \overline{\xi}'^l = \sum_{p,q,u,v} R_{p\bar{q}u\bar{v}} \xi^p \overline{\xi}^q \xi^u \overline{\xi}^v, \tag{5.30}$$

We see that the above expression does not depend on the choice of local frame fields. If the type $(1,0)$ tangent vector $\xi \neq 0$, then we define the quantity

$$R(x,\xi) = \frac{2 \sum_{i,k,j,l} R_{i\bar{k}j\bar{l}} \xi^i \overline{\xi}^k \xi^j \overline{\xi}^l}{\left(\sum_{i,k} h_{i\bar{k}} \xi^i \overline{\xi}^k \right)^2}, \tag{5.31}$$

called the **holomorphic sectional curvature** of the Hermitian manifold M at (x,ξ).

By the third equation in (5.22) we have

$$\mathrm{Tr}\, \Omega' = \mathrm{Tr}\, \Omega. \tag{5.32}$$

Therefore $\Phi = \mathrm{Tr}\, \Omega$ is a differential $(1,1)$-form defined globally on M, called the **Ricci form** of the Hermitian manifold M.

Let $h^{i\bar{k}}$ be an element of the matrix H^{-1}. Then

$$R = \sum_{i,j,k,l} R_{i\bar{k}j\bar{l}} h^{i\bar{k}} h^{j\bar{l}} \tag{5.33}$$

is also independent of the choice of local frame fields, and

$$\begin{aligned}
\overline{R} &= \sum_{i,k,j,l} \overline{R}_{i\bar{k}j\bar{l}} \overline{h^{i\bar{k}}}\, \overline{h^{j\bar{l}}} \\
&= \sum_{i,k,j,l} R_{k\bar{i}l\bar{j}} h^{k\bar{i}} h^{l\bar{j}} \\
&= R.
\end{aligned}$$

Thus (5.33) defines a real function on M called the **scalar curvature** of the Hermitian manifold.

Topologically, a compact Kählerian manifold satisfies very strong restrictions. For instance, the second Betti number of a compact Kählerian manifold cannot be zero. This is because the Kählerian form $\hat{H}$ is a real-valued closed exterior differential form. Thus it determines an element u of the second de Rham cohomology group $H^2(M, \mathbb{R})$. It is easy to see that $u \neq 0$. In fact, according to the local expression for a Kählerian form

$$\hat{H} = \frac{i}{2} \sum_{i,k} h_{i\bar{k}} dz^i \wedge d\bar{z}^k,$$

we have

$$\hat{H} = \left(\frac{i}{2}\right)^m m! (\det H) \bigwedge_k \left(dz^k \wedge d\bar{z}^k\right).$$

Since the matrix H is positive definite, $\det H > 0$, and

$$\int_M \hat{H}^m > 0.$$

$\hat{H}^m$ corresponds to the element u^m in $H^{2m}(M, \mathbb{R})$ (the cup product of m factors of u), and $\int_M \hat{H}^m$ is the value of the cohomology class u^m on the fundamental class M. Therefore $u^m \neq 0$, $u \neq 0$, which proves the above conclusion.

Suppose M and N are m-dimensional and n-dimensional complex manifolds, respectively, and $f : M \longrightarrow N$ is a holomorphic map. If $m \leq n$ and the rank of the Jacobian matrix of f is m everywhere, then we call f an **immersion**. If f is also one to one, that is, for any $x \neq y \in M$, $f(x) \neq f(y)$, then we say f is an **imbedding**.

Theorem 5.3. *Suppose N is a Kählerian manifold, and $f : M \longrightarrow N$ a holomorphic immersion. Then there is an induced Kählerian structure on M from N.*

Proof. Suppose $p \in M$, $(z^1, \ldots, z^n)$ are the complex coordinates of the point $q = f(p)$ in N, and $(w^1, \ldots, w^m)$ the complex coordinates of the point p in M. Then the map f can be expressed locally by

$$z^\alpha = f^\alpha\left(w^1, \ldots, w^m\right).$$

Suppose the Hermitian structure on N is

$$H = \sum_{\alpha,\beta} h_{\alpha\bar{\beta}} dz^\alpha d\bar{z}^\beta.$$

The Kählerian form is then

$$\hat{H} = \frac{i}{2} \sum_{\alpha,\beta} h_{\alpha\overline{\beta}} dz^\alpha \wedge d\overline{z}^\beta,$$

and $d\hat{H} = 0$. Let

$$h'_{i\overline{j}} = \sum_{\alpha,\beta} \left(h_{\alpha\overline{\beta}} \circ f \right) \frac{\partial z^\alpha}{\partial w^i} \frac{\partial \overline{z}^\beta}{\partial \overline{w}^j}.$$

Then the matrix $H' = \left(h'_{i\overline{j}} \right)$ is still positive definite. It gives the positive definite Hermitian structure

$$H' = \sum_{i,j} h'_{i\overline{j}} dw^i d\overline{w}^j,$$

on M, whose Kählerian form is

$$\hat{H}' = \frac{i}{2} \sum_{i,j} h'_{i\overline{j}} dw^i \wedge d\overline{w}^j.$$

Obviously

$$\hat{H}' = f^* \hat{H},$$

and

$$d\hat{H} = d \circ f^* \hat{H} = f^* \left(d\hat{H} \right) = 0.$$

Thus the complex manifold M becomes a Kählerian manifold with respect to the induced Hermitian structure H'.   □

Chapter 8

Finsler Geometry

§8–1 Preliminaries

In Chapter Five we gave a treatment of Riemannian geometry, which is
the metric geometry based on a positive definite (or at least nondegenerate)
quadratic differential form

$$ds^2 = G = g_{ij}(u)du^i \otimes du^j$$

where the u^i are the local coordinates and $g_{ij} = g_{ji}$ are smooth functions on
the Riemannian manifold.

In this chapter we will consider the more general case without the above
quadratic restriction. It is characterized by

$$ds = F(u^1, \ldots, u^m; du^1, \ldots, du^m), \tag{1.1}$$

where $F(x; y)$, known as the **Finsler function** , is a smooth, non-negative
function in the $2m$ variables, and has the value zero only when $y = 0$. $F(x; y)$
is also required to be symmetrically homogeneous of degree one in the y's, that
is,

$$F(x^1, \ldots, x^m; \lambda y^1, \ldots, \lambda y^m) = |\lambda| F(x^1, \ldots, x^m; y^1, \ldots, y^m), \ \lambda \in \mathbb{R}. \tag{1.2}$$

This case was already introduced by Riemann in his historical Habilitation
address of 1854.[a] Hence the geometry based on (1.1) should properly be called
Riemann–Finsler geometry. We will follow the conventional designation and
call it Finsler geometry for brevity, in recognition of Finsler's thesis on the

---

[a] An English translation of as well as a useful commentary on this famous lecture can be
found in Chapter 4A and 4B of Spivak, Vol. II, 1979. For comments on it with respect to
developments in Finsler geometry, see Chern 1996(a).

subject in 1918. The starting point of Finsler geometry is the first notion of the integral calculus, namely, the calculation of arc lengths. The generality of (1.1) is required for diverse applications. For example, solid state physics involves lattices and the geometry is naturally Finsler. In a complex manifold, there are intrinsic metrics, such as the Caratheodory and Kobayashi metrics, which are Finsler and generally not Riemannian. (For a recent survey of the applications of Finsler geometry, see Bao, Chern, and Shen 1996).

Our treatment originates from work carried out by one of the authors in 1948 (see Chern 1948). This work was long neglected but paved the way for some notable progress in the subject in recent years [see Bao and Chern 1993 and Chern 1996(a), (b), (c)]. In this chapter we will present an introductory account of some remarkable results stemming from this progress, and hope to demonstrate that the Finsler setting is the more natural starting point for Riemannian geometry.

The crucial idea, which seems to have been unknown to Riemann, is to consider the projectivised tangent bundle PTM or the sphere bundle SM on a Finsler manifold M, rather than the tangent bundle TM, as was done in most conventional treatments. The issue of the relative merits of PTM versus SM is a rather delicate one whose resolution is beyond the scope of this text. Suffice it to mention here that PTM is the simpler and more natural setting from a geometrical perspective, while SM admits a wider class of interesting Finsler spaces, from both theoretical and applied standpoints. For simplicity, we will use PTM in our presentation, with the understanding that many of our results are applicable to SM as well. (See Bao, Chern, and Shen, to appear, for a detailed treatment of the SM case).

On the canonical pullback bundle $p^*TM \to PTM$ with base manifold PTM, one can introduce moving frames. A distinguished section of the coframe field, known as the Hilbert form, determines a contact structure on PTM. Exterior differentiation of the Hilbert form and the associated coframe sections then leads to a unique, torsion-free, and 'almost metric-compatible' connection, known as the Chern connection, with remarkable properties. This connection is seen to be a generalization of the Christoffel–Levi–Civita connection in Riemannian geometry, and also provides a solution to the equivalence problem in Finsler geometry.

The torsion-freeness, together with 'just the right amount' of metric compatibility of the Chern connection, also ensure that the formulas for the first and second variations of arc length in Finsler geometry retain their well-known Riemannian forms. This fortunate circumstance allows one to generalize easily many important theorems relating curvature and topology in Riemannian geometry to the Finsler setting.

The above ideas will be explained in detail in the following sections. The central results on the existence and remarkable properties of the Chern connection are summarized in Theorems 3.1 and 3.2 (Chern's Theorems), which are the main theorems of this chapter.

§8–2 Geometry on the Projectivised Tangent Bundle (PTM) and the Hilbert Form

Throughout this chapter, except when otherwise stated, lower case Latin indices (except m) run from 1 to m, where m is the dimension of the Finsler manifold; and lower case Greek indices run from 1 to $m - 1$.

Definition 2.1. Let M be an m-dimensional manifold. It is said to be a **Finsler manifold** if the length s of any curve $t \mapsto (u^1(t), \ldots, u^m(t))$, $a \le t \le b$, is given by an integral

$$s = \int_a^b F\left(u^1, \ldots, u^m; \frac{du^1}{dt}, \ldots, \frac{du^m}{dt}\right) dt, \tag{2.1}$$

where the function F has the properties specified immediately following Eq. (1.1) in the preceding section.

Remark. The requirement of first-degree homogeneity is imposed on F so that the arc length s is invariant under reparametrization.

A Finsler manifold M has a tangent bundle $\pi : TM \to M$ and a cotangent bundle $\pi^* : T^*M \to M$. From TM we obtain the **projectivised tangent bundle** of M, PTM, by identifying the non-zero vectors differing from each other by a real factor. Geometrically PTM is the space of line elements on M. Let u^i, $1 \le i \le m$, be local coordinates on M. Then a non-zero tangent vector can be expressed as

$$X = X^i \frac{\partial}{\partial u^i}, \qquad X^i \text{ not all zero.}$$

The u^i, X^i are local coordinates on TM. They are also local coordinates on PTM, with X^i being homogeneous coordinates (determined up to a real factor). Our fundamental idea is to consider PTM as the base manifold of the vector bundle $p^* TM$, pulled back with the canonical projection map $p : PTM \to M$ defined by

$$p(u^i, X^i) = (u^i).$$

The fibers of $p^* TM$ are then vector spaces of dimension m and the base manifold PTM is of dimension $2m - 1$. (See Fig. 14)   Since the function $F(u^i, X^i)$

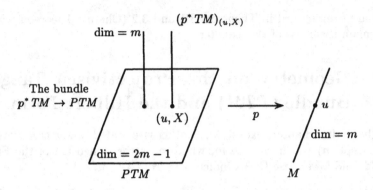

FIGURE 14.

is homogeneous of degree 1 in the X^i, **Euler's theorem** on homogeneous functions gives

$$X^i \frac{\partial F}{\partial X^i} = F, \tag{2.2}$$

and by differentiation

$$X^i \frac{\partial^2 F}{\partial X^i \partial X^j} = 0. \tag{2.3}$$

The latter implies that $\frac{\partial F}{\partial X^i}$ are homogeneous functions of degree zero in the X^i's. Such functions are functions on PTM.

Let (v^j, Y^j) be another local coordinate system on PTM. Then we have

$$Y^i = \frac{\partial v^i}{\partial u^j} X^j, \tag{2.4}$$

whence

$$\frac{\partial F}{\partial X^i} = \frac{\partial F}{\partial Y^j} \frac{\partial v^j}{\partial u^i}. \tag{2.5}$$

It follows that the one-form

$$\omega \equiv \frac{\partial F}{\partial X^i} du^i = \frac{\partial F}{\partial Y^i} dv^i \tag{2.6}$$

is independent of the choice of local coordinates, and is thus intrinsically defined on PTM. We will call it the **Hilbert form**. By Euler's theorem [(2.2)], we can rewrite the arc length integral (2.1) on M as Hilbert's invariant integral

$$s = \int_a^b \omega. \tag{2.7}$$

The Hilbert form is a powerful piece of data, and we will see in the next section how, on exterior differentiation, it yields a connection with remarkable properties. The geometrical significance of this one-form in the calculus of variations was already recognized by Hilbert in the last of his famous twenty three problems enunciated in 1900.[b]

We now wish to calculate the exterior derivative $d\omega$, and will use moving frames. In this calculation and similar ones following, it is crucial to remember that the base manifold of p^*TM is PTM, and thus all differential forms introduced will be forms on PTM. A differential form on PTM can be represented as one on TM provided the latter is invariant under rescaling in the X^i and yields zero when contracted with $X^i \frac{\partial}{\partial X^i}$.[c] Our differential forms on PTM will be so represented, and exterior differentiation on PTM will be obtained by formal differentiation on TM.

---

[b]For comments on Hilbert's 23rd problem in relation to Finsler geometry, see Chern 1996(c).

[c]Let (u^i, X^i), $i = 1, \ldots, m$, be local coordinates of PTM, where the X^i are homogeneous coordinates. Suppose

$$\omega = f_i dX^i + g_i du^i$$

is a 1-form on TM which yields 0 on contraction with $X^i \frac{\partial}{\partial X^i}$ and is invariant under rescaling in the X^i. Then

$$f_i X^i = 0,$$

$$f(\lambda X^i) = \frac{1}{\lambda} f(X^i), \ \lambda \neq 0,$$

and

$$g_i(u^j, \lambda X^k) = g_i(u^j, X^k), \ \lambda \neq 0.$$

The first of the above three equations implies

$$f_m = -\frac{f_\alpha X^\alpha}{X^m}.$$

Then

$$\omega = f_\alpha dX^\alpha + f_m dX^m + g_i du^i$$

$$= f_\alpha \left(dX^\alpha - \frac{X^\alpha}{X^m} dX^m \right) + g_i du^i$$

$$= X^m f_\alpha(u^i, X^i) d\left(\frac{X^\alpha}{X^m} \right) + g_i(u^j, X^k) du^i$$

$$= f_\alpha \left(u^i, \frac{X^\beta}{X^m} \right) d\left(\frac{X^\alpha}{X^m} \right) + g_i \left(u^j, \frac{X^\beta}{X^m} \right) du^i, \ X^m \neq 0,$$

which is a 1-form on PTM.

Let

$$e_i = p_i^j \frac{\partial}{\partial u^j} \tag{2.8}$$

be an orthonormal frame field on the bundle $p^* TM$, and

$$\omega^j = q_k^j du^k \tag{2.9}$$

its dual coframe field, so that

$$(e_i, e_j) \equiv p_i^l g_{lk} p_j^k = \delta_{ij}, \tag{2.10}$$
$$\langle e_i, \omega^j \rangle = \delta_i^j.$$

The former is the orthonormality condition and the latter the duality condition, which is equivalent to

$$p_i^j q_j^k = \delta_i^k, \tag{2.11}$$

that is, the matrices (p_i^j) and (q_j^k) are inverse to each other. The orthonormality is with respect to the following symmetric covariant 2-tensor (the fundamental tensor)

$$\begin{aligned}
G &= g_{ij} du^i \otimes du^j \\
&\equiv \frac{\partial^2(\frac{1}{2}F^2)}{\partial X^i \partial X^j} du^i \otimes du^j \\
&= \left(F \frac{\partial^2 F}{\partial X^i \partial X^j} + \frac{\partial F}{\partial X^i} \frac{\partial F}{\partial X^j} \right) du^i \otimes du^j
\end{aligned} \tag{2.12}$$

defined intrinsically on PTM. We shall suppose this to be positive definite (strong convexity hypothesis), and call it the **Finsler metric**.

Remark. In the Riemannian case

$$F^2(u^i, X^i) = g_{ij}(u) X^i X^j \tag{2.13}$$

where the g_{ij} are functions of the u^i only. In the Finsler setting, the g_{ij} [in (2.12)] are functions of both u^i and X^i in general, and are homogeneous of degree zero in the X^i. Hence the g_{ij} are functions on PTM.

We now distinguish the global sections

$$e_m = \frac{X^i}{F} \frac{\partial}{\partial u^i}, \tag{2.14}$$

and

$$\omega^m = \frac{\partial F}{\partial X^i} du^i = \omega \tag{2.15}$$

on $p^* TM$ and $p^* T^* M$, respectively. These choices imply

$$q_k^m = \frac{\partial F}{\partial X^k},$$
$$p_m^l = \frac{X^l}{F}, \tag{2.16}$$

which, together with (2.11), lead to

$$q_k^\alpha X^k = 0,$$
$$\frac{\partial F}{\partial X^k} p_\alpha^k = 0. \tag{2.17}$$

Note that according to the discussion in the paragraph immediately preceding (2.8), and the fact that $\frac{\partial F}{\partial X^i}$ is homogeneous of degree 0 in X, ω^i, $i = 1, \ldots, m$, are indeed one forms on PTM provided the $q_k^\alpha(u^i, X^j)$ are also homogeneous of degree 0 in X.

Taking the exterior derivative of the Hilbert form ω^m on PTM, we obtain

$$
\begin{aligned}
d\omega^m &= \frac{\partial^2 F}{\partial u^i \partial X^k} du^i \wedge du^k + \frac{\partial^2 F}{\partial X^j \partial X^k} dX^j \wedge du^k \\
&= \frac{\partial^2 F}{\partial u^i \partial X^k} p_j^i p_l^k \omega^j \wedge \omega^l + \frac{\partial^2 F}{\partial X^j \partial X^k} p_l^k dX^j \wedge \omega^l.
\end{aligned}
\tag{2.18}
$$

In the last expression the coefficient of $dX^j \wedge \omega^m$ vanishes by Euler's theorem in the form of (2.3). It follows that

$$d\omega^m = \omega^\alpha \wedge \omega_\alpha^m, \tag{2.19}$$

where the most general expression for the one-forms w_α^m is

$$
\begin{aligned}
\omega_\alpha^m &= -p_\alpha^i \frac{\partial^2 F}{\partial X^i \partial X^j} dX^j + \frac{p_\alpha^i}{F}\left(\frac{\partial F}{\partial u^i} - X^j \frac{\partial^2 F}{\partial X^i \partial u^j}\right) \omega^m \\
&\quad + p_\alpha^i p_\beta^j \frac{\partial^2 F}{\partial u^i \partial X^j} \omega^\beta + \lambda_{\alpha\beta} \omega^\beta; \quad \alpha = 1, \ldots, m - 1.
\end{aligned}
\tag{2.20}
$$

The coefficients $\lambda_{\alpha\beta}$ introduced in the last equation must satisfy $\lambda_{\alpha\beta} = \lambda_{\beta\alpha}$, but are otherwise arbitrary. They will be determined in the next section [Eq. (3.26)] in our quest for the Chern connection.

We digress to establish an important property of the Hilbert form.

Lemma 1. *The Hilbert form on PTM given by*

$$\omega = \frac{\partial F}{\partial X^i} du^i$$

satisfies the condition

$$\omega \wedge (d\omega)^{m-1} \neq 0. \tag{2.21}$$

Proof. Let $A = \omega \wedge (d\omega)^{m-1}$. We use (2.19) for $d\omega$, (2.20) for ω_α^m, and the choice $\omega^m = \omega$ to obtain

$$A = \pm(m-1)! \bigwedge_i \omega^i \bigwedge_\alpha \omega_\alpha^m$$

$$= \pm(m-1)! \bigwedge_i \omega^i \bigwedge_\alpha p_\alpha^j \frac{\partial^2 F}{\partial X^j \partial X^k} dX^k. \tag{2.22}$$

Note that the terms in (2.20) for ω_α^m involving ω^m and ω^β do not enter due to the symmetry properties of the exterior product. By (2.12), A can be rewritten as

$$A = \pm \frac{(m-1)!}{F} \bigwedge_i \omega^i \bigwedge_\alpha p_\alpha^j \left(g_{jk} - \frac{\partial F}{\partial X^j} \frac{\partial F}{\partial X^k} \right) dX^k$$

$$= \pm \frac{(m-1)!}{F} \bigwedge_i \omega^i \bigwedge_\alpha p_\alpha^j g_{jk} dX^k, \tag{2.23}$$

where in the last equality we have used (2.17). Now by the strong convexity hypothesis, that is, g_{ij} is positive definite, $g_{jk} dX^k$, $j = 1, \ldots, m$, must be linearly independent one-forms on PTM. Since p_i^j is invertible, the same must be true of $p_i^j g_{jk} dX^k$. Hence the $(m-1)$ one-forms $p_\alpha^j g_{jk} dX^k$, $\alpha = 1, \ldots, m-1$, as a subset, are also linearly independent. Finally, we recall that ω^i does not involve dX^k [c.f. (2.9)]. Thus the $(2m-1)$ one-forms ω^i and $p_\alpha^j g_{jk} dX^k$ are linearly independent. The lemma then follows from Theorem 3.3 of Chapter Two.   □

Eq. (2.21) remains satisfied if ω is multiplied by a non-zero smooth function. In general, a manifold of dimension $2m-1$ is said to have a **contact structure** if there is a one-form ω, defined up to a factor, which satisfies (2.21). The one-form ω is then called a **contact form**. Thus our manifold PTM has the important property of possessing a contact structure.

Remark. A contact structure on an odd dimensional manifold of dimension $2m-1$ is closely related to a **symplectic structure** on an even dimensional manifold of dimension $2m$, which is given by a closed, nondegenerate 2-form on the even dimensional manifold. Given the Hilbert form ω as a contact form on PTM, we can construct a line bundle with PTM as the base manifold, and a fiber at $p \in PTM$ given by the set $\lambda \omega_p$, $\lambda \neq 0$. This bundle, of dimension $2m$, is called the **symplectification** S of PTM. On S we can define a one-form ω', called the canonical one-form, by

$$\omega'_{p'}(T') = p'(\pi_* T'), \qquad p' \in S, \; T' \in T_{p'}S,$$

where $\pi : S \to PTM$ is the projection $(p, \lambda \omega_p) \mapsto p$, $\lambda \neq 0$. The unique symplectic structure defined on S induced by ω is then given by the nondegenerate two-form $d\omega'$, the property of non-degeneracy being assured by the condition (2.21) of ω.

§8–3 The Chern Connection

Among the spaces associated with a Finsler Manifold M we have the diagram

$$
\begin{array}{ccccc}
q^* TM & \longrightarrow & p^* TM & \longrightarrow & TM \\
\downarrow & & \downarrow & & \pi \downarrow \\
TM & \xrightarrow{\;h\;} & PTM & \xrightarrow{\;p\;} & M
\end{array}
$$

where $q = p \circ h$. We will mainly be concerned with the bundle in the middle column, although we will also use the bundle $q^* TM$ in some of the calculations. [See discussion in the paragraph immediately preceding (2.8), and §8–3.3].

As pointed out in §8–2, the fibers of the bundle $p^* TM \to PTM$ are vector spaces of dimension m and are provided with a scalar product given by the functions g_{ij} on PTM such that $\det(g_{ij}) \neq 0$. With a choice of an orthonormal frame field $\{e_i\}$ (2.8) and it's dual coframe field $\{\omega^i\}$ (2.9) such that the orthonormality and duality conditions (2.10) are satisfied, a connection in the bundle is given by

$$De_i = \omega_i^j e_j, \tag{3.1}$$

where ω_i^j is a matrix of one-forms on PTM. Making use of the intrinsically defined tensor field $e_i \otimes \omega^i$, the connection is then said to be torsion-free if the Cartan covariant derivative on it vanishes:

$$D(e_i \otimes \omega^i) = \omega_i^j e_j \wedge \omega^i + e_i d\omega^i = 0, \tag{3.2}$$

that is,

$$d\omega^i = \omega^j \wedge \omega_j^i. \tag{3.3}$$

[c.f. Eq.(1.37) in Chapter Five].

In this section we will establish the remarkable fact that with a Finsler structure on M, there is a uniquely defined torsion-free connection on the bundle $p^*TM \to PTM$. This connection generalizes the Christoffel–Levi–Civita connection in Riemannian geometry, thus making the latter a special case of Finsler geometry.

§8–3.1 Determination of the Connection

To determine the desired torsion-free connection is to determine the connection forms satisfying (3.3). For $i = m$, $d\omega^m$ is already given by (2.19). The expressions for $d\omega^m$ in (2.19) and (3.3) then become identical if ω_α^m is given by (2.20) and we choose

$$\omega_m^m = 0. \tag{3.4}$$

We continue by exteriorly differentiating the one-forms ω^α as given by (2.9). We have

$$
\begin{aligned}
d\omega^\alpha &= dq_k^\alpha \wedge du^k = p_i^k dq_k^\alpha \wedge \omega^i \\
&= -q_k^\alpha dp_\beta^k \wedge \omega^\beta - q_k^\alpha dp_m^k \wedge \omega^m \\
&= \omega^\beta \wedge (q_k^\alpha dp_\beta^k) + \omega^m \wedge \left(\frac{1}{F} q_k^\alpha dX^k \right),
\end{aligned}
\tag{3.5}
$$

where (2.16) and (2.17) have been used in the last equality. Thus we can write

$$d\omega^\alpha = \omega^\beta \wedge \omega_\beta^\alpha + \omega^m \wedge \omega_m^\alpha, \tag{3.6}$$

where the most general expression for the one-forms ω_β^α and ω_m^α are given by

$$\omega_\beta^\alpha = q_k^\alpha dp_\beta^k + \xi_\beta^\alpha \omega^m + \mu_{\beta\gamma}^\alpha \omega^\gamma, \tag{3.7}$$

$$\omega_m^\alpha = \frac{1}{F} q_k^\alpha dX^k + \xi_i^\alpha \omega^i. \tag{3.8}$$

The coefficients $\mu_{\beta\gamma}^\alpha$ must be symmetric in the two lower indices but are otherwise arbitrary, while the ξ_i^α are completely arbitrary.

Equations (2.19) for $d\omega^m$ and (3.6) for $d\omega^\alpha$ together reduce to (3.3), provided $\omega_m^m = 0$ [Eq. (3.4)]. The existence of a torsion-free connection on $p^*TM \to PTM$ is thus established. We summarize our results thus far by the following theorem.

Theorem 3.1 (Chern). *Let M be an m-dimensional Finsler manifold with Finsler function F. Suppose $e_i = p_i^j \frac{\partial}{\partial u^j}$, $i = 1, \ldots, m$, is an orthonormal* *frame field on the bundle* $p^* TM \to PTM$ *and* $\omega^i = q_j^i du^j$ *its dual coframe field on* $p^* T^* M \to PTM$ *with the distinguished global sections*

$$e_m = \frac{X^i}{F} \frac{\partial}{\partial u^i} \quad , \quad \omega^m = \frac{\partial F}{\partial X^i} du^i,$$

where u^i *and* X^i *are the local coordinates on* M *and* TM, *respectively. Then there exists a torsion-free connection on* $p^* TM \to PTM$:

$$D : \Gamma(p^* TM) \to \Gamma(p^* TM \otimes T^*(PTM)),$$

given by

$$De_i = \omega_i^j e_j, \quad \omega_m^m = 0.$$

The connection forms ω_i^j *satisfy the torsion-free structure equation*

$$d\omega^i = \omega^j \wedge \omega_j^i.$$

We will now fix the ξ_i^α introduced in (3.7) and (3.8) by requiring that ω_α^m and ω_m^α are negatives of each other, that is,

$$\omega_\alpha^m + \omega_m^\alpha = 0. \tag{3.9}$$

Note that this condition, together with the condition $\omega_m^m = 0$, imply that the matrix of one-forms ω_i^j is skew-symmetric in the last row and last column.

Remark. As will be seen later [see Eq. (3.46) and subsequent discussion], Eq. (3.9) has the geometrical meaning that $G(e_i, e_m)$ remains constant under parallel translations, i.e., $DG(e_i, e_m) = 0$.

Using (2.20) and (3.8), (3.9) can be recast in the form

$$\left(\frac{\delta_{\alpha\sigma} q_j^\sigma}{F} - p_\alpha^i \frac{\partial^2 F}{\partial X^i \partial X^j} \right) dX^j$$

$$+ \left\{ \delta_{\alpha\sigma} \xi_m^\sigma + \frac{p_\alpha^i}{F} \left(\frac{\partial F}{\partial u^i} - X^j \frac{\partial^2 F}{\partial X^i \partial u^j} \right) \right\} \omega^m$$

$$+ \left(\delta_{\alpha\sigma} \xi_\beta^\sigma + p_\alpha^i p_\beta^j \frac{\partial^2 F}{\partial u^i \partial X^j} + \lambda_{\alpha\beta} \right) \omega^\beta = 0. \tag{3.10}$$

In the above, the coefficient of dX^j vanishes. Indeed

$$\delta_{\alpha\sigma} q_j^\sigma - p_\alpha^i F \frac{\partial^2 F}{\partial X^i \partial X^j} = \delta_{\alpha\sigma} q_j^\sigma - p_\alpha^i \left(g_{ij} - \frac{\partial F}{\partial X^i} \frac{\partial F}{\partial X^j} \right) = 0 \tag{3.11}$$

where in the first equality we have used the definition of g_{ij} in (2.12), and in the second we have used (2.17) and the orthonormality condition in (2.10). Since ω^m and the ω^β are linearly independent one-forms, their coefficients in (3.10) can be separately equated to zero. It follows that

$$\xi^\alpha_m = -\frac{\delta^{\alpha\sigma} p^i_\sigma}{F}\left(\frac{\partial F}{\partial u^i} - X^j \frac{\partial^2 F}{\partial X^i \partial u^j}\right), \tag{3.12}$$

$$\xi^\alpha_\beta = -\delta^{\alpha\sigma}\left(p^i_\sigma p^j_\beta \frac{\partial^2 F}{\partial X^j \partial u^i} + \lambda_{\sigma\beta}\right). \tag{3.13}$$

The connection ω^i_j will be completely determined upon fixing the tensors $\lambda_{\alpha\beta}$ and $\mu^\alpha_{\beta\gamma}$ introduced in (2.20) and (3.7), respectively. To this end, we impose the further condition, besides (3.9), that the $(m-1) \times (m-1)$ block in the matrix (ω^β_α) is 'almost-symmetric'. Specifically, we require that

$$\omega^\alpha_\rho \delta_{\alpha\sigma} + \omega^\alpha_\sigma \delta_{\alpha\rho} = 0 \mod \omega^\beta_m. \tag{3.14}$$

This condition implies that, among the ω^β_α, we can choose at most $(m-1)(m-2)/2$ of them, say ω^β_α $(\alpha < \beta)$, to be linearly independent of each other, and also of the ω^α_m. We thus have:

Lemma 1. *The* $(2m-1)+(m-1)(m-2)/2$ *Pfaffian forms* ω^i $(i = 1,\ldots,m)$, ω^α_m $(\alpha = 1,\ldots,m-1)$, *and* ω^β_α $(\alpha,\beta = 1,\ldots,m-1; \alpha < \beta)$ *are linearly independent and constitute a multiplicative basis of the algebra of exterior differential forms in the space of preferred coframes on PTM specified by (2.9), (2.10) and (2.15). The* $(2m-1)$ *Pfaffian forms* ω^i $(i = 1,\ldots,m)$, ω^α_m $(\alpha = 1,\ldots,m-1)$ *at each point* $p \in PTM$ *constitute a basis for* $T^*_p(PTM)$.

Remark 1. The number of independent Pfaffian forms in the above Lemma is equal to the total number of essential variables in the Finsler setting: $2m-1$ from the local coordinates u^i, X^i of PTM (remember that the X^i are homogeneous coordinates), and $(m-1)(m-2)/2$ from the freedom to specify the orthonormal moving frame field $(e_1,\ldots,e_m)$, with e_m specified by (2.14). This degree of freedom results from the following considerations. In the $m \times m$ matrix p^j_i, the last row (p^j_m) is fixed by (2.16). The remaining $m(m-1)$ elements (p^i_α) satisfy the $(m-1)$ conditions $p^i_\alpha \frac{\partial F}{\partial X^i} = 0$ [Eq. (2.17)] and the $(m-1)+(m-1)(m-2)/2 = m(m-1)/2$ orthonormality conditions $p^i_\alpha g_{lk} p^k_\beta = \delta_{\alpha\beta}$ [Eq. (2.10)]. Thus the required number is $m(m-1) - (m-1) - m(m-1)/2 = (m-1)(m-2)/2$.

Remark 2. The linearly independent Pfaffian forms in Lemma 1 provide a solution to the so-called **equivalence problem** in Finsler geometry: to decide when two Finsler metrics are locally equivalent. Let two Finsler metrics be specified by the Finsler functions $F(u^i, X^i)$ and $F'(u'^i, X'^i)$, with the respective sets of linearly independent Pfaffian forms (ω^i, ω^i_j) and (ω'^i, ω'^j_i). Then they are locally equivalent if and only if there is a diffeomorphism between the spaces of preferred coframes such that

$$\omega^i = \omega'^i \quad , \quad \omega^j_i = \omega'^j_i.$$

(For details, see Chern 1948).

To implement the condition (3.14) we first apply the exterior derivative (on PTM) to the equation

$$q^\alpha_j q^\beta_k \delta_{\alpha\beta} = F \frac{\partial^2 F}{\partial X^j \partial X^k}, \tag{3.15}$$

which follows from the definition of the Finsler metric g_{ij} [Eq. (2.12)] and the orthonormality condition of our preferred frame field in (2.10). We then use (2.11) to contract the result twice with p to obtain

$$\delta_{\alpha\sigma} q^\alpha_i dp^i_\rho + \delta_{\alpha\rho} q^\alpha_i dp^i_\sigma = -p^j_\sigma p^i_\rho d \left(\frac{\partial^2 F}{\partial X^i \partial X^j} \right). \tag{3.16}$$

Using this equation and (3.13) in (3.7), the left hand side of (3.14) can then be written as

$$\omega^\alpha_\rho \delta_{\alpha\sigma} + \omega^\alpha_\sigma \delta_{\alpha\rho}$$
$$= -p^j_\sigma p^i_\rho \left\{ d \left(F \frac{\partial^2 F}{\partial X^i \partial X^j} \right) + \left(\frac{\partial^2 F}{\partial u^i \partial X^j} + \frac{\partial^2 F}{\partial u^j \partial X^i} \right) \omega^m \right\}$$
$$- 2\lambda_{\rho\sigma} \omega^m + (\delta_{\alpha\sigma} \mu^\alpha_{\rho\gamma} + \delta_{\alpha\rho} \mu^\alpha_{\sigma\gamma}) \omega^\gamma. \tag{3.17}$$

Next we write, according to Lemma 3.1,

$$d \left(F \frac{\partial^2 F}{\partial X^i \partial X^j} \right) = S^\alpha_{ij} \omega^m_\alpha + G_{ijl} \omega^l, \tag{3.18}$$

where both S^α_{ij} and G_{ijl} are symmetric in i, j. These quantities can be calculated as follows. Contracting (3.18) with $Fp^k_\beta \frac{\partial}{\partial X^k}$ and using (2.20) for ω^m_α, we find that

$$Fp^k_\beta \frac{\partial}{\partial X^k} \left(F \frac{\partial^2 F}{\partial X^i \partial X^j} \right) = -S^\alpha_{ij} p^s_\alpha p^k_\beta F \frac{\partial^2 F}{\partial X^s \partial X^k} \tag{3.19}$$
$$= -S^\alpha_{ij} \delta_{\alpha\beta},$$

where in the second equality we have used (3.15) and (2.11). Thus orthonormality of the q's and duality between the q's and p's imply

$$S_{ij}^\alpha = -F q_i^\alpha g^{lk} \frac{\partial}{\partial X^k} \left(F \frac{\partial^2 F}{\partial X^i \partial X^j} \right). \tag{3.20}$$

To find G_{ijm}, we use (2.20) again to contract (3.18) with e_m and obtain

$$G_{ijm} = \frac{-p_\alpha^s S_{ij}^\alpha}{F} \left(\frac{\partial F}{\partial u^s} - X^r \frac{\partial^2 F}{\partial X^s \partial u^r} \right) + \frac{X^s}{F} \frac{\partial}{\partial u^s} \left(F \frac{\partial^2 F}{\partial X^i \partial X^j} \right). \tag{3.21}$$

On using (3.20) for S_{ij}^α, the first term on the right hand side can be expressed as

$$
\begin{aligned}
p_\alpha^s q_i^\alpha g^{kl} & \frac{\partial}{\partial X^k} \left(F \frac{\partial^2 F}{\partial X^i \partial X^j} \right) \left(\frac{\partial F}{\partial u^s} - X^r \frac{\partial^2 F}{\partial X^s \partial u^r} \right) \\
&= \left(\delta_l^s - \frac{X^s}{F} \frac{\partial F}{\partial X^l} \right) g^{kl} \frac{\partial}{\partial X^k} \left(F \frac{\partial^2 F}{\partial X^i \partial X^j} \right) \left(\frac{\partial F}{\partial u^s} - X^r \frac{\partial^2 F}{\partial X^s \partial u^r} \right) \\
&= g^{ks} F \frac{\partial^3 F}{\partial X^k \partial X^i \partial X^j} \left(\frac{\partial F}{\partial u^s} - X^r \frac{\partial^2 F}{\partial X^s \partial u^r} \right).
\end{aligned} \tag{3.22}
$$

In the second equality above we have used the facts

$$
\begin{aligned}
g^{kl} \frac{\partial F}{\partial X^l} &= p_c^k \delta^{cd} p_d^l \frac{\partial F}{\partial X^l} = p_m^k p_m^l \frac{\partial F}{\partial X^l} \\
&= \frac{X^k X^l}{F^2} \frac{\partial F}{\partial X^l} = \frac{X^k}{F},
\end{aligned} \tag{3.23}
$$

and

$$\frac{X^s}{F} \left(\frac{\partial F}{\partial u^s} - X^r \frac{\partial^2 F}{\partial X^s \partial u^r} \right) = 0, \tag{3.24}$$

the last on account of Euler's theorem [Eq. (2.2)]. Eqs. (3.21) and (3.22) result in

$$
\begin{aligned}
G_{ijm} = g^{ks} F & \frac{\partial^3 F}{\partial X^k \partial X^i \partial X^j} \left(\frac{\partial F}{\partial u^s} - X^r \frac{\partial^2 F}{\partial X^s \partial u^r} \right) \\
&+ X^s \frac{\partial^3 F}{\partial u^s \partial X^i \partial X^j} + \frac{X^s}{F} \left(\frac{\partial F}{\partial u^s} \right) \left(\frac{\partial^2 F}{\partial X^i \partial X^j} \right).
\end{aligned} \tag{3.25}
$$

The $\lambda_{\alpha\beta}$ can now be determined in terms of G_{ijm} by plugging (3.18) into the right hand side of (3.17) and then, as per (3.14), setting the coefficient of ω^m equal to zero. We obtain

$$\lambda_{\rho\sigma} = -\frac{1}{2} p_\rho^i p_\sigma^j \left(G_{ijm} + \frac{\partial^2 F}{\partial X^j \partial u^i} + \frac{\partial^2 F}{\partial X^i \partial u^j} \right). \tag{3.26}$$

Requiring the coefficient of ω^γ to be zero also in the same expression, there follows

$$\delta^{\alpha\beta} p^i_\rho p^j_\sigma G_{ij\beta} = \delta^{\alpha\beta} \delta_{\nu\sigma} \mu^\nu_{\rho\beta} + \delta^{\alpha\beta} \delta_{\nu\rho} \mu^\nu_{\sigma\beta}. \tag{3.27}$$

Two similar equations can be obtained by commuting the index set (ρ, σ, β) in cyclic order. Adding these and (3.27), we have

$$\mu^\alpha_{\rho\sigma} = \frac{1}{2} \delta^{\alpha\beta} \left(p^i_\beta p^j_\rho G_{ij\sigma} - p^i_\rho p^j_\sigma G_{ij\beta} + p^i_\sigma p^j_\beta G_{ij\rho} \right). \tag{3.28}$$

It remains to determine $G_{ij\beta}$. This is achieved by contracting (3.18) with e_β to first obtain

$$G_{ij\beta} = -S^\alpha_{ij} \langle \omega^m_\alpha, e_\beta \rangle + p^k_\beta \frac{\partial}{\partial u^k} \left(F \frac{\partial^2 F}{\partial X^i \partial X^j} \right). \tag{3.29}$$

Next use (3.20) for S^α_{ij} and (3.26) for $\lambda_{\alpha\beta}$. Remembering (3.23), the identity

$$q^\alpha_i p^r_\alpha = \delta^r_i - q^m_i p^r_m = \delta^r_i - \left(\frac{\partial F}{\partial X^i} \right) \left(\frac{X^r}{F} \right),$$

and Euler's theorem in the form

$$X^k \frac{\partial^3 F}{\partial X^i \partial X^j \partial X^k} = -\frac{\partial^2 F}{\partial X^i \partial X^j},$$

we have

$$-S^\alpha_{ij} \langle \omega^m_\alpha, e_\beta \rangle = \frac{F}{2} g^{lk} p^s_\beta \left(\delta^r_l - \frac{X^r}{F} \left(\frac{\partial F}{\partial X^l} \right) \right)$$
$$\times \left(\frac{\partial^2 F}{\partial u^r \partial X^s} - \frac{\partial^2 F}{\partial X^r \partial u^s} - G_{rsm} \right) \frac{\partial}{\partial X^k} \left(F \frac{\partial^2 F}{\partial X^i \partial X^j} \right)$$
$$= \frac{1}{2} p^s_\beta \left(\frac{\partial^2 F}{\partial u^r \partial X^s} - \frac{\partial^2 F}{\partial X^r \partial u^s} - G_{rsm} \right)$$
$$\times \left(F^2 g^{rk} \frac{\partial^3 F}{\partial X^i \partial X^j \partial X^k} + X^r \frac{\partial^2 F}{\partial X^i \partial X^j} \right). \tag{3.30}$$

Finally,

$$G_{ij\beta} = p^s_\beta \left\{ \frac{1}{2} \left(\frac{\partial^2 F}{\partial u^r \partial X^s} - \frac{\partial^2 F}{\partial X^r \partial u^s} - G_{rsm} \right) \left(F^2 g^{rk} \frac{\partial^3 F}{\partial X^i \partial X^j \partial X^k} \right. \right. \tag{3.31}$$
$$\left. \left. + X^r \frac{\partial^2 F}{\partial X^i \partial X^j} \right) + \left(\frac{\partial F}{\partial u^s} \right) \left(\frac{\partial^2 F}{\partial X^i \partial X^j} \right) + F \frac{\partial^3 F}{\partial u^s \partial X^i \partial X^j} \right\}.$$

This completes the determination of the connection ω^j_i. It will be called the **Chern connection**.

§8–3.2   The Cartan Tensor and Characterization of Riemannian Geometry

We can write (3.14) as

$$\omega_\rho^\alpha \delta_{\alpha\sigma} + \omega_\sigma^\alpha \delta_{\alpha\rho} = 2A_{\rho\sigma}^\alpha \omega_\alpha^m$$
$$= -2A_{\rho\sigma\alpha}\omega_m^\alpha, \tag{3.32}$$

where

$$A_{\rho\sigma\alpha} = A_{\rho\sigma}^\beta \delta_{\beta\alpha} \tag{3.33}$$

and, according to (3.17) and (3.18),

$$A_{\rho\sigma}^\alpha \equiv -\frac{1}{2}p_\rho^i p_\sigma^j S_{ij}^\alpha. \tag{3.34}$$

If we extend the definition of A_{ijk} to include m as a possible value assumed by its indices such that

$$A_{ijk} = 0 \quad \text{whenever any index has the value } m, \tag{3.35}$$

then (3.32) can be written as

$$\omega_{ik} + \omega_{ki} = -2A_{ikj}\omega_m^j \tag{3.36}$$

where

$$\omega_{ik} = \omega_i^j \delta_{jk}. \tag{3.37}$$

The $(0,3)$ tensor

$$A = A_{ijk}\omega^i \otimes \omega^j \otimes \omega^k \tag{3.38}$$

is called the **Cartan tensor**.

A useful expression for A_{ijk} in terms of the Finsler function F can be straightforwardly obtained by using (3.20) for S_{ij}^α in (3.34). Recalling from (2.10) that

$$g^{lk} = p_r^l \delta^{rs} p_s^k,$$

we have

$$A_{\rho\sigma\alpha} = \frac{F}{2}p_\sigma^j p_\rho^i p_\alpha^k \left[\left(\frac{\partial F}{\partial X^k} \right) \left(\frac{\partial^2 F}{\partial X^i \partial X^j} \right) + F\frac{\partial^3 F}{\partial X^i \partial X^j \partial X^k} \right]. \tag{3.39}$$

The first term inside the square brackets does not contribute on account of (2.17). Hence we can add two similar non-contributing terms and rewrite the above equation as

$$
\begin{aligned}
A_{\rho\sigma\alpha} &= \frac{F}{2} p_\sigma^j p_\rho^i p_\alpha^k \left[\left(\frac{\partial F}{\partial X^k} \right) \left(\frac{\partial^2 F}{\partial X^i \partial X^j} \right) + \left(\frac{\partial F}{\partial X^j} \right) \left(\frac{\partial^2 F}{\partial X^i \partial X^k} \right) \right. \\
&\quad \left. + \left(\frac{\partial F}{\partial X^i} \right) \left(\frac{\partial^2 F}{\partial X^j \partial X^k} \right) + F \frac{\partial^3 F}{\partial X^i \partial X^j \partial X^k} \right] \\
&= \frac{F}{2} \frac{\partial^3 \left(\frac{F^2}{2} \right)}{\partial X^i \partial X^j \partial X^k} p_\sigma^j p_\rho^i p_\alpha^k = \frac{F}{2} \frac{\partial g_{ij}}{\partial X^k} p_\sigma^j p_\rho^i p_\alpha^k .
\end{aligned}
$$

(3.40)

Since

$$
X^j \frac{\partial g_{lk}}{\partial X^j} = X^l \frac{\partial g_{lk}}{\partial X^j} = X^k \frac{\partial g_{lk}}{\partial X^j} = 0,
\tag{3.41}
$$

which follows from (2.12) and Euler's theorem, we can allow all indices in (3.40) to range from 1 to m and finally write

$$
A_{ijk} = \frac{F}{2} \frac{\partial^3 \left(\frac{F^2}{2} \right)}{\partial X^r \partial X^s \partial X^t} p_i^r p_j^s p_k^t .
\tag{3.42}
$$

Note that the above equation gives the components of the Cartan tensor with respect to the ω^i. Eqs. (2.9), (2.11) and (2.12) then yield

$$
\begin{aligned}
A_{abc} &= \frac{F}{2} \frac{\partial^3 \left(\frac{F^2}{2} \right)}{\partial X^a \partial X^b \partial X^c} \\
&= \frac{F}{2} \frac{\partial g_{bc}}{\partial X^a} = \frac{F}{2} \frac{\partial g_{ab}}{\partial X^c} = \frac{F}{2} \frac{\partial g_{ac}}{\partial X^b},
\end{aligned}
\tag{3.43}
$$

where

$$
A = A_{abc} du^a \otimes du^b \otimes du^c
\tag{3.44}
$$

with the components given with respect to the natural basis du^i. Both A_{ijk} and A_{abc} are clearly totally symmetric. By virtue of (2.16) and (3.41), A_{ijk} [as given by (3.42)] also vanishes whenever any of its indices assumes the value m, as required by (3.35).

Since

$$
G = g_{ij} du^i \otimes du^j = \delta_{ij} \omega^i \otimes \omega^j,
\tag{3.45}
$$

it follows from (1.47) in Chapter Four and (3.36) that, for any given $v \in TM$,

$$
\begin{aligned}
(D_v G)(e_i, e_j) &= -\langle v, \omega_i^j \rangle - \langle v, \omega_j^i \rangle \\
&= 2 A_{ijk} \langle v, \omega_m^k \rangle .
\end{aligned}
\tag{3.46}
$$

Hence $G(e_m, e_j)$ and $G(e_i, e_m)$ are covariantly constant in all directions whereas $G(e_\alpha, e_\beta)$ is covariantly constant only along those directions v which satisfy $\langle v, \omega_m^k \rangle = 0$, that is, along which e_m is parallelly displaced. Under these circumstances we say that the Chern connection is **almost metric-compatible**. It can be viewed as a natural analog, in the Finsler setting, of the Christoffel–Levi–Civita connection of Riemannian geometry, which is entirely metric-compatible.

Eqs. (3.42) or (3.43) leads to a most important property of the Cartan tensor: it vanishes if and only if the Finsler metric G [c.f. (2.12)] is Riemannian, that is, g_{ij} is independent of X^k. The developments in this section culminate in the following theorem, which is a generalization of Theorem 1.3 in Chapter Five to the Finsler setting.

Theorem 3.2 (Chern). *The Chern connection forms ω_i^j on the Finsler bundle $p^* TM \to PTM$ are unique solutions of the structure equations*

$$d\omega^i = \omega^j \wedge \omega_j^i, \qquad \text{(torsion-freeness)}$$

and
$$\omega_{ij} + \omega_{ji} = -2A_{ijk}\omega_m^k, \qquad \text{(almost metric-compatibility)}$$

where $\omega_{ij} = \omega_i^k \delta_{kj}$, and the Cartan tensor

$$A = A_{ijk}\omega^i \otimes \omega^j \otimes \omega^k$$

is given by

$$A_{ijk} = \frac{F}{2} \frac{\partial^3 \left(\frac{F^2}{2} \right)}{\partial X^r \partial X^s \partial X^t} p_i^r p_j^s p_k^t.$$

The Finsler metric g_{ij} is Riemannian if and only if the Cartan tensor vanishes. Furthermore, no torsion free connection can be entirely metric-compatible at the same time unless the Finsler structure is Riemannian.

Thus our development includes Riemannian geometry as a special case, and the Chern connection is a natural generalization of the Christoffel–Levi–Civita connection.

§8-3.3 Explicit Formulas for the Connection Forms in Natural Coordinates

We begin by recasting the connection forms ω_α^m in terms of the natural basis du^i and dX^i. It will be convenient to define the following quantities:

$$\mathcal{G} \equiv \frac{1}{2} F^2 \tag{3.47}$$

$$\mathcal{G}_l \equiv \frac{1}{2} \left(X^s \frac{\partial^2 \mathcal{G}}{\partial X^l \partial u^s} - \frac{\partial \mathcal{G}}{\partial u^l} \right)$$

$$= \frac{1}{2} \left(X^s \frac{\partial F}{\partial u^s} \frac{\partial F}{\partial X^l} + X^s F \frac{\partial^2 F}{\partial X^l \partial u^s} - F \frac{\partial F}{\partial u^l} \right), \tag{3.48}$$

$$\mathcal{G}^i \equiv g^{il} \mathcal{G}_l. \tag{3.49}$$

First we use (2.9), (2.15) and (3.26)(for $\lambda_{\rho\sigma}$) to rewrite (2.20) for ω_α^m in the form

$$\omega_\alpha^m = -p_\alpha^i \frac{\partial^2 F}{\partial X^i \partial X^j} dX^j$$

$$+ p_\alpha^i \left[\frac{X^j}{F} \frac{\partial F}{\partial X^k} \left(G_{ijm} + \frac{\partial^2 F}{\partial X^j \partial u^i} - \frac{\partial^2 F}{\partial X^i \partial u^j} \right) \right. \tag{3.50}$$

$$\left. - \frac{1}{2} \left(G_{ikm} + \frac{\partial^2 F}{\partial X^i \partial u^k} - \frac{\partial^2 F}{\partial X^k \partial u^i} \right) \right] du^k.$$

With the help of (3.25) for G_{ijm}, the expression within the square brackets in the above equation can be written, after some tedious but straightforward algebra, as

$$\frac{-g_{it}}{F} \frac{\partial \mathcal{G}^t}{\partial X^k} + \frac{1}{2F} \left(\frac{\partial F}{\partial X^i} \right) \left\{ F \frac{\partial^2 F}{\partial X^k \partial X^i} g^{tl} \left(\frac{\partial F}{\partial u^l} - X^r \frac{\partial^2 F}{\partial X^l \partial u^r} \right) \right.$$

$$\left. + \frac{\partial F}{\partial u^k} + X^r \frac{\partial^2 F}{\partial X^k \partial u^r} \right\}. \tag{3.51}$$

This allows ω_α^m to be finally expressed in the compact form:

$$\omega_\alpha^m = -p_\alpha^i \left[\frac{g_{ij}}{F} \frac{\partial \dot{\mathcal{G}}^j}{\partial X^k} du^k + \frac{\partial^2 F}{\partial X^i \partial X^k} dX^k \right], \tag{3.52}$$

where we have used the duality condition given by the second equation of (2.17). A simpler expression for G_{ijk} in terms of natural coordinates [which combines (3.25) and (3.31)] can now be obtained by plugging (3.52) for ω_α^m and (3.20) for S_{ij}^α into (3.18), and then contracting the resulting equation with

e_k:

$$G_{ijk} = p_k^l \left[\frac{\partial}{\partial u^l} \left(F \frac{\partial^2 F}{\partial X^i \partial X^j} \right) - \frac{\partial \mathcal{G}^r}{\partial X^l} \frac{\partial}{\partial X^r} \left(F \frac{\partial^2 F}{\partial X^i \partial X^j} \right) \right]. \tag{3.53}$$

The above equation can then be used in (3.26) and (3.28) to give $\lambda_{\rho\sigma}$ and $\mu_{\rho\sigma}^\alpha$ in natural coordinates.

We will now construct canonical basis sets on $T(TM\backslash\{0\})$ and $T^*(TM\backslash\{0\})$ using the natural coordinates u^i, X^i, $i = 1, \ldots, m..$ By (3.9), (2.12) and the fact that $g_{ij} = q_i^k \delta_{kl} q_j^l$ [which follows from (2.10)], (3.52) is equivalent to

$$\omega_m^\alpha = q_j^\alpha \left(\frac{dX^j}{F} + N_k^j du^k \right) = q_j^\alpha \delta X^j, \tag{3.54}$$

where

$$N_j^i \equiv \frac{1}{F} \frac{\partial \mathcal{G}^i}{\partial X^j}, \tag{3.55}$$

and

$$\delta X^j \equiv \frac{dX^j}{F} + N_k^j du^k. \tag{3.56}$$

Note the similarity of (3.54) to the expression for ω^i in terms of the natural basis given by (2.9). Thus the dual orthonormal vectors [in $T(PTM)$] to the basis set $\omega^i = q_j^i du^j$ and ω_m^α [in $T^*(PTM)$] are given by

$$\hat{e}_i = p_i^j \frac{\delta}{\delta u^j}, \quad i = 1, \ldots, m, \tag{3.57}$$

and

$$\hat{e}_{m+\alpha} = p_\alpha^j \frac{\delta}{\delta X^j}, \quad \alpha = 1, \ldots, m-1, \tag{3.58}$$

where

$$\frac{\delta}{\delta u^i} \equiv \frac{\partial}{\partial u^i} - F N_i^j \frac{\partial}{\partial X^j}, \tag{3.59}$$

and

$$\frac{\delta}{\delta X^i} \equiv F \frac{\partial}{\partial X^i}. \tag{3.60}$$

The set $\{ \frac{\delta}{\delta u^i}, \frac{\delta}{\delta X^i} \}$ is naturally dual to the set $\{ du^i, \delta X^i \}$. These form local bases for $T(TM \backslash \{0\})$ and $T^*(TM \backslash \{0\})$, respectively.

PTM	$\omega^i = q^i_j du^j$ $\omega^\alpha_m = q^\alpha_j \delta X^j$	$\hat{e}_i = p^j_i \frac{\delta}{\delta u^j}$ $\hat{e}_{m+\alpha} = p^j_\alpha \frac{\delta}{\delta X^j}$
TM \ {0}	du^i $\delta X^i = \frac{dX^i}{F} + N^i_j du^j$	$\frac{\delta}{\delta u^i} = \frac{\partial}{\partial u^i} - F N^j_i \frac{\partial}{\partial X^j}$ $\frac{\delta}{\delta X^i} = F \frac{\partial}{\partial X^i}$

TABLE 1.

We summarize the relationship between our frame fields and coframe fields in orthonormal and natural bases in Table 1, where the canonical objects in the two columns are dual to each other. The designations on the left refer to the base manifolds on which the various one-forms and vectors live.

We are now ready to derive explicit formulas for the Chern connection

$$D : \Gamma(p^* TM) \to \Gamma(p^* TM \otimes T^*(TM \setminus \{0\}))$$

in natural coordinates, given by

$$D \frac{\partial}{\partial u^i} = \theta^j_i \frac{\partial}{\partial u^j} \tag{3.61}$$

It will be useful to give first the relationship (gauge transformation) between the θ^j_i and the ω^b_a. Applying D to (3.1) we obtain

$$De_i = \omega^j_i e_j = D \left(p^k_i \frac{\partial}{\partial u^k} \right),$$

which gives

$$\omega^j_i p^l_j \frac{\partial}{\partial u^l} = dp^l_i \frac{\partial}{\partial u^l} + p^k_i \theta^l_k \frac{\partial}{\partial u^l}.$$

The above implies

$$\omega^j_i p^l_j = dp^l_i + p^k_i \theta^l_k.$$

Contraction with q then yields

$$\omega^j_i = q^j_l (dp^l_i + p^k_i \theta^l_k). \tag{3.62}$$

Inversion of the above produces

$$\theta^j_i = p^j_l (dq^l_i + q^k_i \omega^l_k). \tag{3.63}$$

Eqs. (3.62) and (3.63) are equivalent to (1.12) of Chapter Four. Again, note that these equations display explicitly the non-tensorial nature of the connection forms.

Using (3.62) and the duality condition (2.11), the torsion-free condition (3.3) is seen to be equivalent to

$$du^i \wedge \theta_i^j = 0. \tag{3.64}$$

This immediately implies that the Chern connection forms θ_i^j cannot have dX^k terms, that is,

$$\theta_i^j = \Gamma_{il}^j du^l \tag{3.65}$$

Eq. (3.64) also implies the symmetry property

$$\Gamma_{il}^j = \Gamma_{li}^j \tag{3.66}$$

[c.f. Def. 2.2 and (2.31) in Chapter Four]. Eq. (3.61) can then be written as

$$D\frac{\partial}{\partial u^i} = \Gamma_{il}^j du^l \otimes \frac{\partial}{\partial u^j}. \tag{3.67}$$

Note that objects of the form $D\frac{\partial}{\partial X^i}$ are undefined, since the covariant differentiation D acts on sections of the tensor products of p^*TM and p^*T^*M.

To obtain the other structure equation reflecting almost metric-compatibility of the Chern connection in natural coordinates, we apply D to the equation

$$g_{ij} = G\left(\frac{\partial}{\partial u^i}, \frac{\partial}{\partial u^j}\right).$$

Thus

$$dg_{ij} = (DG)\left(\frac{\partial}{\partial u^i}, \frac{\partial}{\partial u^j}\right) + G\left(D\frac{\partial}{\partial u^i}, \frac{\partial}{\partial u^j}\right) + G\left(\frac{\partial}{\partial u^i}, D\frac{\partial}{\partial u^j}\right). \tag{3.68}$$

Writing

$$G = \delta_{ij}\omega^i \otimes \omega^j$$

and using (3.36), we have

$$\begin{aligned}
DG &= -(\omega_{ik} + \omega_{ki})\omega^i \otimes \omega^k \\
&= 2A_{ikl}\omega_m^j p_j^l du^i \otimes du^k \\
&= 2A_{ijk}\delta X^k du^i \otimes du^j
\end{aligned} \tag{3.69}$$

where the Cartan tensor components are with respect to the natural basis du^i [given by (3.43)], and in the third equality we have used (3.54). Hence

$$DG\left(\frac{\partial}{\partial u^i}, \frac{\partial}{\partial u^j}\right) = 2A_{ijk}\delta X^k. \tag{3.70}$$

Meanwhile, by (3.67),

$$G\left(D\frac{\partial}{\partial u^i}, \frac{\partial}{\partial u^j}\right) = \Gamma_{il}^k g_{kj}du^l = \Gamma_{ijl}du^l, \tag{3.71}$$

$$G\left(\frac{\partial}{\partial u^i}, D\frac{\partial}{\partial u^j}\right) = \Gamma_{jl}^k g_{ik}du^l = \Gamma_{jil}du^l. \tag{3.72}$$

Plugging the above equations in (3.68), we obtain the almost metric-compatibility condition in natural coordinates:

$$dg_{ij} = g_{kj}\theta_i^k + g_{ik}\theta_j^k + 2A_{ijk}\delta X^k. \tag{3.73}$$

Comparing coefficients of du^k in the above equation, we have, on recalling (3.56),

$$\Gamma_{ijk} + \Gamma_{jik} = \frac{\partial g_{ij}}{\partial u^k} - 2A_{ijl}N_k^l. \tag{3.74}$$

Comparison of coefficients of dX^k confirms the expression (3.43) for the Cartan tensor.

Let us now compute the combination

$$(\Gamma_{ijk} + \Gamma_{jik}) - (\Gamma_{jki} + \Gamma_{kji}) + (\Gamma_{kij} + \Gamma_{ikj}).$$

Using (3.66) and (3.74), we find

$$\begin{aligned}
\Gamma_{ijk} = {} & \frac{1}{2}\left(\frac{\partial g_{ij}}{\partial u^k} - \frac{\partial g_{ki}}{\partial u^j} + \frac{\partial g_{jk}}{\partial u^i}\right) \\
& - \frac{F}{2}\left(\frac{\partial g_{ij}}{\partial X^l}N_k^l - \frac{\partial g_{ki}}{\partial X^l}N_j^l + \frac{\partial g_{jk}}{\partial X^l}N_i^l\right)
\end{aligned} \tag{3.75}$$

where N_j^i is given by (3.55) [and also (3.79) below]. The corresponding expression for Γ_{ik}^j can then be obtained from the above by noting that

$$\Gamma_{ik}^j = g^{jl}\Gamma_{ilk}. \tag{3.76}$$

As in the Riemannian case, the formal expressions

$$\gamma_{ijk} \equiv \frac{1}{2}\left(\frac{\partial g_{ij}}{\partial u^k} - \frac{\partial g_{ki}}{\partial u^j} + \frac{\partial g_{jk}}{\partial u^i}\right) \tag{3.77}$$

and

$$\gamma_{ik}^j \equiv \frac{g^{il}}{2}\left(\frac{\partial g_{il}}{\partial u^k} - \frac{\partial g_{ki}}{\partial u^l} + \frac{\partial g_{lk}}{\partial u^i}\right) \tag{3.78}$$

are called the Christoffel symbols of the first and second kinds, respectively. For future use, it will be convenient to express the N_j^i in terms of the Christoffel symbols and the Cartan tensor. Using (3.55), (3.49) and (3.78), it can be established, after some tedious algebra, that

$$N_j^i = \gamma_{jk}^i \frac{X^k}{F} - A_{jk}^i \gamma_{rs}^k \frac{X^r X^s}{F^2}, \tag{3.79}$$

where $A_{jk}^i = g^{il}A_{ljk}$, with A_{ljk} given by (3.43). It then follows from (3.41) and (3.75) that

$$\Gamma_{jk}^i \frac{X^j}{F} = N_k^i, \tag{3.80}$$

or

$$N_k^i du^k = \theta_j^i \frac{X^j}{F}. \tag{3.81}$$

Eq. (3.75) should be compared with (1.34) in Chapter Five, which gives an expression for the Christoffel symbols of the first kind in Riemannian geometry. It is clear that in the Riemannian case, when g_{ij} is a function of the u^i only, the tensor

$$M_{ijk} \equiv F\frac{\partial g_{ij}}{\partial X^l}N_k^l \tag{3.82}$$

vanishes. Hence in that case, the Γ_{ijk} in (3.75) reduces to the Christoffel symbols of the first kind, and the Chern connection reduces to the Christoffel–Levi–Civita connection. Thus M_{ijk} can be viewed as a measure of the departure from metric-compatibility of the Chern connection.

§8–4   Structure Equations and the Flag Curvature

In this section we will explore the properties of the curvature tensor Ω of the Chern connection. These are completely determined by the structure equations describing torsion-freeness and almost metric-compatibility [Eqs. (3.3) and (3.36), respectively]. We will see that Ω splits into two parts, R and P, where the R-part is the generalization of the Riemann curvature tensor introduced in §4–2. Both R and P, as well as the Cartan tensor, play a fundamental role in the classification of special Finsler spaces, some examples of which will be given in §8–4.3.

§8–4.1　The Curvature Tensor

The structure equation describing the torsion-freeness of the Chern connection [Eq. (3.3)]:

$$dw^i = \omega^j \wedge \omega_j^i$$

can be exteriorly differentiated to yield

$$\omega^k \wedge \left(dw_k^i - \omega_k^j \wedge \omega_j^i \right) = 0. \tag{4.1}$$

Recalling Def. 1.2 in Chapter Four, the matrix of two-forms

$$\Omega_k^i \equiv dw_k^i - \omega_k^j \wedge \omega_j^i \tag{4.2}$$

is called the curvature form of the Chern connection. Being a two-form on PTM, it must be expressible as a linear superposition of only two-forms of type $\omega^i \wedge \omega^j$ and $\omega^i \wedge \omega_m^\alpha$. Two-forms of the type $\omega_m^\alpha \wedge \omega_m^\beta$ are ruled out because they contain terms proportional to $dX^i \wedge dX^j$ [c.f. (3.52)], which would violate torsion-freeness [as expressed by (4.1)]. Thus the most general expression for Ω_k^i is given by

$$\Omega_k^i = \frac{1}{2} R_{kjl}^i \omega^j \wedge \omega^l + P_{kj\alpha}^i \omega^j \wedge_m^\alpha, \tag{4.3}$$

where, without loss of generality, we can set

$$R_{kjl}^i + R_{klj}^i = 0. \tag{4.4}$$

Note that Eq. (4.3) without the P-part is formally identical to the expression for the Riemannian curvature [see (2.22) in Chapter Four]. The R-part of the curvature tensor is called the horizontal-horizontal (h-h) part, and the P-part the horizontal-vertical (h-v) part. We will refer to these as **the first and second Chern curvature tensors**, respectively. These separate parts of the curvature tensor, as well as the Cartan tensor, will be shown to provide signatures in the classification of special Finsler spaces. We have already seen in §8–3.2 that the vanishing of the Cartan tensor is equivalent to the Finsler structure being Riemannian. In §8–4.3, we will see two more important examples of Finsler spaces characterized by properties of the curvature tensor.

Putting (4.3) in (4.1) we have

$$\omega^k \wedge \left(\frac{1}{2} R_{kjl}^i \omega^j \wedge \omega^l + P_{kj\alpha}^i \omega^j \wedge \omega_m^\alpha \right) = 0. \tag{4.5}$$

Two other similar expressions can be obtained on permuting the indices k, j, l in cyclic order. Addition of these to (4.5) yields

$$\frac{1}{2}\left(R^i_{kjl} + R^i_{jlk} + R^i_{lkj}\right) \omega^k \wedge \omega^j \wedge \omega^l + 3P^i_{kja}\omega^k \wedge \omega^j \wedge \omega^\alpha_m = 0. \qquad (4.6)$$

The coefficient of each term in (4.6) separately vanishes. Hence we have the **Bianchi identity**

$$R^i_{kjl} + R^i_{jlk} + R^i_{lkj} = 0, \qquad (4.7)$$

which is again formally the same as that for the Riemannian case [compare with (1.57) in Chapter Five]; and the following symmetry property for P:

$$P^i_{kj\alpha} = P^i_{jk\alpha}. \qquad (4.8)$$

To obtain more information on the curvature tensors R and P we exteriorly differentiate the structure equation of the Chern connection reflecting the property of almost metric-compatibility [Eq. (3.36)]. It is a straightforward procedure, making use of (4.2) and the condition $\omega^m_m = 0$, to establish the following fundamental identity for the curvature tensor:

$$\begin{aligned}
\Omega_{ik} + \Omega_{ki} &= \Omega^j_i \delta_{jk} + \Omega^j_k \delta_{ji} \\
&= -2A_{kij}\Omega^j_m - 2(DA)_{ki\alpha} \wedge \omega^\alpha_m,
\end{aligned} \qquad (4.9)$$

where the covariant derivative DA of the Cartan tensor is given (according to the discussion in §4–2) by

$$\begin{aligned}
(DA)_{ki\alpha} &= dA_{ki\alpha} - A_{si\alpha}\omega^s_k - A_{ks\alpha}\omega^s_i - A_{kis}\omega^s_\alpha \\
&\equiv \frac{1}{2}\left(L_{ki\alpha\beta}\omega^\beta_m + Q_{ki\alpha s}\omega^s\right).
\end{aligned} \qquad (4.10)$$

The last equality defines the tensors L and Q, called the vertical and horizontal parts, respectively, of the covariant derivative of the Cartan tensor. On interchanging the indices k and i in the above equation, the left hand side remains unchanged due to the total symmetry of A. Hence

$$L_{ki\alpha\beta} = L_{ik\alpha\beta} \qquad (4.11)$$

and

$$Q_{ki\alpha s} = Q_{ik\alpha s}. \qquad (4.12)$$

Furthermore, letting $k = m$ in (4.10) and using the fact that $A_{ijk} = 0$ whenever any of its indices has the value m, we have

$$Q_{mi\alpha s} = 0. \qquad (4.13)$$

Substituting (4.3) and (4.10) into (4.9) we have

$$\frac{1}{2} \left(\delta_{ji} R^j_{kab} + \delta_{jk} R^j_{iab} \right) \omega^a \wedge \omega^b + \left(\delta_{ji} P^j_{ksa} + \delta_{jk} P^j_{isa} \right) \omega^s \wedge \omega^\alpha_m$$

$$= -2A_{kij} \left(\frac{1}{2} R^j_{mab} \omega^a \wedge \omega^b + P^j_{msa} \omega^s \wedge \omega^\alpha_m \right) \tag{4.14}$$

$$- L_{ki\alpha\beta} \omega^\beta_m \wedge \omega^\alpha_m - Q_{ki\alpha s} \omega^s \wedge \omega^\alpha_m.$$

The term on the right hand side involving $\omega^\beta_m \wedge \omega^\alpha_m$ must be zero. Consequently

$$L_{ki\alpha\beta} = L_{ki\beta\alpha}. \tag{4.15}$$

Comparing coefficients of $\omega^a \wedge \omega^b$ in (4.14) we have

$$R_{kisl} + R_{iksl} = R^j_{ksl} \delta_{ji} + R^j_{isl} \delta_{jk}$$
$$= -2A_{kij} R^j_{msl}. \tag{4.16}$$

This generalizes property 1) of the curvature tensor for a Riemannian manifold given in Theorem 1.4 of Chapter Five. In particular, on setting $k = m$ in the above equation, we obtain

$$R_{misl} + R_{imsl} = 0, \tag{4.17}$$

which in turn implies

$$R_{mmsl} = 0. \tag{4.18}$$

We also have

$$R_{kmml} = R_{lmmk} \tag{4.19}$$

from (4.4) and Bianchi's identity (4.7). On account of (4.17), (4.19) is equivalent to

$$R_{mkml} = R_{mlmk}. \tag{4.20}$$

An analogous equation to (4.16), which can be obtained from (4.14) on comparison of the coefficients of $\omega^s \wedge \omega^\alpha_m$, is

$$P_{kis\alpha} + P_{iks\alpha} = -2A_{kij} P^j_{ms\alpha} - Q_{kis\alpha}. \tag{4.21}$$

Thus

$$(P_{kis\alpha} + P_{iks\alpha}) - (P_{skis\alpha} + P_{ksi\alpha}) + (P_{isk\alpha} + P_{sik\alpha})$$
$$= 2P_{kis\alpha} \tag{4.22}$$
$$= -2A_{kij} P^j_{ms\alpha} + 2A_{skj} P^j_{mi\alpha} - 2A_{isj} P^j_{mk\alpha} - Q_{kias} + Q_{skai} - Q_{isak},$$

where in the first equality we have used (4.8). Setting $k = s = m$, and using (4.12), (4.13), and (3.40), we have

$$P_{mim\alpha} = 0. \tag{4.23}$$

Setting only $k = m$ in (4.22) then yields

$$P_{mis\alpha} = -P_{ims\alpha} = -\frac{1}{2}Q_{is\alpha m}. \tag{4.24}$$

Consequently

$$A_{ijk}P^j_{ms\alpha} = A_{ijk}\delta^{jl}P_{mls\alpha} = \frac{1}{2}A^j_{ik}Q_{js\alpha m}. \tag{4.25}$$

Using (4.25) and (4.12) on the right hand side of (4.22), we then find

$$P_{kis\alpha} = \frac{1}{2}\left(A^j_{ik}Q_{js\alpha m} - A^j_{ks}Q_{ji\alpha m} + A^j_{si}Q_{jk\alpha m}\right)$$
$$- \frac{1}{2}\left(Q_{ik\alpha s} - Q_{ks\alpha i} + Q_{si\alpha k}\right). \tag{4.26}$$

Thus we have expressed the second Chern curvature tensor P in terms of the Cartan tensor A and the Q (horizontal)-part of its covariant derivative. The symmetry properties of P are given by (4.8), (4.21) and (4.24). Eqs. (4.21) and (4.26) immediately imply the following useful fact.

Lemma 1. *The second Chern curvature tensor $P^j_{ik\alpha}$ vanishes if and only if the horizontal part of the covariant derivative of the Cartan tensor, Q_{kias}, vanishes.*

To conclude thus subsection we will express the first Chern curvature tensor R in terms of natural coordinates. Recalling Table 1 in §8–3.3, and Eqs. (3.61) and (3.64), (4.3) is equivalent to

$$\Omega^i_k = d\theta^i_k - \theta^j_k \wedge \theta^i_j$$
$$= \frac{1}{2}R^i_{kjl}du^j \wedge du^l + P^i_{kjl}du^j \wedge \delta X^l, \tag{4.27}$$

where δX^l is given by (3.56). In view of the torsion-free condition (3.64),

$$d\theta^i_k = (d\Gamma^i_{kj}) \wedge du^j. \tag{4.28}$$

Introduce the quantity

$$\frac{\delta\Gamma^i_{kj}}{\delta u^l} \equiv \left\langle d\Gamma^i_{kj}, \frac{\partial}{\partial u^l} \right\rangle \tag{4.29}$$

with $\frac{\delta}{\delta u^l}$ given by (3.59). We can then operate both sides of (4.27) on the pair $\left(\frac{\delta}{\delta u^j}, \frac{\delta}{\delta u^l}\right)$ to obtain

$$R^i_{kjl} = \frac{\delta \Gamma^i_{kl}}{\delta u^j} - \frac{\delta \Gamma^i_{kj}}{\delta u^l} + \Gamma^h_{kl}\Gamma^i_{hj} - \Gamma^h_{kj}\Gamma^i_{hl}. \tag{4.30}$$

This formula should be compared with the structurally identical one for Riemannian geometry, (in which the $\frac{\delta \Gamma^i_{kl}}{\delta u^j}$ are replaced by $\frac{\partial \Gamma^i_{kl}}{\partial u^j}$) [Eq. (2.23) of Chapter Four]. Similarly, operating (4.27) on the pair $\left(\frac{\delta}{\delta u^j}, \frac{\delta}{\delta X^l}\right)$ yields

$$P^i_{kjl} = -\frac{\delta \Gamma^i_{kj}}{\delta X^l} = -F\frac{\partial \Gamma^i_{kj}}{\partial X^l}. \tag{4.31}$$

We summarize the symmetry properties of $R_{ijkl} = R^h_{ikl}g_{hj}$ by the following theorem, [d] which is the generalization of Theorem 1.4 of Chapter Five to the Finsler setting.

Theorem 4.1. *This first Chern curvature tensor R_{ijkl} (the h-h part) on a Finsler manifold satisfies the following relations:*

1) $R_{ijkl} + R_{jikl} = -2A_{ija}R^a_{bkl}\frac{X^b}{F} \equiv 2B_{ijkl}$,
2) $R_{ijkl} + R_{kjli} + R_{ljik} = 0$ *(Bianchi identity)*,
3) $R_{ijkl} - R_{klij} = (B_{ijkl} - B_{klij}) + (B_{iljk} + B_{jkil}) + (B_{ljki} + B_{kilj})$,
4) $R_{ijkl} = -R_{ijlk}$.

[In property 1) A_{ija} are the components of the Cartan tensor with respect to the natural basis.]

Proof. Property 1) is the counterpart of (4.16) and follows directly from it on recalling (2.14). Properties 2) and 4) are counterparts of (4.7) and (4.4), respectively. To obtain property 3), we cyclically permute the set $(ijkl)$ to obtain three more versions of the Bianchi identity in addition to 2) and then add the four resulting equations. Using 1) and 4), and relabelling the indices, we get 3). $\qquad\square$

§8–4.2 The Flag Curvature and the Ricci Curvature

Entirely analogous to the Riemannian case [c.f. (3.6) and (3.7) in Chapter Five], we can define the curvature operator

$$R(X,Y) : \Gamma(p^* TM) \to \Gamma(p^* TM)$$

---

[d]Statement 3) of Theorem 4.1 is first given in Bao, Chern, and Shen, to appear.

given by

$$R(X,Y)Z = R^j_{ikl} Z^i X^k Y^l \frac{\partial}{\partial u^j};$$
(4.32)

and the quadriliniear function

$$R(X,Y,Z,W) \equiv G\left(R(Z,W)X,Y\right),$$
(4.33)

where $X,Y,Z,W \in p^* TM$. We caution that, unlike the Riemannian case, g_{ij} is a function on PTM, and depends in general on the choice of a 'reference vector' at which it is evaluated. Corresponding to the symmetry properties of R_{ijkl} given by Theorem 4.1 we have the following symmetries for $R(X,Y,Z,W)$:

1)

$$R(X,Y,Z,W) + R(Y,X,Z,W) = -2A\left(X,Y,R(Z,W)e_m\right)$$
$$\equiv 2B(XYZW),$$
(4.34)

2)

$$R(X,Y,Z,W) + R(Z,Y,W,X) + R(W,Y,X,Z) = 0,$$
(4.35)

3)

$$R(X,Y,Z,W) - R(Z,W,X,Y) = [B(XYZW) - B(ZWXY)]$$
$$+ [B(XWYZ) + B(YZXW)]$$
$$+ [B(WYZX) + B(ZXWY)],$$
(4.36)

4)

$$R(X,Y,Z,W) = -R(X,Y,W,Z).$$
(4.37)

The above generalize properties 1) to 3) of the Riemannian curvature function given in §5-3.

We recall that the sectional curvature of a Riemannian manifold M is given, at any point $p \in M$ and for any two linearly independent tangent vectors $X,Y \in T_p M$, by

$$K(X,Y) = \frac{-R(X,Y,X,Y)}{G(X,X)G(Y,Y) - (G(X,Y))^2},$$

[c.f. §5-3]. This can be thought of as a quantity defined on a 'flag' based at p, with 'flagpole' X and a transverse edge Y. $K(X,Y)$ is then a function only of the two-dimensional subspace in T_pM containing the flag, and is independent of the particular choice of the flagpole and the transverse edge. In the Finsler setting, we define a generalization of the sectional curvature in Riemannian geometry, called the **flag curvature**, as follows:

$$K(e_m, Y) \equiv \frac{-R(e_m, Y, e_m, Y)}{G(e_m, e_m)G(Y, Y) - (G(e_m, Y))^2}. \tag{4.38}$$

Based at each $p \in PTM$, the flagpole is always chosen to be the distinguished $e_m = \frac{X^i}{F}\frac{\partial}{\partial u^i}$, while the transverse edge can be any vector independent of e_m. Though $G(e_m, e_m) = 1$, we have left it in the denominator of the above equation for conceptual clarity. Note that the flag curvature is a function on $(u^i, X^i) \in PTM$ and $Y = Y^i\frac{\partial}{\partial u^i} \in (p^*TM)_{(u,X)}$. By inspection it is seen to be invariant under rescaling in Y.

A fundamental quantity associated with the flag curvature is the **Ricci curvature** Ric_p, $p \in PTM$, defined on PTM. For an orthonormal frame field $\{e_i\}$, $i = 1,\ldots,m$, on p^*TM with e_m given by (2.14), Ric_p is defined as the average of the flag curvatures $K(e_m, e_\alpha)$, $\alpha = 1,\ldots,m-1$:

$$\mathrm{Ric}_p \equiv \frac{1}{m-1}\sum_{\alpha=1}^{m-1} K(e_m, e_\alpha)$$

$$= -\frac{1}{m-1}\sum_{\alpha=1}^{m-1} R(e_m, e_\alpha, e_m, e_\alpha). \tag{4.39}$$

As in Riemannian geometry, the flag and Ricci curvatures contain important information regarding global properties of a Finsler manifold. Some examples will be given in §8-8.

§8-4.3   Special Finsler Spaces

i) Riemannian Spaces

This is historically the most important example of Finsler spaces, and has already been characterized in §8-3.2. A Finsler manifold (M, F) is said to be **Riemannian** if the Finsler metric g_{ij} is a function of u^i only. The Chern Theorem (3.2) then stipulates that M is Riemannian if and only if the Cartan tensor A vanishes.

ii) Berwald Spaces

A Finsler manifold (M, F) is called a **Berwald space** if the second Chern (h-v) curvature tensor P vanishes. By virtue of Lemma 1 and (4.31) we have the following result.

Lemma 2. *Let (M, F) be a Finsler manifold. The following statements are equivalent.*

1) *M is a Berwald space;*
2) *Q (the horizontal part of the covariant derivative of the Cartan tensor) vanishes;*
3) *The Chern connection coefficients Γ^i_{kj} are independent of X^l.*

In view of property 3) and (4.29), we have

$$\frac{\delta \Gamma^i_{kl}}{\delta u^j} = \frac{\partial \Gamma^i_{kl}}{\partial u^j} \tag{4.40}$$

and

$$R^i_{kjl} = \frac{\partial \Gamma^i_{kl}}{\partial u^j} - \frac{\partial \Gamma^i_{kj}}{\partial u^l} + \Gamma^h_{kl}\Gamma^i_{hj} - \Gamma^h_{kj}\Gamma^i_{hl} \tag{4.41}$$

for a Berwald space.

Remark. Eq. (4.41) is formally identical to its Riemannian counterpart [Eq. (2.23) in Chapter Four], but the Chern connection coefficients Γ^j_{ik} [as given by (3.75) and (3.76)] are not the same as the Christoffel symbols of the second kind [as given by (3.78)].

For specific examples of Berwald spaces involving positively homogeneous Finsler functions, where the relevant Finsler bundle is $p^*TM \to SM$, the reader should consult Bao, Chern, and Shen, to appear.

iii) Locally Minkowskian Spaces

A Finsler manifold (M, F) is said to be **locally Minkowskian** if the first and second Chern curvature tensors both vanish, that is, if $R = P = 0$.

We have the following useful characterization of locally Minkowskian spaces.

Lemma 3. *A Finsler manifold (M, F) is locally Minkowskian if and only if, at each $p \in M$, there exist local coordinates (u^i, X^i) on TM such that F is a function of X^i only.*

The proof of this lemma makes use of the following result on general affine connection spaces which we will state without proof. [See, for example, Spivak 1979 (Vol II)].

Lemma 4. *Let D be a torsion-free connection on a finite dimensional manifold M. If the curvature of this connection vanishes in a neighborhood of some $p \in M$, then there exists a local coordinate system (u^i) about p in which all the connection coefficients Γ^j_{ik} vanish.*

Proof of Lemma 3. Suppose there always exist (u^i, X^i) such that $\frac{\partial F}{\partial u^i} = 0$. Hence $\frac{\partial g_{ij}}{\partial u^k} = 0$ by (2.12). It then follows from (3.78) that $\gamma^j_{ik} = 0$, which in turn implies, by (3.79), that $N^i_j = 0$. By (3.75) and (3.76), the Chern connection coefficients Γ^j_{ik} vanish also. Then $R = P = 0$ from (4.30) and (4.31).

Conversely, suppose $R = P = 0$. By Lemma 2 $\frac{\partial \Gamma^j_{ik}}{\partial X^l} = 0$, and by (4.41)

$$R^i_{kjl} = \frac{\partial \Gamma^i_{kl}}{\partial u^j} - \frac{\partial \Gamma^i_{kj}}{\partial u^l} + \Gamma^h_{kl}\Gamma^i_{hj} - \Gamma^h_{kj}\Gamma^i_{hl} = 0. \tag{4.42}$$

Thus Γ^j_{ik} specify a torsion-free connection on M with zero curvature. By Lemma 4, there exists a local coordinate system about any point $p \in M$ in which $\Gamma^j_{ik} = 0$. Eq. (3.80) then implies that $N^i_j = 0$, and consequently, by (3.74), $\frac{\partial g_{ij}}{\partial u^k} = 0$. Thus g_{ij} is independent of u^k, and so is F, by virtue of the fact

$$X^i X^j g_{ij} = F^2,$$

which follows from (2.12) and Euler's theorem (2.2). $\square$

Remark. A locally Minkowskian manifold may be intuitively constructed by putting the same Minkowski norm on each tangent space $T_p M$ of M. However there are typically topological obstructions to this procedure. For an in depth discussion on the relationship between geometry and topology on a locally Minkowskian manifold, the reader may consult Bao, Chern, and Shen, to appear.

Example 1. A useful family of Minkowski norms on $\mathbb{R}^2$ is given by

$$F(X^1, X^2) = \sqrt{(X^1)^2 + (X^2)^2 + \alpha\sqrt{(X^1)^4 + (X^2)^4}} \, ; \, \alpha \geq 0.$$

§8–5   The First Variation of Arc Length and Geodesics

The flag curvature introduced in the last section plays an important role in variational problems in Finsler geometry. This is evidenced by the formulas for the first and second variations of arc length. In this and the next section, we will demonstrate the remarkable fact that, using the Chern connection, both of these formulas are formally identical to those in Riemannian geometry, with the sectional curvature in the Riemannian case replaced by the flag curvature in the Finsler case in the second variation formula. Consequently, many of the classical theorems in Riemannian geometry which follow from these formulas generalize to the Finsler setting also. Some important examples will be presented in §8–7 and §8–8. We will see that the variational formulas [given by (5.24) and (6.8) below] are direct consequences of the structure equations (3.3) and (3.36) of the Chern connection (c.f. Theorem 3.2).

We first define the notion of a **variation** of a smooth curve in a manifold M.

Definition 5.1. Let $c : [0, a] \to M$ be a smooth curve in M. A variation of c is a differentiable mapping $\sigma : [0, a] \times (-\epsilon, \epsilon) \to M$ such that $\sigma(t, 0) = c(t)$, $t \in [0, a]$. $c(t)$ is called the **base curve** of the variation. For each $u \in (-\epsilon, \epsilon)$, the parametrized curve $\sigma_u : [0, a] \to M$ defined by $\sigma_u(t) = \sigma(t, u)$ is called a **curve in the variation** (t-curve); while the parametrized curves $\sigma_t(u) = \sigma(t, u)$, t fixed, are called **transversal curves** (u-curves) of the variation.

A variation $\sigma(t, u)$ gives rise to two vector fields on M:

$$T \equiv \sigma_* \frac{\partial}{\partial t} = \frac{\partial \sigma}{\partial t} \tag{5.1}$$

and

$$U \equiv \sigma_* \frac{\partial}{\partial u} = \frac{\partial \sigma}{\partial u}, \tag{5.2}$$

which are velocity fields on the t- and u-curves, respectively. In particular, the field

$$U(t, 0) = \left. \frac{\partial \sigma(t, u)}{\partial u} \right|_{u=0}$$

is called the **variation field** of the variation σ along the base curve $c(t)$.

In our Finsler setting, the length of a t-curve is given by

$$L(u) = \int_0^a F(\sigma(t, u), T) dt = \int_0^a \sqrt{G_T(T, T)} dt, \tag{5.3}$$

where F is the Finsler function and the subscript T in G_T means that the Finsler metric G is evaluated at the **canonical lift**

$$\tilde{\sigma}(t, u) \equiv (\sigma(t, u), T(t, u)) \in PTM. \tag{5.4}$$

The **first variation of arc length** is thus

$$L'(u) = \int_0^a \frac{\partial}{\partial u} \sqrt{G_T(T, T)} dt. \tag{5.5}$$

Define the following vector fields on PTM:

$$\tilde{T} \equiv \tilde{\sigma}_* \frac{\partial}{\partial t}, \tag{5.6}$$

and

$$\tilde{U} \equiv \tilde{\sigma}_* \frac{\partial}{\partial u}, \tag{5.7}$$

and choose the preferred orthonormal frame field $e_i = 1, \ldots, m$, on $p^* TM$ such that e_m at the lift (5.4) is the unit vector along T:

$$e_m = \frac{X^i}{F} \frac{\partial}{\partial u^i} = \frac{T}{\sqrt{G_T(T,T)}}. \tag{5.8}$$

Let the dual coframe field to e_i be $\omega^i = 1, \ldots, m$. It will be more convenient and intuitive to work with the pullback forms on the rectangle $[0, a] \times (-\epsilon, \epsilon)$:

$$\tilde{\sigma}^* \omega^i = a^i dt + b^i du, \tag{5.9}$$

and

$$\tilde{\sigma}^* \omega_i^j = a_i^j dt + b_i^j du, \tag{5.10}$$

where

$$a^i = \omega^i(\tilde{T}), \tag{5.11}$$
$$b^i = \omega^i(\tilde{U}), \tag{5.12}$$
$$a_i^j = \omega_i^j(\tilde{T}), \tag{5.13}$$

and

$$b_i^j = \omega_i^j(\tilde{U}). \tag{5.14}$$

Since ω^i has no dX^j terms,

$$a^i = \omega^i(\tilde{T}) = \omega^i(T) = T^i \tag{5.15}$$

and

$$b^i = \omega^i(\tilde{U}) = \omega^i(U) = U^i. \tag{5.16}$$

The choice for e_m in (5.8) then implies

$$a^\alpha = \omega^\alpha(T) = 0, \tag{5.17}$$
$$a^m = \omega^m(T) = \sqrt{G_T(T,T)} \; ; \tag{5.18}$$

and from (5.5) it follows that

$$L'(u) = \int_0^a \frac{\partial a^m}{\partial u} dt. \tag{5.19}$$

To calculate the derivative in the above integrand, we use $\tilde{\sigma}^*$ to pull back the torsion-free condition [(3.3)]

$$d\omega^i = \omega^i \wedge \omega_j^i$$

of the Chern connection. Since $\tilde{\sigma}^*$ commutes with the exterior derivative and the exterior product [c.f. Eqs. (2.43) and (2.42) in Chapter Three, respectively], we obtain

$$d(\tilde{\sigma}^*\omega^i) = \tilde{\sigma}^*\omega^i \wedge \tilde{\sigma}^*\omega_j^i. \tag{5.20}$$

Eq. (5.9) implies

$$d(\tilde{\sigma}^*\omega^i) = \left(-\frac{\partial a^i}{\partial u} + \frac{\partial b^i}{\partial t} \right) dt \wedge du. \tag{5.21}$$

Substituting (5.21), (5.9) and (5.10) in (5.20), we have

$$-\frac{\partial a^i}{\partial u} + \frac{\partial b^i}{\partial t} = a^j b_j^i - b^j a_j^i. \tag{5.22}$$

Thus, on recalling $\omega_m^m = 0$ [(3.4)], and applying (5.13), (5.14) and (5.17) in (5.22) with $i = m$, the torsion-free condition becomes

$$\frac{\partial a^m}{\partial u} = \frac{\partial b^m}{\partial t} + b^\alpha a_\alpha^m. \tag{5.23}$$

Using this result in (5.19), we obtain the following formula for the first variation of arc length in Finsler geometry:

$$L'(u) = b^m \Big|_0^a + \int_0^a dt\, b^\alpha a_\alpha^m. \tag{5.24}$$

We note that , on using (5.11) to (5.16), the torsion-free condition (5.22) can also be written as

$$D_T U = D_U T, \tag{5.25}$$

where the covariant derivative is with respect to the Chern connection on $p^* TM \to PTM$. If we then recall that the almost metric-compatibility condition (3.36) is equivalent to

$$\frac{\partial}{\partial t} G_T(X,Y) = G_T(D_T X,Y) + G_T(X, D_T Y)$$
$$+ 2X^i Y^j A_{ij\alpha} \omega_m^\alpha(\tilde{T}), \tag{5.26}$$

and observe that

$$D_T \left(\frac{T}{\sqrt{G_T(T,T)}} \right) = D_T e_m$$

(5.27)

$$= -a_k^m \delta^{ki} e_i,$$

the formula (5.24) for the first variation of arc length can also be written in the intrinsic form:

$$L'(u) = \int_0^a \left[\frac{\partial}{\partial t} G_T \left(U, \frac{T}{\sqrt{G_T(T,T)}} \right) - G_T \left(U, D_T \left(\frac{T}{\sqrt{G_T(T,T)}} \right) \right) \right] dt.$$

(5.28)

For variations with fixed end-points, the variation field U satisfies $U(0) = U(a) = 0$, and thus the boundary term in (5.28) vanishes. We now formalize the notion of a geodesic curve in a Finsler manifold.

Definition 5.2. A **geodesic curve** $c : [0, a] \to M$ in a Finsler manifold M is a **critical curve** of the arc length functional $L(u)$ [(5.3)] with respect to all smooth variations σ with fixed end-points, that is c is a geodesic if $L'(0) = 0$ for all σ satisfying $\sigma(0, u) = c(0)$ and $\sigma(a, u) = c(a)$.

According to (5.28), the equation for a geodesic in Finsler space is thus

$$D_T \left(\frac{T}{\sqrt{G_T(T,T)}} \right) = 0.$$

(5.29)

If we impose the condition of constant speed parametrization, that is, $G_T(T,T) = $ constant, (5.29) reduces to the familiar condition of "auto-parallels" in Riemannian geometry [Eq. (2.20) in Chapter Four]:

$$D_T T = 0.$$

(5.30)

In terms of the coframe field ω^i and the Chern connection forms ω_i^j, the condition for a geodesic [(5.29)] can also be restated as follows: A geodesic in a Finsler space M is a curve $c(t)$ whose canonical lift $\tilde{c}(t) = (c(t), T(t)) \in PTM$ is a solution of the differential system:

$$a^\alpha = \omega^\alpha (\tilde{T}) = 0, \qquad a_m^\alpha = \omega_m^\alpha (\tilde{T}) = 0$$

(5.31)

where $\tilde{T} \equiv \tilde{c}_* \frac{d}{dt}$ [c.f. (5.13) and (5.17)].

We will now focus on some issues concerning geodesics in the Finsler versus Riemannian settings. It is remarkable that all the results in §5–2 for the

Riemannian case are still valid in the Finsler setting. There are, however, some important differences in the arguments leading to these results which distinguish the Finsler case. Space does not permit a full treatment here, and we can only give a brief discussion of the essential points. For more in-depth accounts, we refer the reader to Bao and Chern 1993, and Bao, Chern and Shen, to appear. Our discussion will be based on constant speed geodesics.

According to (5.30) a constant speed geodesic curve $c(t)$ in a Finsler manifold satisfies an ordinary differential equation which is formally identical to that for a Riemannian manifold:

$$\frac{d^2 u^i}{dt^2} + \frac{du^k}{dt}\frac{du^l}{dt}(\Gamma^i_{kl})_{\tilde{c}} = 0, \tag{5.32}$$

or equivalently,

$$\frac{d^2 u^i}{dt^2} + \frac{du^k}{dt}\frac{du^l}{dt}\left\{\frac{g^{ij}}{2}\left(\frac{\partial g_{kj}}{\partial u^l} + \frac{\partial g_{lj}}{\partial u^k} - \frac{\partial g_{kl}}{\partial u^j}\right)\right\} = 0, \tag{5.33}$$

where $\tilde{c}$ is the canonical lift of $c(t)$ and Γ^i_{kl} are the Chern connection coefficients with respect to the natural coordinates [c.f. (3.75)]. Note that in the second term in (5.32) the M-terms [(3.82)] do not contribute in the double contraction, due to (3.41).

Remark. Even though the geodesic equation (5.33) is formally identical to that in Riemannian geometry, the metric tensor g_{ij} depends on both u^i and $T^i(t) = \frac{du^i}{dt}$. Hence the nonlinear dependence on the velocity $\frac{du^i}{dt}$ in the second term on the left hand side of the equation may be more than quadratic. Yet the left hand side scales in the expected manner under an affine transformation $t \to \alpha t + \beta$. More precisely, let $w^i = u^i(\alpha s + \beta)$, $\tilde{w}(s) = \left(w^i(s), \frac{dw^i}{ds}\right)$. Then

$$\frac{d^2 w^i}{ds^2} + \frac{dw^k}{ds}\frac{dw^l}{ds}(\Gamma^i_{kl})_{\tilde{w}} = \alpha^2\left[\frac{d^2 u^i}{dt^2} + \frac{du^k}{dt}\frac{du^l}{dt}(\Gamma^i_{kl})_{\tilde{u}}\right]_{t=\alpha s+\beta}, \tag{5.34}$$

which follows simply from the chain rule and the fact that $\tilde{w}^i(s)$ and $\tilde{u}^i(t)$ are identical points on PTM. This last statement would no longer be true in general if the base manifold for the Finsler bundle p^*TM were either TM, or even the sphere bundle SM, instead on PTM. Eq. (5.34) also implies that if $c(t) = (u^i(t))$ is a geodesic, so is its reverse curve

$$w^i(s) = u^i(a - s), \qquad 0 \le s \le a. \tag{5.35}$$

This is not necessarily true if the base manifold of p^*TM is SM.

Analogous to the development in §5–2 for the Riemannian case, the theory of second order ordinary differential equations applied to (5.33) allows us to define geodesic normal coordinates and normal coordinate neighborhoods in a Finsler manifold. For the Finsler setting we need to introduce the notions of the **Finsler sphere**:

$$S_p(r) = \{X \in TM | F(p, X) = r\}, \tag{5.36}$$

and the **Finsler ball**:

$$B_p(r) = \{X \in TM | F(p, X) < r\}, \tag{5.37}$$

of radius r in $T_p M$. Analogous to (2.9) in Chapter Five, one can define the **exponential map** $\exp_p : T_p M \to M$, which maps rays through $0 \in T_p M$ to unique geodesics passing through p, for small enough $B_p(r)$. Specifically, in the notation of §5–2,

$$u^i = (\exp_x X)^i \equiv f^i(1, x^k, X^k). \tag{5.38}$$

The exponential map is smooth for $X \neq 0$ and C^1 at $X = 0$. (See Bao, Chern and Shen, to appear.) Furthermore $(\exp_x)_*$ is the identity map at $0 \in T_x M$ [c.f. (2.11) in Chapter Five]. Consequently, $\exp_x$ is a diffeomorphism for some small $B_x(r)$.

The generalization of the hyperspheres Σ_σ centered at $x \in M$ (**geodesic spheres**) in Theorem 2.5 of Chapter Five is obtained by exponentiating the Finsler sphere $S_x(\delta)$ in $T_x M$:

$$\Sigma_\sigma = \exp_x(S_x(\delta)). \tag{5.39}$$

Each $X \in S_x(\delta)$ yields a **radial geodesic** $\sigma(t) = \exp_x(tX)$, $0 \leq t \leq 1$, which intersects all geodesic spheres of radii $\leq \delta$ and centered at x. We have (analogous to the Corollary of Theorem 2.3 in Chapter Five):

Lemma 1 (The Gauss Lemma). *A radial geodesic $\sigma(t)$ with velocity $T(t)$ intersects the geodesic spheres orthogonally with respect to the scalar product G_T.*

Proof. For $X \in S_x(\delta)$ and $\tau \in [0, 1]$, consider the radial geodesic

$$\sigma(t) = \exp_x(t\tau X), \qquad 0 \leq t \leq 1. \tag{5.40}$$

Note that $\tau X \in S_x(\tau \delta)$. Let $Y(u)$ be any curve on $S_x(\tau \delta)$ such that $Y(0) = \tau X$. Now define the variation $\sigma : [0, 1] \times (-\epsilon, \epsilon) \to M$ on the radial geodesic $\sigma(t)$ by

$$\sigma(t, u) = \exp_x(tY(u)). \tag{5.41}$$

Denote the velocity of $Y(u)$ at $u = 0$ by V. The variation field $U(t, 0)$ of $\sigma(t, u)$ then satisfies $U(0, 0) = 0$ and $U(1, 0) = \exp_{x*} V$, and the latter is tangent to the geodesic sphere $\exp_x(S_x(\tau\delta))$. Every t-curve of the variation (5.41) is a geodesic emanating from x with constant speed $\tau\delta$. Thus their lengths are all equal and the first variation of arc length $L'(0)$ vanishes. Denoting the velocity field of $\exp_x(tX)$ by T, that of $\sigma(t) = \exp_x(t\tau X)$ is then τT. Eq. (5.29) and the formula for the first variation of arc length (5.28) then imply that

$$G_T(\exp_{x*} V, \tau T) = 0, \tag{5.42}$$

which proves the lemma.   □

Using the Gauss Lemma, we can prove the following important result, which is the generalization of result 2) of Theorem 2.4 in Chapter Five to the Finsler setting:

Theorem 5.1. *Geodesics in a Finsler manifold are locally minimal.*

To prove this theorem we will also need the following fundamental result on the Finsler function $F(u, X)$. (This result has been adapted from Rund 1959. For an extensive discussion on it, refer to Bao, Chern and Shen, to appear).

Lemma 2. *For* (u, X), $(u, V) \in PTM$,

$$\left(\frac{\partial F}{\partial X^i}\right)_{(u,X)} V^i \leq F(u, V). \tag{5.43}$$

Proof. It follows from the positive-definiteness of the Finsler metric that, for $u \in M$ and any $X, V, W \in T_u M$,

$$G_{(u,X)}(V, W) \leq \sqrt{G_{(u,X)}(V, V) G_{(u,X)}(W, W)}, \tag{5.44}$$

where the equality holds if and only if V is proportional to W. Eq. (5.44) is known as the **Cauchy-Schwarz inequality**.

Using (2.12) for the Finsler metric and Euler's theorem, we have

$$g_{ij}(u, X)V^i V^j = F(u, X)V^i V^j \frac{\partial^2 F}{\partial X^i \partial X^j} + V^i V^j \frac{\partial F}{\partial X^i} \frac{\partial F}{\partial X^j}, \tag{5.45}$$

$$X^i g_{ij}(u, X) = F(u, X) = \frac{\partial F}{\partial X^j}, \tag{5.46}$$

and

$$X^i X^j g_{ij}(u, X) = F^2. \tag{5.47}$$

On using (5.46) for $\frac{\partial F}{\partial X^i}$, the Cauchy-Schwarz inequality (5.44) with $W = X$, and then (5.47), the second term on the right hand side of (5.45) is seen to satisfy

$$V^i V^j \frac{\partial F}{\partial X^i} \frac{\partial F}{\partial X^j} \leq g_{ij}(u, X)V^i V^j. \tag{5.48}$$

Thus (5.45) implies

$$\frac{\partial^2 F}{\partial X^i \partial X^j} V^i V^j \geq 0, \tag{5.49}$$

where the equality holds if and only if V is proportional to X. Now expand $F(u, V)$ as

$$
\begin{aligned}
F(u, V) = F(u, X) &+ (V^i - X^i)\frac{\partial F(u, X)}{\partial X^i} \\
&+ (V^i - X^i)(V^j - X^j)\frac{\partial^2 F(u, X + \epsilon(V - X))}{\partial X^i \partial X^j}, \, 0 < \epsilon < 1.
\end{aligned} \tag{5.50}
$$

It then follows from (5.49) that, for all $V \in T_u M$,

$$F(u, X) + (V^i - X^i)\frac{\partial F(u, X)}{\partial X^i} \leq F(u, V). \tag{5.51}$$

Application of Euler's theorem to the second term on the left hand side of the above inequality then yields the lemma.   □

We now return to the proof of Theorem 5.1.

Proof of Theorem 5.1. We follow the proof first given in Bao and Chern 1993. For $p \in M$, choose $\delta > 0$ small enough so that the exponential map $(t, X) \mapsto \exp_p(tX)$, $0 \leq t \leq \delta$, $X \in T_p M$, $F(p, X) = 1$, is a diffeomorphism. This map sends unit vectors X to radial geodesics, and for $t = \delta$, the Finsler sphere $S_p(1)$ onto the geodesic sphere $\exp_p(S(\delta))$. Now consider the radial geodesic of length δ, $\exp_p(tY)$, $0 \leq t \leq \delta$, corresponding to the unit vector $Y \in T_p M$. This geodesic begins and $p \in M$ and ends at a point $q \in \exp_p(S(\delta))$. Construct a neighboring comparison curve from p to q,

$$c(u) \doteq \exp_p(t(u)X(u)), \tag{5.52}$$

where $0 \leq u \leq 1$, $F(p, X(u)) = 1$, $X(0) = X(1) = Y$, $t(0) = 0$, and $t(1) = \delta$. Our objective is to show that the length of $c(u)$ is at least δ.

At each u, the curve $c(u)$ intersects a radial geodesic emanating from p at a point with geodesic normal coordinates $(t(u)X^i(u))$. Thus

$$c(u) = \sigma(t(u), u), \tag{5.53}$$

where $\sigma(t, u)$ is the variation of $\exp_p(tY)$ defined by

$$\sigma(t, u) = \exp_p(tX(u)), \, 0 \leq t \leq \delta, 0 \leq u \leq 1, \tag{5.54}$$

whose *t*-curves are all unit speed radial geodesics of length δ. By the chain rule,

$$\frac{dc}{du} = \frac{\partial \sigma}{\partial t}\frac{dt}{du} + \frac{\partial \sigma}{\partial u} = T\frac{dt}{du} + U. \tag{5.55}$$

Now the length of $c(u)$ is given by

$$L(c(u)) = \int_0^1 du F\left(c(u), \frac{dc}{du}\right). \tag{5.56}$$

On applying Lemma 2 with $V = \frac{dc}{du}$ and $X = T$, we have

$$L(c(u)) \geq \int_0^1 du \left[\left(\frac{\partial F}{\partial X^i}\right)_{(c,T)} T^i \frac{dt}{du} + \left(\frac{\partial F}{\partial X^i}\right)_{(c,T)} U^i\right]$$

$$= \int_0^1 du \left(\frac{dt}{du} + \frac{1}{F}(g_{ij})_{(c,T)} U^i T^j\right). \tag{5.57}$$

The equality obtains since $\left(\frac{\partial F}{\partial X^i}\right)_{(c,T)} T^i = 1$ by virtue of Euler's theorem, the fact that T has unit length $[F(c(u), T) = 1]$, and $F\frac{\partial F}{\partial X^i} = \frac{X^j}{F} g_{ij}$, which follows from (2.12). But Gauss Lemma implies that

$$(g_{ij})_{(c,T)} U^i T^j = 0. \tag{5.58}$$

Thus

$$L(c(u)) \geq t(1) - t(0) = \delta. \tag{5.59}$$

$\square$

Remark. The usual proof of Theorem 5.1 for the Riemannian case (such as that given in the proof of Theorem 2.4 in Chapter Five) does not carry over to the Finsler setting. This is due to the fact that, in Finsler geometry, the metric function used to compute the velocity along the curve $c(u)$ depends on both $c(u)$ and the velocity $\frac{dc}{du}$ itself, whereas in the Riemannian case, it is independent of the velocity. (For details, consult Bao and Chern 1993).

§8–6   The Second Variation of Arc Length and Jacobi Fields

For clarity we will work with the pull-back forms on the variation rectangle again. The **second variation of arc length** is given according to (5.19) by

$$L''(u) = \int_0^a \frac{\partial^2 a^m}{\partial u^2} dt. \tag{6.1}$$

Differentiating (5.23) with respect to u we have

$$\frac{\partial^2 a^m}{\partial u^2} = \frac{\partial^2 b^m}{\partial u \partial t} + a_\alpha^m \frac{\partial b^\alpha}{\partial u} + b^\alpha \frac{\partial a_\alpha^m}{\partial u}. \tag{6.2}$$

The derivatives of the Chern connection coefficients a_α^m can be calculated by using $\tilde{\sigma}^*$ to pull back the structure equations (4.2) and (4.3) for the Chern curvature tensor. We have, on using (5.9), (5.10), (5.17), (5.18), and the symmetry property of the first Chern curvature tensor (4.4),

$$\frac{\partial b_k^i}{\partial t} - \frac{\partial a_k^i}{\partial u} = a_k^j b_j^i - b_k^j a_j^i + R_{kml}^i a^m b^l \tag{6.3}$$
$$+ P_{km\beta}^i a^m b_m^\beta - P_{kj\beta}^i b^j a_m^\beta.$$

Set $i = m$ and $k = \alpha$ in the above equation. Then the first P-term involves $P_{\alpha m \beta}^m$, which vanishes on account of (4.26), (4.12), (4.13), and (3.35). Thus

$$\frac{\partial a_\alpha^m}{\partial u} = \frac{\partial b_\alpha^m}{\partial t} - a_\alpha^j b_j^m + b_\alpha^j a_j^m \tag{6.4}$$
$$- R_{\alpha m l}^m a^m b^l + P_{\alpha j \beta}^m b^j a_m^\beta.$$

On using this equation in (6.2) we obtain

$$\frac{\partial^2 a^m}{\partial u^2} = \frac{\partial}{\partial t} \left(\frac{\partial b^m}{\partial u} + b^\alpha b_\alpha^m \right) - b_\alpha^m \frac{\partial b^\alpha}{\partial t} - b^\alpha a_\alpha^\beta b_\beta^m$$
$$- a^m b^\alpha b^l R_{\alpha m l}^m \tag{6.5}$$
$$+ a_\alpha^m \left(\frac{\partial b^\alpha}{\partial u} + b^\beta b_\beta^\alpha - b^\beta b^j P_{\beta j}^{m\alpha} \right),$$

where $P_{\beta j}^{m\alpha} = \delta^{\alpha\gamma} P_{\beta j\gamma}^m$. Now set $i = \alpha$ in (5.22) and recall $a^\alpha = 0$ [c.f. (5.17)], to obtain

$$\frac{\partial b^\alpha}{\partial t} = a^m b_m^\alpha - b^j a_j^\alpha. \tag{6.6}$$

With this result, (6.5) assumes the form

$$\frac{\partial^2 a^m}{\partial u^2} = \frac{\partial}{\partial t} \left(\frac{\partial b^m}{\partial u} + b^\alpha b_\alpha^m \right)$$
$$+ a_\alpha^m \left(\frac{\partial b^\alpha}{\partial u} + b^\beta b_\beta^\alpha - b^\beta b^j P_{\beta j}^{m\alpha} - \delta^{\alpha\beta} b^m b_\beta^m \right) \tag{6.7}$$
$$- a^m (b_m^\alpha b_\alpha^m + b^\alpha b^l R_{\alpha m l}^m).$$

Using the symmetry property (4.17) of the first Chern curvature tensor, the term involving R in the above equation can be written

$$-a^m b^\alpha b^l R^m_{\alpha m l} = -a^m b^\alpha b^l R_{\alpha i m l} \delta^{im}$$
$$= a^m b^\alpha b^l R_{i \alpha m l} \delta^{im}$$
$$= a^m b^\alpha b^l R_{m \alpha m l}$$
$$= a^m b^i b^j R_{m i m j},$$

where the last equality holds on account of (4.18). Further, due to (3.4) and (3.9),

$$-a^m b^\alpha_m b^m_\alpha = a^m b^\alpha_m \delta_{\alpha\beta} b^\beta_m$$
$$= a^m \delta_{ij} b^i_m b^j_m.$$

We finally restrict (6.1) to $u = 0$ and assume that the base curve is a geodesic. Thus it follows from (5.31) and (3.9) that the term in (6.7) proportional to a^m_α vanishes. The formula for the second variation of arc length of a geodesic in Finsler geometry is therefore:

$$L''(0) = \left(\frac{\partial b^m}{\partial u} - \delta_{ij} b^j b^j_m \right) \Big|_0^a$$
$$+ \int_0^a a^m (\delta_{ij} b^i_m b^j_m + R_{mimj} b^i b^j) dt. \tag{6.8}$$

The term involving the first Chern curvature tensor R_{mimj} can be seen to be proportional to the flag curvature. Indeed, from (3.6) in Chapter Five, the action of the curvature operator $R(T, U)$ on T is given by

$$R(T, U)T = R^j_{ikl} T^k U^l T^i e_j$$
$$= (F(T))^2 R^j_{mml} U^l e_j. \tag{6.9}$$

It follows that

$$G_T(R(T, U)T, U) = g_{ij} F^2 R^j_{mml} U^i U^l$$
$$= F^2 R_{mimj} U^i U^j. \tag{6.10}$$

According to the definition of the flag curvature $K(e_m, U)$ given by (4.38), we then have

$$a^m R_{mimj} b^i b^j = -FK(e_m, U) [G_T(U, U) G_T(e_m, e_m) - \{G_T(e_m, U)\}^2]$$
$$= -a^m K(e_m, U) [\delta_{ij} b^i b^j - (b^m)^2]. \tag{6.11}$$

Note that the term in (6.7) involving the second Chern curvature tensor P does not appear in the result (6.8). This remarkable fact makes the formula for the second variation of arc length in Finsler geometry formally identical to that in Riemannian geometry, with the sectional curvature in the Riemannian case replaced by the flag curvature in the Finsler case.

It is not difficult to see that, using the intrinsic form for the structure equations of the Chern connection [(5.25) and (5.26)], the second variation formula (6.8) can also be written in the intrinsic form

$$
L''(0) = G_T\left(D_U U, \frac{T}{F(T)}\right)\Big|_0^a
$$
$$
+ \int_0^a \frac{1}{F(T)}[G_T(D_T U, D_T U) + G_T(R(T, U)T, U)]dt \qquad (6.12)
$$
$$
- \int_0^a \frac{1}{F(T)}\left(\frac{\partial F(T)}{\partial u}\right)^2 dt,
$$

The reader is asked to supply the details.

We now define the **index form** on smooth vector fields V and W along a geodesic with tangent field T:

$$
I(V, W) \equiv \int_0^a \frac{1}{F(T)}[G_T(D_T V, D_T W) + G_T(R(T, V)T, W)]dt. \qquad (6.13)
$$

The second variation formula (6.12) then assumes the form

$$
L''(0) = I(U, U) + G_T\left(D_U U, \frac{T}{F(T)}\right)\Big|_0^a - \int_0^a \frac{1}{F(T)}\left(\frac{\partial F(T)}{\partial u}\right)^2 dt. \qquad (6.14)
$$

This can be simplified further by introducing the component of U that is G_T-orthogonal to T:

$$
U_\perp \equiv U - G_T\left(U, \frac{T}{F(T)}\right)\frac{T}{F(T)}. \qquad (6.15)
$$

It the follows from the structure equation (5.26) and the geodesic condition (5.29) that

$$
G_T(D_T U_\perp, D_T U_\perp) = G_T(D_T U, D_T U) - \left(\frac{\partial F}{\partial u}\right)^2. \qquad (6.16)
$$

On the other hand, the properties of the flag curvature imply that

$$
G_T(R(T, U_\perp)T, U_\perp) = G_T(R(T, U)T, U). \qquad (6.17)
$$

Using (6.16) and (6.17) in (6.13), we have the following compact form for the second variation expression:

$$L''(0) = I(U_\perp, U_\perp) + G_T\left(D_U U, \frac{T}{F(T)}\right)\Big|_0^a. \qquad (6.18)$$

Let us return to the index form defined by (6.13). Under a constant speed parametrization $[F(T) = \text{constant}]$, $I(J, V)$ can be rewritten in the following form on using integration by parts, the structure equation (5.26), and the geodesic condition (5.31):

$$I(J,V) = \frac{1}{F}G_T(D_T J, V)\Big|_0^a - \frac{1}{F}\int_0^a dt\, G_T(D_T D_T J + R(J,T)T, V). \qquad (6.19)$$

A vector field J along a geodesic is said to be a **Jacobi field** if it satisfies the equation

$$D_T D_T J + R(J,T)T = 0. \qquad (6.20)$$

The above equation is called the **Jacobi equation**, which is formally identical to that in Riemannian geometry. The theory of second order ordinary differential equations implies that for a geodesic $\sigma(t)$, $0 \le t \le a$, with velocity field T, and given $V, W \in T_{\sigma(0)}M$, there exists a unique Jacobi field $J(t)$ along σ such that $J(0) = V$ and $D_T J(0) = W$. We have the following result.

Theorem 6.1. *Given any smooth variation (not necessarily with fixed endpoints) in which all the t-curves are geodesics on a Finsler manifold, the variation field U is necessarily a Jacobi field. Conversely, a Jacobi field $J(t)$ along a geodesic $\sigma(t)$ determines a variation field U of σ.*

Proof. We will consider constant speed geodesics ($D_T T = 0$) and apply D_T to the torsion-free condition $D_T U = D_U T$ [(5.25)] of the Chern connection. With the help of (5.25), (5.25) and (6.3) we obtain

$$\begin{aligned}
D_T D_T U &= D_T D_U T = (D_T D_U - D_U D_T)T \\
&= -(F(T))^2(R^i_{mjm}U^j - P^i_{mm\alpha}\omega^\alpha_m(\tilde{U}))e_i \\
&= (F(T))^2 R^i_{mmj}U^j e_i \\
&= R(T,U)T \\
&= -R(U,T)T,
\end{aligned} \qquad (6.21)$$

where in the third equality we have used (4.23) to eliminate the P-term, in the fourth equality we have used (6.9), and in the last equality, the symmetry property of R given by (4.4) has been used.

Given a Jacobi field $J(t)$ along a geodesic $\sigma(t)$, $0 \leq t \leq a$, with velocity field $T(t)$, we construct a variation, $\sigma(t, u)$, of $\sigma(t)$ as follows. Choose a curve $\gamma(u)$, with $\gamma(0) = \sigma(0)$ and velocity field $V(u)$ such that $V(u = 0) = J(t = 0)$. Next construct vector fields $Y(u)$ and $W(u)$ that are parallel along $\gamma(u)$ $(D_V Y = D_V W = 0)$ and satisfy the initial conditions $Y(0) = T(0)$ and $W(0) = (D_T J)|_{t=0}$. The sought-for variation is then given by

$$\sigma(t, u) = \exp_{\gamma(u)}\{t(Y(u) + uW(u))\}. \tag{6.22}$$

Note that $\sigma(0, u) = \gamma(u)$ and $\sigma(t, 0)$ is the base geodesic $\sigma(t)$. Consider the variation field $U(t) = \left.\frac{\partial \sigma(t, u)}{\partial u}\right|_{u=0}$ of $\sigma(t)$. By the first part of the theorem, $U(t)$ is a Jacobi field. We wish to show that $U(t)$ is in fact $J(t)$, which will follow from the uniqueness of a Jacobi field with given initial conditions provided we can establish $U(0) = J(0)$ and $(D_T U)|_{t=0} = (D_T J)|_{t=0}$. Indeed, we have

$$U(0) = \left.\frac{\partial \sigma(0, u)}{\partial u}\right|_{u=0} = \frac{d}{du}\gamma(u) = V(0) = J(0), \tag{6.23}$$

and

$$\begin{aligned}
(D_T U)|_{t=0} &= (D_U T)\big|_{t=0} \\
&= D_U \frac{\partial}{\partial t} \exp_{\gamma(u)}\{t(Y(u) + uW(u))\}\bigg|_{t=0, u=0},
\end{aligned} \tag{6.24}$$

where in the first equality we have used the torsion-free condition. Due to the fact that $\exp_{\gamma(u)*}$ is the identity at the origin of $T_{\gamma(u)}M$,

$$\frac{\partial}{\partial t} \exp_{\gamma(u)}\{t(Y(u) + uW(u))\}\bigg|_{t=0, u} = Y(u) + uW(u). \tag{6.25}$$

Thus

$$\begin{aligned}
(D_T U)_{t=0} &= D_U(Y(u) + uW(u))\big|_{t=0, u=0} \\
&= D_{U(0)}(Y(u) + uW(u))\big|_{u=0} \\
&= D_{V(0)}(Y(u) + uW(u))\big|_{u=0} \\
&= W(0) = (D_T J)\big|_{t=0}.
\end{aligned} \tag{6.26}$$

This proves the converse part of the theorem.   □

As in Riemannian geometry, Jacobi fields constitute an important tool in the study of the properties of geodesics in Finsler spaces. In particular the relationship between the flag curvature and geodesics can be studied through the notion of **conjugate points**, which can be characterized by the behavior of Jacobi fields at the points in question. (See Def. 6.1 below.) The methods of analysis in the Finsler setting follow closely those employed in Riemannian geometry. This is possible due, again, to the remarkable properties of the Chern connection. In the remainder of this section, we will identify a few important properties of the index form in relation to conjugate points and Jacobi fields (which are also standard results in Riemannian geometry). Space does not permit a presentation of complete proofs, which are very similar to the corresponding ones in the Riemannian case. The reader may consult, for example, Spivak, Vol. III, 1975, and do Carmo, 1993, for the Riemannian case, and Bao, Chern and Shen, to appear, for the Finsler case.

Definition 6.1. Suppose the derivative map of the exponential map, $\exp_{p*}$, is degenerate at $V \in T_p M$, that is, there exists some nonzero $W \in T_V(T_p M)$ such that $\exp_{p*} W = 0 \in T_{\exp_p V} M$, then the point $\sigma(1) = \exp_p V$ on the geodesic $\sigma(t) = \exp_p tV$, $0 \leq t \leq 1$, is said to be **conjugate** to the point $\sigma(0) = p$.

Construct the variation to the geodesic $\sigma(t)$ in the above definition according to

$$\sigma(t, u) = \exp_p(t(V + uW)). \tag{6.27}$$

Each t-curve in this variation is a geodesic. By Theorem 6.1, the variation field $U(t)$ of this variation is a Jacobi field, and (6.27) implies that $U(0) = U(1) = (\exp_p)_* W = 0$. Thus we have

Lemma 1. $\sigma(t)$ *is conjugate to* $\sigma(0)$ *on a geodesic* $\sigma(t) = \exp_p tV$, $0 \leq t \leq 1$, *or equivalently,* $\exp_{p*}$ *is degenerate at* $V \in T_{\sigma(0)} M$, *if and only if there exists a nonzero Jacobi field* $J(t)$ *along* $\sigma(t)$ *such that* $J(0) = J(1) = 0$.

For variations with fixed end-points, that is, $U(0) = U(1) = 0$, $D_U U$ also vanishes at the end-points. It then follows from (6.14) that $L''(0) < 0$ if $I(U, U) < 0$. Thus we have

Lemma 2. *If a variation field* $U(t)$ *of a geodesic* $\sigma(t)$ *keeps the end-points fixed, then* $I(U, U) < 0$ *implies that the geodesic cannot be a minimal geodesic.*

The following are important results concerning the relationship between conjugate points on a geodesic and the sign of the index form on fields vanishing at the end-points. For the remainder of this section, let $\mathcal{V}_0$ denote the vector space of all smooth vector fields $W(t)$ on a geodesic $\sigma(t)$, $0 \leq t \leq a$, joining p and q such that $W(0) = W(a) = 0$. Then

Lemma 3. *(1) If no point along the geodesic* $\sigma(t)$, $0 \leq t \leq a$, *is conjugate to* $\sigma(0)$, *then*

$$I(W, W) \geq 0 \qquad \text{for all} \quad W(t) \in \mathcal{V}_0,$$

and the equality holds only if $W(t) = 0$.

(2) If $\sigma(a)$ is conjugate to $\sigma(0)$, but for all $0 < \tau < a$, $\sigma(\tau)$ is not conjugate to $\sigma(0)$, then

$$I(W, W) \geq 0 \qquad \text{for all} \quad W(t) \in \mathcal{V}_0,$$

and the equality holds if W is a Jacobi field.

We also have

Lemma 4. *For a geodesic $\sigma(t)$, $0 \leq t \leq a$, a necessary and sufficient condition for a point $\sigma(b)$, $0 < b < a$, to be conjugate to $\sigma(0)$ is that there exists a $W \in \mathcal{V}_0$ and G_T-orthogonal to T such that $I(W, W) < 0$.*

The proofs of the above two lemmas can be found in Bao and Chern 1993. As a corollary of Lemma 3, we have

Lemma 5. *Suppose $\sigma(t)$, $0 \leq t \leq a$, is a geodesic containing no conjugate points. Let W and J be smooth fields on σ such that $W(0) = J(0)$, $W(a) = J(a)$, and J is a Jacobi field. Then*

$$I(W, W) \geq I(J, J),$$

and the equality holds if and only if $W = J$.

Proof. By assumption $J - W \in \mathcal{V}_0$. Lemma 3 then implies

$$\begin{aligned}
0 &\leq I(J - W, J - W) \\
&= I(J, J) - 2I(J, W) + I(W, W) \\
&= \frac{1}{F} G_T(D_T J, J) \Big|_0^a - \frac{2}{F} G_T(D_T J, W) \Big|_0^a + I(W, W) \\
&= -\frac{1}{F} G_T(D_T J, J) \Big|_0^a + I(W, W) \\
&= -I(J, J) + I(W, W).
\end{aligned}$$

In the first inequality we have used the bilinearity and symmetry properties of the index form. In the second equality we have used (6.19). Applying Lemma 3 again, we see that the equality holds if and only if $J - W = 0$. □

Finally we state

Lemma 6. $I(V, W) = 0$ *for all $W \in \mathcal{V}_0$ if and only if V is a Jacobi field.*

Proof. If V is a Jacobi field, $I(V, W) = 0$ for all $W \in \mathcal{V}_0$ follows directly from (6.19). Conversely, assume $I(V, W) = 0$ for all $W \in \mathcal{V}_0$. Let $f(t)$ be a smooth function on $[0, a]$ satisfying $f(0) = f(a) = 0$ and $f(t) > 0$ for $0 < t < a$. Choose W to be $f(t)\{D_T D_T V + R(V, T)T\}$. Then, on using (6.19) for $I(V, W)$, we have

$$0 = -\frac{1}{F} \int_0^a dt f(t) G_T(D_T D_T V + R(V, T)T, D_T D_T V + R(V, T)T),$$

which implies that V must be a Jacobi field. $\qquad \square$

§8–7 Completeness and the Hopf-Rinow Theorem

An obvious question on the global properties of a Finsler manifold is whether it is a proper open submanifold of another Finsler manifold. An important concept related to this question is completeness, which is a property of metric spaces. Hence we will first review some general results of metric spaces, and then proceed to study complete Finsler spaces. Out results will automatically apply to the Riemannian case also.

Definition 7.1. A continuous image of the closed interval $0 \le t \le 1$ to a metric space M is called an **arc** in M. A continuous image of the half-open interval $0 \le t < 1$ to M is called a **path** in M.

Let $p(t)$, $0 \le t \le 1$, be an arc in M. If for any $0 \le t_1 \le t_2 \le t_3 \le 1$ we have

$$\rho(p(t_1), p(t_2)) + \rho(p(t_2), p(t_3)) = \rho(p(t_1), p(t_2)), \tag{7.1}$$

where ρ is the distance function of the metric space M, then we call $p(t)$ a **line segment** in M. A line segment in M that has p, q as its initial and end points, respectively, will be denoted by pq.

Since all closed intervals (or half-open intervals) are homeomorphic, the length of any interval in the above definition is not restricted to be 1. The restriction of an arc (or path) to a closed subinterval of the domain is called a **subarc**.

Let $p(t)$, $0 \le t < 1$ be a path in M. If

1) $p(t)$ is a closed subset of M, and
2) every subarc of $p(t)$ is a line segment,

then $p(t)$ is called a **ray** in M. If the function $\rho(p(0), p(t))$, $0 \le t < 1$, is bounded, then its supremum is called the length of the ray.

Example 1. Suppose M is the space obtained by deleting one point from the plane $\mathbb{R}^2$, say, $M = \mathbb{R}^2 - \{0\}$. It is a metric space with respect to the distance function induced from $\mathbb{R}^2$. Let

$$p_1(t) = (t - 1, 0), \quad 0 \le t < 1.$$

Then $p_1(t)$ is a ray in M whose length is obviously finite, and is in fact one. But the ray

$$p_2(t) = \left(1, \frac{1}{1-t}\right), \quad 0 \le t < 1,$$

does not have a finite length.

Example 2. Not every metric space has rays. For instance, there are no rays in a compact metric space. Indeed, suppose $p(t)$, $0 \le t < 1$, is a ray in a compact metric space M. Then the limit $\lim_{t \to 1} p(t) = p_0 \in M$ exists. Since any ray is a closed subset of M, p_0 is on the ray, i.e., there is a t_0, $0 \le t_0 < 1$, such that $p(t_0) = p_0$. Let $t_1 = \frac{t_0+1}{2}$. Then $t_0 < t_1 < 1$. By the definition of rays we have

$$\rho(p(t_0), p(t_1)) + \rho(p(t_1), p(t)) = \rho(p(t_0), p(t))$$

when $t_1 < t < 1$. Hence

$$\rho(p(t_0), p(t)) \ge \rho(p(t_0), p(t_1)) > 0.$$

But as $t \to 1$, the left hand side approaches zero. This is a contradiction. Hence there can be no rays in M.

It is obvious that the unit disc $D = \{(x, y) \in \mathbb{R}^2 | x^2 + y^2 \le 1\}$ in $\mathbb{R}^2$ is a compact metric space, so there cannot be any rays in D.

Lemma 1. *Suppose there is a sequence of points $a_1, \ldots, a_n$ in M which satisfies the condition*

$$\sum_{i=1}^{n-1} \rho(a_i, a_{i+1}) = \rho(a_1, a_n). \tag{7.2}$$

Then for any set of integers $1 \le i_1 \le \cdots \le i_k \le n$, we have

$$\sum_{r=1}^{k-1} \rho(a_{i_r}, a_{i_{r+1}}) = \rho(a_{i_1}, a_{i_k}). \tag{7.3}$$

Proof. Suppose the lemma is false. Then there exists a set of integers $1 \leq i_1 \leq \cdots \leq i_k \leq n$ such that

$$\sum_{r=1}^{k-1} \rho(a_{i_r}, a_{i_{r+1}}) > \rho(a_{i_1}, a_{i_k}).$$

Thus

$$\sum_{i=1}^{n-1} \rho(a_i, a_{i+1}) \geq \rho(a_1, a_{i_1}) + \sum_{r=1}^{k-1} \rho(a_{i_r}, a_{i_{r+1}}) + \rho(a_{i_k}, a_n)$$
$$> \rho(a_1, a_{i_1}) + \rho(a_{i_1}, a_{i_k}) + \rho(a_{i_k}, a_n)$$
$$\geq \rho(a_1, a_n),$$

which contradicts (7.2). Hence (7.3) must be true.   □

Lemma 2. *Suppose $a_k a_{k+1}$ $(1 \leq k \leq n-1)$ are line segments in a metric space M, and*

$$\sum_{k=1}^{n-1} \rho(a_k, a_{k+1}) = \rho(a_1, a_n). \qquad (7.4)$$

Then $\gamma = \sum_{k=1}^{n-1} a_k a_{k+1}$ is a line segment.

Proof. Suppose $n = 3$. Obviously $a_1 a_2 + a_2 a_3$ is an arc. We need to show that it is also a line segment. Suppose x, y, z are any three points in the arc, and y lies between x and z. If x and z both belong to $a_1 a_2$ or $a_2 a_3$, then by the definition of line segments we have

$$\rho(x, y) + \rho(y, z) = \rho(x, z).$$

Now assume that x, $y \in a_1 a_2$ and $z \in a_2 a_3$. Then

$$\rho(a_1, x) + \rho(x, y) + \rho(y, a_2) = \rho(a_1, a_2),$$
$$\rho(a_2, z) + \rho(z, a_3) = \rho(a_2, a_3).$$

Hence, by (7.4), we have

$$\rho(a_1, x) + \rho(x, y) + \rho(y, a_2) + \rho(a_2, z) + \rho(z, a_3) = \rho(a_1, a_2) + \rho(a_2, a_3)$$
$$= \rho(a_1, a_3).$$

By Lemma 1 we then have

$$\rho(x, y) + \rho(y, z) = \rho(x, z).$$

Thus $a_1 a_2 + a_2 a_3$ is a line segment.

By a similar analysis, it is not difficult to prove this lemma for any n by induction.   □

Lemma 3. *Suppose $\beta : p(t)$, $0 \le t < 1$, is a path in a metric space M, and every subarc is a line segment. If β is a closed subset of M, then β is a ray; if it is not a closed subset of M, then $\lim_{t\to1} p(t) = p$ exists, and $\beta + p$ is a line segment.*

Proof. The first conclusion follows from the definition of rays. Now assume that β is not a closed subset of M. Then there must be a limit point p of β such that $p \notin \beta$, and hence a sequence $t_\nu \to 1$, $0 \le t_\nu < 1$, such that $a_\nu = p(t_\nu) \to p(\nu \to +\infty)$. We wish to show that $\lim_{t\to1} p(t) = p$.

We may assume $\{t_\nu\}$ to be a monotonically increasing sequence. Since $p(t_\nu) \to p(\nu \to +\infty)$, for any give positive number ϵ there exists a positive integer N such that when $\nu > N$

$$\rho(p(t_\nu), p) < \frac{\epsilon}{4}. \tag{7.5}$$

Thus for ν, $\nu' > N$ we always have

$$\rho(p(t_\nu), p(t_{\nu'})) \le \rho(p(t_\nu), p) + \rho(p(t_{\nu'}), p) < \frac{\epsilon}{2}. \tag{7.6}$$

Choose $\delta = 1 - t_{N+1} > 0$. Then for $0 < 1 - t < \delta$, we have $t_{N+1} < t < 1$. Since $t_\nu \to 1$, for any such t we can always find an index $\nu' > N$ such that $t_{N+1} < t < t_{\nu'} < 1$. Since any subarc of β is a line segment, we have

$$\begin{aligned}
\rho(p(t_{N+1}), p(t)) &\le \rho(p(t_{N+1}), p(t)) + \rho(p(t), p(t_{\nu'})) \\
&= \rho(p(t_{N+1}), p(t_{\nu'})) < \frac{\epsilon}{2}.
\end{aligned} \tag{7.7}$$

Thus, as long as $0 < 1 - t < \delta$, we have

$$\rho(p(t), p) \le \rho(p(t), p(t_{N+1})) + \rho(p(t_{N+1}), p) < \epsilon,$$

that is,

$$\lim_{t\to1} p(t) = p.$$

Let $p(1) = p$. We will show that $p(t)$, $0 \le t \le 1$, is a line segment M. Suppose x, y, z are any three points on this arc, and y lies between x and z. If $z \ne p(1)$, then by assumption

$$\rho(x, y) + \rho(y, z) = \rho(x, z).$$

If $z = p(1)$, we may assume that $y \ne p(1)$. Since $a_\nu = p(t_\nu) \to p$, y must lie between x and a_ν for sufficiently large ν. Since xa_ν is a line segment, we have

$$\rho(x, y) + \rho(y, a_\nu) = \rho(x, a_\nu).$$

Letting $\nu \to +\infty$, the above result gives,

$$\rho(x, y) + \rho(y, z) = \rho(x, z).$$

Hence $\beta + p$ is a line segment.   □

Now assume that M is a connected Finsler manifold. Under the assumption that the Finsler function F is symmetrically homogeneous, it is also a metric space with respect to the distance function induced by the Finsler metric [c.f. Eq. (2.47) in Chapter Five]. The following lemma shows that a curve segment whose arc length is equal to the distance between its two endpoints is precisely a line segment, as defined in Definition 7.1.

Lemma 4. *Suppose γ is a curve in M connecting points p and q with measurable arc length equal to $\rho(p, q)$. Then γ is a geodesic curve in the Finsler manifold M, and is also a line segment in the metric space M. Conversely, if γ is a line segment in M that connects p and q, then γ is a geodesic curve, and the distance between any two points on γ is equal to the arc length of the part of γ between these two points.*

If a geodesic connecting two points has length equal to the distance between them, then it is called a **minimal geodesic**. The essence of the above lemma is that the concepts of a minimal geodesic and a line segment in a Finsler manifold are equivalent.

Proof of Lemma 4. Let r be a point on the curve γ. Choose a normal coordinate neighborhood U of r such that it satisfies the requirements of Theorem 2.4 in Chapter Five. Then choose a point $r_1 \in \gamma \cap U$ such that the part of γ between r and r_1, denoted by γ_1, lies in U. By Theorem 5.1, there exists a unique geodesic g in U connecting r and r_1 that is at the same time the shortest line connecting these two points. If $g \neq \gamma$, then the length of γ_1 must be greater than that of g, which contradicts the assumption that γ is a curve whose length is equal to the distance between p and q. Therefore γ is a geodesic.

Choose any three points x, y, $z \in \gamma$, with y lying between x and z. Then

$$\widehat{px} + \widehat{xy} + \widehat{yz} + \widehat{zq} = \widehat{pq} = \rho(p, q),$$

where $\widehat{pq}$ denotes the arc length of the curve between p and q. By the definition of the distance between two points in a Finsler manifold, we have

$$\rho(p, q) \leq \rho(p, x) + \rho(x, y) + \rho(y, z) + \rho(z, q) \leq \widehat{px} + \widehat{xy} + \widehat{yz} + \widehat{zq} = \rho(p, q).$$

Hence the equalities must hold. By Lemma 1, we then have

$$\rho(x, y) + \rho(y, z) = \rho(x, z).$$

Therefore γ is a line segment.

Conversely, suppose γ is a line segment connecting p and q. Choose a partition of γ, say $p = r_0, r_1, \ldots, r_n = q$, such that

$$\rho(r_i, r_{i+1}) = \frac{1}{n}\rho(p, q).$$

Then

$$\text{length of } \gamma = \sum_{i=1}^{n-1} r_i r_{i+1}.$$

When n is sufficiently large the difference between the arc length $\widehat{r_i r_{i+1}}$ and $\rho(r_i, r_{i+1})$ is small, of order higher than $\frac{1}{n}$. Therefore

$$\text{length of } \gamma = \sum_{i=0}^{n-1} \left(\rho(r_i, r_{i+1}) + o\left(\frac{1}{n}\right) \right)$$
$$= \rho(p, q) + o(1),$$

that is, length of $\gamma = \rho(p, q)$. γ is thus a geodesic curve. $\qquad\square$

Lemma 5. *For every point p in M there exists a positive number $\rho(p)$ such that*

1) *for any point $q \in M$, if $\rho(p, q) \le \rho(p)$, then there exists a line segment that connects p and q;*
2) *if $\rho(p, q) > \rho(p)$,*

then there is a point $x \in M$ such that $\rho(p, x) = \rho(p)$, and

$$\rho(p, x) + \rho(x, q) = \rho(p, q).$$

Proof. Choose a geodesic sphere W centered at p with radius ϵ such that ϵ has the properties stated in Theorem 2.5 of Chapter Five. We need only choose the positive number $\rho(p)$ such that $\rho(p) < \epsilon$. If $\rho(p, q) < \rho(p)$, then $q \in W$, so there exists a unique geodesic in W that connects p and q, whose length is exactly $\rho(p, q)$ (see the discussion immediately following the proof of Theorem 2.6 of Chapter Five). By Lemma 4, this geodesic is precisely the line segment pq. Hence 1) holds.

Suppose $q \in M$, $\rho(p, q) > \rho(p)$. By the definition of distance, there must exist a sequence $\{\beta_\nu\}$ of rectifiable arcs such that the lengths s_ν of the β_ν approach $\rho(p, q)$ as $\nu \to \infty$. Choose a point x_ν on each β_ν such that $\rho(p, x_\nu) = \rho(p)$. Then the x_ν lie on the geodesic sphere $\sum_{\rho(p)}$ centered at p with radius $\rho(p)$. Due to the compactness of $\sum_{\rho(p)}$, there exists a subsequence of $\{x_\nu\}$ that converges to a point x in $\sum_{\rho(p)}$. We may assume that $\{x_\nu\}$ itself is convergent. Since

$$\rho(x_\nu, q) \leq \widehat{x_\nu q} \leq s_\nu - \rho(p, x_\nu) = s_\nu - \rho(p),$$

we have

$$\rho(p, q) \leq \rho(p, x_\nu) + \rho(x_\nu, q) \leq s_\nu.$$

Let $\nu \to \infty$. Then $s_\nu \to \rho(p, q)$. Hence

$$\rho(p, q) = \rho(p, x) + \rho(x, q).$$

$\square$

Theorem 7.1. *Suppose M is a connected Finsler manifold, and p, q are two points in M. Then one of the following statements must hold:*

1) *there exists a line segment connecting p and q;*
2) *there exists a ray γ starting from p such that the condition*

$$\rho(p, x) + \rho(x, q) = \rho(p, q)$$

holds for any point $x \in \gamma$.

Proof. Suppose no line segment connects p and q. Let $a_i = p$. We need to construct a sequence $a_i a_{i+1}$, $i = 1, 2, \ldots$, of line segments such that the following holds for any positive integer k:

$$\rho(a_1, a_2) + \cdots + \rho(a_{k-1}, a_k) + \rho(a_k, q) = \rho(p, q). \tag{C_k}$$

If there already exist line segments $a_1 a_2, \ldots, a_{k-1} a_k$ such that (C_k) holds, then it follows from Lemma 1 that

$$\rho(a_1, a_2) + \cdots + \rho(a_{k-1}, a_k) = \rho(a_1, a_k), \tag{7.8}$$
$$\rho(a_1, a_k) + \rho(a_k, p) = \rho(a_1, q). \tag{7.9}$$

By Lemma 2, $\gamma_k = \sum_{i=1}^{k-1} a_i a_{i+1}$ is the line segment $a_1 a_k$. Therefore, by assumption, a_k and q cannot be connected by a line segment. Consider the point set

$$S_k = \{x \in M | \ x \text{ and } a_k \text{ can be connected by a line}$$
$$\text{segment and } \rho(a_k, x) + \rho(x, q) = \rho(a_k, q)\}.$$

By Lemma 5, $S_k \neq \varnothing$, and

$$T_{k+1} = \sup_{x \in S_k} \rho(a_k, x) > 0. \tag{7.10}$$

Obviously $T_{k+1} \leq \rho(a_k, q)$. Choose any point $a_{k+1} \in S_k$ with

$$\rho(a_k, a_{k+1}) \geq \frac{1}{2} T_{k+1}. \tag{7.11}$$

Then the line segments $a_1 a_2, \ldots, a_k a_{k+1}$ satisfy condition (C_{k+1}). Let

$$\gamma = \sum_{i=1}^{\infty} a_i a_{i+1}. \tag{7.12}$$

Then γ is a path. Any subarc of γ must be a subarc of some line segment $\sum_{i=1}^{k-1} a_i a_{i+1}$ (as long as k is sufficiently large), and hence must be a line segment. By Lemma 3, we only need to show that $\{a_k\}$ has no limit points in order to show that γ is a ray.

For this, we assume that $\lim_{k \to \infty} a_k = r \in M$. Then by Lemma 3, $\gamma + r$ is a line segment. Letting $k \to \infty$ in (7.9), we have

$$\rho(a_1, r) + \rho(r, q) = \rho(a_1, q). \tag{7.13}$$

Similarly,

$$\rho(a_k, r) + \rho(r, q) = \rho(a_k, q). \tag{7.14}$$

Since, by assumption, no line segment connects a_1 and q, $r \neq q$. By Lemma 5, there must exist a point $x \neq r$ such that x and r can be connected by a line segment, and

$$\rho(r, x) + \rho(x, q) = \rho(r, q). \tag{7.15}$$

Combining (7.14) and (7.15), we have

$$\rho(a_k, r) + \rho(r, x) + \rho(x, q) = \rho(a_k, q).$$

By Lemma 1,

$$\rho(a_k, r) + \rho(r, x) = \rho(a_k, x).$$

Thus, by Lemma 2, $a_k r + rx$ is a line segment. Hence $x \in S_k$, and

$$T_{k+1} \geq \rho(a_k, x) \geq \rho(r, x) > 0. \tag{7.16}$$

This formula holds for any $k \geq 1$. On the other hand, by (7.11),

$$\frac{1}{2} \sum_{k=1}^{\infty} T_{k+1} \leq \sum_{k=1}^{\infty} \rho(a_k, a_{k+1}) \leq \rho(a_1, q).$$

This implies $\lim_{k \to \infty} T_{k+1} = 0$, which contradicts (7.16). Hence the sequence $\{a_k\}$ cannot have a limit point; and thus γ is a ray.

Now choose any point x on γ. When k is sufficiently large, x lies on the line segment $\sum_{i=1}^{k-1} a_i a_{i+1}$. Thus

$$\rho(a_1, x) + \rho(x, a_k) = \rho(a_1, a_k).$$

Combining this with (5.17) we obtain

$$\rho(a_1, x) + \rho(x, a_k) + \rho(a_k, q) = \rho(a_1, q).$$

It follows from Lemma 1 that

$$\rho(a_1, x) + \rho(x, q) = \rho(a_1, q). \tag{7.17}$$

This proves the theorem. $\qquad\qquad\qquad\qquad\qquad\qquad\qquad\qquad\qquad\square$

Remark. In case 2), since

$$\rho(p, x) \leq \rho(p, q), \qquad x \in \gamma,$$

the ray γ has finite length. Therefore, if there exists no rays of finite length in a connected Finsler manifold, any two points can be connected by a line segment. From Example 2, we see that any two points on a compact connected Finsler manifold can be connected by a line segment.

Definition 7.2. Suppose $\{a_n\}$ is a sequence of points in a metric space M. If for any given positive number ϵ there exists a positive integer $N(\epsilon)$ such that

$$\rho(a_n, a_l) < \epsilon$$

for any $n, l > N(\epsilon)$, then we call $\{a_n\}$ a **Cauchy sequence**.

Definition 7.3. If every Cauchy sequence in a metric space M converges, then we say M is **complete**. If a connected Finsler manifold M is also a complete metric space with respect to the distance function induced by the Finsler metric, then we call M a **complete Finsler manifold**.

By the Cauchy criterion for convergent sequences, the Euclidean space $\mathbb{R}^m$ is naturally a complete metric space, and can also be viewed as a complete Riemannian (or Finsler) manifold. But $\mathbb{R}^m$ is not compact. Hence compactness is not the most appropriate condition from the viewpoint of geometry. For the global study of Finsler manifolds, completeness is considered the most appropriate condition.

Theorem 7.2 (Hopf-Rinow). *Suppose M is a connected Finsler manifold. Then the following statements are equivalent*

1) *M is complete;*
2) *any geodesic curve in M can be infinitely extended;*
3) *every closed and bounded subset of M is compact.*

Proof. 1) $\implies$ 2). Suppose there is a geodesic curve γ in M of finite length L that starts from $p \in M$ and that cannot extended. Then the curve can be expressed as $p(t)$, $0 \leq t < 1$, and $\lim_{t \to 1} p(t)$ does not exist. Indeed, if $\lim_{t \to 1} p(t)$ existed, we can take $p(1) = q$ and obtain a geodesic line segment, with $0 \leq t \leq 1$. But every geodesic line segment can always be extended from the endpoints, thus contradicting our assumption. Now choose any monotonically increasing sequence of numbers t_k, $0 \leq t_k < 1$, such that $t_k \to 1$ as $k \to \infty$. Let $a_k = p(t_{k-1})$, $k = 1, 2, \ldots$. Then

$$\sum_{k=1}^{\infty} (a_k, a_{k+1}) \leq L. \tag{7.18}$$

Thus for any given $\epsilon > 0$, there exists a positive integer N such that

$$\sum_{k=l}^{n-1} (a_k, a_{k+1}) < \epsilon$$

for all $n > l > N$. Hence

$$\rho(a_l, a_n) < \epsilon, \tag{7.19}$$

which implies that $\{a_k\}$ is a Cauchy sequence. On the other hand, if M is complete, then there must exist a point $q \in M$ such that $\lim_{k \to \infty} a_k = q$. Obviously, q is independent of the choice of the sequence $\{t_k\}$. Thus $\lim_{t \to 1} p(t) = q$, which is a contradiction. Therefore any geodesic curve can be extended infinitely.

2) $\implies$ 3). Assume that any geodesic curve in M can be infinitely extended. Then there are no rays in M with finite length. By Theorem 7.1, any two points in M can be connected by a line segment. Let S be a bounded infinite subset of M. Then there is a point $a \in M$ and a positive number K such that S is contained in the open geodesic sphere centered at a with radius K. Choose an infinite sequence $\{x_k\}$ in S of distinct points. Since $\rho(a, x_k) < K$, $\{\rho(a, x_k)\}$ is a bounded infinite sequence with a convergent subsequence. We may assume that the sequence itself is convergent, and let

$$\lim_{k \to \infty} \rho(a, x_k) = l \leq K.$$

Connect a and x_k by a line segment ax_k. Let ν_k be the unit tangent vector to ax_k at the point a. Then ν_k lies on the unit Finsler sphere in $T_a M$ centered at the origin. Due to the compactness of the unit sphere, $\{\nu_k\}$ converges to ν, say, also on the unit sphere. Construct a geodesic curve γ starting at the point a along the tangent direction ν. Since γ can be infinitely extended, there exists a point x_0 on γ such that the length of the part of γ between a and x_0 is l. The line segment ax_k is a geodesic that starts at a, is tangent to ν_k, and has length $\rho(a, x_k)$. Because of the continuous dependence of geodesics on the initial conditions, $\lim_{k \to \infty} x_k = x_0$, which implies that x_0 is a limit point of S. If S is also closed, then $x_0 \in S$. Thus every closed and bounded infinite subset S of M has a limit point belonging to the subset, which implies that the subset is compact.

3) $\implies$ 1). In fact, if $\{a_n\}$ is a Cauchy sequence in M, then $\{a_n\}$ must be a bounded set. By condition 3), the closure of this point set must be compact, hence there exists a convergent subsequence $\{a_{n_k}\} \to a_0$ $(k \to \infty)$. Since

$$|\rho(a_n, a_0) - \rho(a_m, a_0)| \leq \rho(a_n, a_m),$$

$\{\rho(a_n, a_0)\}$ is a Cauchy sequence. Therefore

$$\lim_{n \to \infty} \rho(a_n, a_0) = \lim_{k \to \infty} \rho(a_{n_k}, a_0) = 0,$$

that is, $\lim_{n \to \infty} a_n = a_0$. $\qquad\square$

Remark 1. Statement 2) of the above theorem is also equivalent to the following: At any point $p \in M$, $\exp_p$ is defined on all of $T_p M$.

Remark 2. As remarked earlier, the reverse geodesic in M with respect to the Finslet bundle $p^* TM \to SM$ may not be a geodesic. Hence the notion of infinite extendability needs to be replaced by infinite forward extendability when the above theorem is applied to SM. For a discussion of this subtle feature of geodesics, consult Bao, Chern, and Shen, to appear.

Corollary 1. *In a complete Finsler manifold, any two points can be connected by a minimal geodesic curve.*

Proof. Because geodesic curves can be extended infinitely in a complete Finsler manifold, there are no rays with finite length in such a manifold. By Theorem 7.1, any two points can be connected by a line segment, that is, by a minimal geodesic. $\qquad\square$

Corollary 2. *A compact connected Finsler manifold is complete.*

Proof. Because any infinite subset of a compact metric space has a limit point, a compact connected Finsler manifold is complete by the Hopf-Rinow Theorem. $\qquad\square$

Definition 7.4. Let M be a connected Finsler manifold. If M is not a proper open submanifold of another connected Finsler manifold, then M is said to be **non-extendable**.

Theorem 7.3. *A complete Finsler manifold is non-extendable.*

Proof. Let M be a complete Finsler manifold. If M is a proper open submanifold of another connected Finsler manifold M', then we may choose a boundary point $p \in (M' - M) \cap \overline{M}$. By Lemma 3, there exist an ϵ-ballshaped neighborhood U of p in M' such that any point in U can be connected to p by a line segment in M'. Since $p \in \overline{M}$, there exists a point $q \in M \cap U$ such that the part of the line segment qp in M is a ray emanating from q with length $\leq \rho(p,q)$. This contradicts the completeness property of M. Therefore M is non-extendable. $\square$

The restriction of completeness is in fact stronger than non-extendability. For example, the universal covering manifold Π of $E^2 - \{0\}$ is a connected Riemannian manifold. It is non-extendable, but not complete.

To see this, choose a coordinate system (ρ, θ), $0 \leq \rho < +\infty$, $-\infty < \theta < \infty$, in Π with the Riemannian metric

$$ds^2 = d\rho^2 + \rho^2 d\theta^2.$$

Let the covering map be $\pi : \Pi \to E^2 - \{0\}$, such that

$$\pi(\rho, \theta) = (\rho \cos\theta, \rho \sin\theta).$$

π preserves the Riemannian metric locally. If the Riemannian manifold Π were extendable, then the point $\rho = 0$ may be added. But the resulting manifold is no longer a two-dimensional Riemannian manifold. It is obvious that Π is not complete, since the distance between the two points

$$(\rho_1, \theta_1) = (1, 0) \quad , \quad (\rho_2, \theta_2) = (1, \pi)$$

in Π is 2, but no line segment in Π connects them.

§8–8 The Theorems of Bonnet-Myers and Synge

These two classical theorems in Riemannian geometry are important global applications of the second variation of arc length formula and demonstrate beautifully the close relationship between curvature and topology. We will see that, using the Chern connection, they also hold in the Finsler setting.

Let T be an arbitrary vector in T_pM and $e_m = \frac{T}{F(T)}$ be the unit vector along T. Construct a G_T-orthonormal basis $\{e_i\}$, $i = 1,\ldots,m$. Recall the Ricci curvature in the direction e_m at p as defined by (4.39) in §8–4.2. We have

Theorem 8.1 (Bonnet-Myers). *Let M be a complete connected Finsler manifold. If the Ricci curvature along any direction in T_pM for all $p \in M$ has a positive lower bound:*

$$\mathrm{Ric}_p \geq \frac{1}{r^2} > 0,$$

then

a) *M is compact with diameter $\leq \pi r$, and*
b) *the fundamental group $\pi_1(M)$ is finite.*

Before proceeding to the proof of this theorem, we illustrate it using the simple and intuitive example of S^2 embedded in $\mathbb{R}^3$, with the induced metric and radius r. Recall that for any metric space M, the diameter is defined by

$$\mathrm{diam}\, M \equiv \sup\{\rho(p,q); \, p,q \in M\}, \tag{8.1}$$

where ρ is the distance function. For S^2 with radius r, the Ricci curvature $\mathrm{Ric}_p = \frac{1}{r^2}$ and the diameter is πr. It is clear that S^2 is simply connected and complete. Furthermore its fundamental group $\pi_1(S^2) = 1$, and is hence finite.

Let us also use this example to derive an interesting fact which will be useful in the proof of Theorem 8.1, namely: If $a > \pi r$, $r > 0$, then there exists a smooth function $f : [0, a] \to \mathbb{R}$ such that

$$\int_0^a \left\{ \left(\frac{df}{dt}\right)^2 - \frac{f^2}{r^2} \right\} dt < 0, \tag{8.2}$$

and $f(0) = f(a) = 0$. Indeed, consider a unit speed geodesic $\sigma(t)$ of length a in S^2 (with radius r), where $a > \pi r$. Let e_1, e_2 be an orthonormal and parallel frame field along σ, with $e_2 = T$. This geodesic contains a conjugate point to $\sigma(0)$, namely, $\sigma(\pi r)$. Hence by Lemma 4 in section §8-6, there exists a $W(t)$ along σ that is orthogonal to e_2, that satisfies $W(0) = W(a) = 0$, and such that $I(W,W) < 0$. If we write $W(t) = f(t)e_1$, then $f(0) = f(a) = 0$; and the fact that $I(W,W) < 0$ implies, by (6.15) and on recalling that $F = 1$ and e_1 is a parallel field along σ,

$$I(W,W) = \int_0^a \left\{ \left(\frac{df}{dt}\right)^2 - \frac{f^2}{r^2} \right\} dt < 0. \tag{8.3}$$

We will now proceed to the proof of Theorem 8.1.

Proof of Theorem 8.1. a) Let p, q be any two points in M. By Corollary 1 to the Hopf-Rinow Theorem (Theorem 7.2), there exists a minimal geodesic connecting p and q. We shall suppose this to be a unit speed geodesic and call it $\gamma : [0, a] \to M$, $\gamma(0) = p$ and $\gamma(a) = q$. Then $\rho(p, q) = L(\gamma) = a$. We would like to show that $a \leq \pi r$. Assume the contrary, that is, $a > \pi r$. Then by the discussion in the last paragraph, there exists a smooth function $f : [0, a] \to \mathbb{R}$, $f(0) = f(a) = 0$, that satisfies (8.2). Construct a G_T-orthonormal, parallel, frame field e_i, $i = 1 \ldots, m$, along $\gamma(t)$ with $e_m = T$. The almost metric-compatibility condition of the Chern connection guarantees that such a frame field exists. [See the discussion following (3.46)]. Let $W_\alpha(t) = f(t)e_\alpha$, $\alpha = 1, \ldots, m - 1$. Then $W_\alpha(0) = W_\alpha(a) = 0$. We then have, according to (6.15) and (4.39),

$$\sum_{\alpha=1}^{m-1} I(W_\alpha, W_\alpha) = (m - 1) \int_0^a dt \left\{ \left(\frac{df}{dt} \right)^2 - f^2 \operatorname{Ric}(e_m) \right\}. \qquad (8.4)$$

By assumption, $\operatorname{Ric}(e_m) \geq \frac{1}{r^2}$. Hence

$$\sum_{\alpha=1}^{m-1} I(W_\alpha, W_\alpha) \leq (m - 1) \int_0^a dt \left\{ \left(\frac{df}{dt} \right)^2 - \frac{f^2}{r^2} \right\} < 0, \qquad (8.5)$$

where the last inequality follows from (8.2). There must then exists a particular α for which $I(W_\alpha, W_\alpha) < 0$. By construction, this W_α is G_T-orthonormal to $e_m = T$ and satisfies $W_\alpha(0) = W_\alpha(a) = 0$. Hence it is a variation field of the geodesic γ with fixed end-points. By (6.18) we have

$$L''(0) = I(W_\alpha, W_\alpha) < 0.$$

Therefore γ cannot be a minimal geodesic containing p and q. Thus we arrive at a contradiction, and it must be the case that $a \leq \pi r$. Since p and q are arbitrary, diam $M \leq \pi r$. M itself is closed, and by hypothesis, complete. We have just shown that it is also bounded. It follows from statement 3) of the Hopf-Rinow Theorem (Theorem 7.2) that M must be compact.

b) Let $\tilde{M}$ be the universal covering space of M with covering projection $\pi : \tilde{M} \to M$, and the pull-back Finsler structure on $\tilde{M}$ be $\tilde{F} = \pi^* F$. Since the projection π is a **local isometry**, $\tilde{M}$ also satisfies the conditions of the theorem for M. By part a) of the theorem, $\tilde{M}$ is compact. Hence $\pi : \tilde{M} \to M$ must be a finite cover. Now since M is connected, all its fundamental groups $\pi_1(M, p)$ with different base points p are isomorphic. Finally, since $\tilde{M}$ is simply connected, there is a bijection between $\pi_1(M, p)$ and the discrete finite set $\pi^{-1}(p)$. Thus $\pi_1(M)$ is finite. $\qquad \square$

Theorem 8.2 (Synge). *If a Finsler manifold M is compact, orientable, even-dimensional, and has positive flag curvatures, then M is simply connected.*

To prove this theorem, we need the following lemma.

Lemma 1. *In a compact, connected Finsler manifold M, every free homotopy class of loops has a minimal, closed geodesic.*

Proof. Suppose $\pi_1(M) \neq 1$. As in the proof of part b) of Theorem 8.1, introduce the universal covering space $\tilde{M}$ of M with covering projection $\pi : \tilde{M} \to M$ and pull-back Finsler structure $\tilde{F} = \pi^* F$. Then π is a local isometry. Since M is compact, by Corollary 2 to Theorem 7.2, it must be complete. Thus $\tilde{M}$ is also complete. Let $\pi^{-1}(p) = \{\tilde{p}_1, \tilde{p}_2, \tilde{p}_3, \ldots\}$. For $p \in M$, we seek to find a minimal geodesic $\tilde{\gamma}$ in $\tilde{M}$ joining $\tilde{p}_1$ to $\tilde{p}_n$, $n \neq 1$, such that

$$\rho(\tilde{p}_1, \tilde{p}_n) = \inf_{k \geq 2} \rho(\tilde{p}_1, \tilde{p}_k) \equiv \lambda > 0.$$

Such a geodesic will be mapped under π to a minimal closed geodesic in M. Let $\{\tilde{p}_{n_i}\}$ be a sequence of points in $\pi^{-1}(p)$ such that $\rho(\tilde{p}_1, \tilde{p}_{n_i}) \to \lambda$. Such a sequence is obviously bounded. According to the arguments in the proof of 2) $\implies$ 3) in the Hopf-Rinow Theorem (Theorem 7.2), completeness of $\tilde{M}$ implies that $\{\tilde{p}_{n_i}\}$ has a limit point $\tilde{p}$. By the continuity of π, $\tilde{p} \in \pi^{-1}(p)$. Hence $\tilde{p} = \tilde{p}_n$ for some $n \geq 2$, and $\rho(\tilde{p}_1, \tilde{p}_n) = \lambda$. The sought-for curve $\tilde{\gamma}$ is the minimal geodesic joining $\tilde{p}_1$ and $\tilde{p}_n$, which must exist, according to Corollary 1 of the Hopf-Rinow Theorem (Theorem 7.2), by virtue of the completeness of $\tilde{M}$. □

We are now ready to prove the Synge Theorem.

Proof of Theorem 8.2. Suppose there exists a non-trivial $\alpha \in \pi_1(M)$, that is, α is not homotopic to zero. Then by the above lemma, α contains a closed minimal geodesic. Suppose this to be of unit speed and denote it by $\sigma(t)$, $0 \leq t \leq L$. We note that the length $L > 0$. Let the parallel transport by the Chern connection once around σ staring at $\sigma(0) = p$ be $P : T_p M \to T_p M$. Due to the almost metric-compatibility of the Chern connection, P preserves G_T lengths and G_T angles. Thus it is an orientation-preserving isomorphism. Let $\mathcal{W}$ be the G_T-orthogonal complement of T in $T_p M$. Then $\mathcal{W}$ is odd-dimensional since, by hypothesis, M is even-dimensional. Denote the restriction of P to $\mathcal{W}$ by $Q : \mathcal{W} \to \mathcal{W}$. Q is also orientation-preserving and thus must have determinant $+1$ as an orthogonal transformation on $\mathcal{W}$. Since the coefficients of the characteristic polynomial of Q are all real, complex eigenvalues must occur as complex conjugate pairs, and there must be an odd number of real eigenvalues whose product is positive. Hence at least one of these real eigenvalues must be positive. Since Q is G_T length-preserving, all its eigenvalues must have norm one. Thus we conclude that the positive real eigenvalue identified above is in fact equal to one. Consequently, there exist a unit vector $U_\perp \in T_pM$ which is G_T orthogonal to T and which is left invariant by P.

The parallelly transported $U_\perp$ along the closed minimal geodesic σ generates a variation of σ with a variation field $U_\perp(t)$, $0 \leq t \leq L$, which satisfies $D_T U_\perp = 0$ and $G_T(U_\perp(t), U_\perp(t)) = 1$ all along σ. It follows from (6.18) [for $L''(0)$], (6.13) [for the index form], and (4.38) [for the flag curvature] that

$$L''(0) = -\int_0^L K(T, U_\perp)dt.$$

By hypothesis of the theorem, all flag curvatures of M are bounded below by a positive number λ. Thus

$$L''(0) \leq -\lambda L < 0.$$

This implies that σ cannot be a minimal geodesic, and we arrive at a contradiction. It follows that the original supposition in this proof of the existence of a non-trivial $\alpha \in \pi_1(M)$ must be invalid. Hence M must be simply connected. $\qquad\square$

Appendix A

Historical Notes

(by S.S. Chern)

§A–1 Classical Differential Geometry

Differential geometry is a natural outgrowth of the infintesimal calculus. In fact, differentiation is the same as the construction of the tangent line of a curve and integration is the study of areas and volumes. Already in the works of Newton, Liebnitz, and the Bernoulli brothers, the calculus has been found to be an effective tool in the treatment of geometrical and physical problems.

The first contributions to surface theory were made by Euler and Monge. The latter wrote the first book on differential geometry.

§A–2 Riemannian Geometry

Riemann's historical Habilitation lecture in 1854 introduced Riemannian geometry. Its two-dimensional case was already developed by Gauss in 1827, and is the core of differential geometry. The subject was enhanced by Einstein's theory of general realtivity (1915).

The fundamental problem is the form problem: to decide when two Riemannian metrics differ only by a change of coordinates. This problem was solved in the same year, 1870, by E. B. Christoffel and R. Lipschitz by different methods.

Christoffel's solution involved the notion of covariant differentiation and led to the founding, by G. Ricci and T. Levi-Civita, of tensor analysis. The latter plays a fundamental role in differential geometry.

§A–3   Manifolds

The fundamental objects of study in differential geometry are manifolds. These are spaces whose properties are described by coordinates which are defined up to certain transformations but are themselves devoid of meaning. Their utilization creates difficulties. It took Einstein seven years to pass from his special relativity in 1908 to his general relativity in 1915. He explained the long delay in the following words: "Why were another seven years required for the construction of the general theory of relativity? The main reason lies in the fact that it is not so easy to free oneself from the idea that coordinates have an immediate metrical meaning" (Einstein 1949).

The technical and philosophical difficulties were overcome by the development of differential topology. The notion of a topological manifold is subtle. With the introduction of differentiation, however, the powerful tool of the calculus can be applied for analytical treatment, and differential manifolds became accessible objects.

An important analytical tool is the exterior differential calculus (cf. Chapter Three). It was developed by Elie Cartan in 1922.

§A–4   Global Geometry

Classical differential geometry is local, that is, the study is generally restricted to a neighborhood, in which a system of coordinates is valid. In recent years, most efforts have been expended on global geometry, that is, the geometry of a whole manifold. While great progress has been made in this direction, I wish to take this occasion to remark that local differential geometry is interesting in its own right and has many important and difficult problems. Geometry is one topic, whether it is local or global.

Finally, I wish to list a few important global results in differential geometry.

1) *Characteristic classes.*

   This is a generalization of the Euler characteristic. Among many applications they provide a ring homomorphism of the K-ring to the cohomology ring.

2) *Hodge's harmonic integrals.*

   By this analytic approach Hodge gave an analysis of the cohomology of algebraic varieties (see, for example, G. de Rham, 1984). More generally Eells and Sampson introduced the notion of harmonic maps (Eells and Sampson 1964). It is clearly a fundamental notion in differential geometry.

3) *Atiyah–Singer Index Theorem.*

A simple but deep theorem in mathematics is the Riemann–Roch theorem on compact Riemann surfaces. It expresses the extent of the space of meromorphic functions in terms of their zeros, poles, and geometric invariants on the surface, the most significant of which is the Euler characteristic. This result has a far-reaching generalization to an arbitrary manifold. The culmination is the so-called Atiyah–Singer index theorem, which expresses the index of an elliptic operator on a manifold as a geometric invariant in the form of an integral (see Berline, Getzler, and Vergne 1991).

Appendix B

Differential Geometry and Theoretical Physics

(by S.S. Chern)

It was in the fall of 1930 when I first met Professor Zhou Pei-Yuan at Qinghua University. That year I graduated from Nankai University and applied to the graduate school in mathematics at Qinghua. Among the subjects in the entrance examination was mechanics, and Professor Zhou was responsible for setting the exam problems as well as grading the papers. The first time he laid eyes on me he greeted me thus, "I have read your exam." By 1937 we were colleagues in the Southwest Associated University, where I also audited his course on electromagnetism.

Differential geometry and theoretical physics both employ the calculus as a tool. One discipline treats geometrical phenomena, while the other treats physical phenomena. Of the two, the latter is naturally broader in scope. Yet all physical phenomena take place in space. Thus differential geometry is also the foundation of theoretical physics. Both disciplines rely on deductive methods; but theoretical physics must also have experimental support. Geometry is free from this constraint. Consequently the choice of its problems is permitted greater freedom, even though its deductive procedures must be accompanied by mathematical rigor. This freedom propelled mathematics to ever new territories. Those who possess mathematical experience and vision can sail the

---

This essay originally appeared in Chinese in *Papers on Theoretical Physics and Mechanics*, Science Press, Beijing (1982). It was dedicated to the late Professor Zhou Pei-Yuan of Peking University. The present translation is carried out by Kai S. Lam.

uncharted seas and reach new domains of great import. For example, Riemannian geometry as required by the general theory of relativity and the theory of connections on fiber spaces required by gauge field theory have been developed by mathematicians prior to the recognition of their applications in physics. This "confluence of divergent paths" of mathematics and physics is indeed a mysterious phenomenon.

The relationship between differential geometry and theoretical physics is beyond simple description. In this essay I offer a few humble perspectives to stimulate discussion.

§B–1   Dynamics and Moving Frames

To describe the motion of a rigid body in dynamics, we attach an orthonormal frame rigidly to the body, and then describe the motion of the frame. By an orthonormal frame in three-dimensional space we mean a point x, together with three mutually perpendicular unit vectors e_i passing through x, $i = 1, 2, 3$. If x also represents the position vector of the point x, we have

$$
\begin{aligned}
\frac{dx}{dt} &= \sum_i p_i(t)\, e_i, \\
\frac{de_i}{dt} &= \sum_j q_{ij}(t)\, e_j, \qquad 1 \le i, j \le 3,
\end{aligned}
\tag{B.1}
$$

where t is time, and

$$
q_{ij}(t) + q_{ji}(t) = 0.
\tag{B.2}
$$

The functions $p_i(t)$ and $q_{ij}(t)$ give a complete description of the motion of the orthonormal frame and the rigid body.

There is an intimate relationship between dynamics and the theory of curves in space. In fact, the latter can be viewed as a special example of the former. In order to exploit this relationship in the theory of surfaces, we need to consider reference frames specified by two parameters. This program was successfully and spectacularly completed by the great French geometer G. Darboux (1842—1917). His magnum opus, the four volumes of *Theorie des Surfaces*, is a classic in differential geometry.

It was Elie Cartan (1869—1951) who further developed this so-called method of moving frames to new heights. Gauss, Riemann, and Cartan are generally recognized to be the three greatest differential geometers in history. At present, the method of moving frames has become a most important tool in differential geometry. We will attempt to describe it in the following paragraphs.

For multi-parameter families of orthonormal frames, the analogous equations to (B.1) are partial differential equations whose coefficients satisfy certain integrability conditions. The best way to represent these conditions is by using exterior derivatives. We rewrite (B.1) and (B.2) in the form

$$dx = \sum_i \omega_i e_i, \qquad de_i = \sum_j \omega_{ij} e_j, \qquad 1 \le i, j \le 3, \tag{B.3}$$

$$\omega_{ij} + \omega_{ji} = 0, \tag{B.4}$$

where ω_i and ω_{ij} are differential one-forms in parameter space. The most general case is when parameter space is the space of all orthonormal frames. This space is six-dimensional, since we need three coordinates to fix x, and an orthonormal frame with origin at x is specified by three parameters. Given an orthonormal frame there is a unique motion which transforms it to another. Thus the space of all orthonormal frames is homeomorphic to the motion group, which we will denote by G.

Exteriorly differentiating (B.3), we obtain

$$\begin{aligned} d\omega_i &= \sum_j \omega_j \wedge \omega_{ji}, \\ d\omega_{ij} &= \sum_k \omega_{ik} \wedge \omega_{kj}, \qquad 1 \le i, j, k \le 3, \end{aligned} \tag{B.5}$$

where "$\wedge$" denotes an exterior product. These are the Maurer–Cartan equations for the group G, which are dual to the multiplication equations for the Lie algebra of G. We thus see that the progression from dynamics to moving frames, and then to the fundamental equations of a Lie group, is one involving a series of natural steps.

This progression can be further extended. After the appearance of Einstein's general theory of relativity, Cartan published an article in 1925 in which the theory of generalized affine spaces and its application to the theory of relativity were developed (see Cartan, 1937). A conclusion of this paper is a generalization of (B.5):

$$\begin{aligned} d\omega_i &= \sum_j \omega_j \wedge \omega_{ji}, \\ d\omega_{ij} &= \sum_k \omega_{ik} \wedge \omega_{kj} + \Omega_{ij}, \qquad 1 \le i, j \le 3, \end{aligned} \tag{B.6}$$

where Ω_{ij} is a differential two-form, called the curvature form. This equation is the fundamental equation for three-dimensional Riemannian geometry.

The usual approach to differential geometry is through tensor analysis. Its basic viewpoint is to use the tangent vectors of local coordinates as coordinate frames. From a contemporary perspective, the disadvantages of this restriction outweigh the advantages. However, the simplicity and clarity of tensor analysis make it undeniably useful in applications to elementary problems.

§B–2   Theory of Surfaces, Solitons and the Sigma Model

Let S be a surface in 3-dimensional Euclidean space E^3. At a point $x \in S$, let x also denote the corresponding position vector, and let ξ denote the normal unit vector. Then the invariants of S are the two differential two-forms:

$$
\begin{aligned}
\mathrm{I} &= (dx, dx) > 0, \\
\mathrm{II} &= -(dx, d\xi),
\end{aligned}
\tag{B.7}
$$

respectively called the first and second fundamental forms; and the former is always positive definite. The two eigenvalues κ_i, $i = 1, 2$, of the second fundamental form are called the principal curvatures of S. The symmetric functions of these eigenvalues:

$$
\begin{aligned}
H &= \frac{1}{2}\left(\kappa_1 + \kappa_2\right), \\
K &= \kappa_1 \kappa_2,
\end{aligned}
\tag{B.8}
$$

are respectively called the mean curvature and the total curvature (Gauss curvature). It is well-known that these curvatures have simple geometric interpretations. For example, a surface with $H = 0$ is a minimal surface.

Surfaces with constant mean curvature or total curvature are obviously worthy of study. If x, y, z are coordinates in E^3, and S is expressed by the equation

$$
z = z(x, y)
\tag{B.9}
$$

then the equations

$$
H = \text{constant} \qquad \text{or} \qquad K = \text{constant}
$$

can be expressed as second-order nonlinear partial differential equations of the function $z(x, y)$. The problem of finding surfaces with constant curvatures is equivalent to the solution of the corresponding differential equations. For example, the equation for the minimal surface $H = 0$ is

$$
\left(1 + z_y^{\,2}\right) z_{xx} - 2 z_x z_y z_{xy} + \left(1 + z_x^{\,2}\right) z_{yy} = 0.
\tag{B.10}
$$

This equation is of non-linear elliptic type.

Another important example is when K is a negative constant. We may assume $K = -1$. The asymptotic curves on such surfaces are non-overlapping real curves. Let φ be the angle between two such curves. Then we can choose parameters u, t on S such that

$$\varphi_{ut} = \sin(\varphi). \tag{B.11}$$

This is the famous Sine-Gordon (SG) equation. Conversely, if a solution to the SG-equation is known, then we can construct a $K = -1$ surface from it.

According to the above discussion, the transformation theory of surfaces has important applications to the theory of partial differential equations. The basis of this claim is the following theorem, known as the Bäcklund theorem. Let the surface S and S^* be paired in such a way that the straight line joining the corresponding paired points $x \in S$ and $x^* \in S^*$ is a common tangent line of the two surfaces. Let r be the distance between the paired points, and ν be the angle between the normals to the surfaces at the paired points. If $r = $ constant and $\nu = $ constant, then the total curvatures of S and S^* are both equal to the negative constant $-\sin\nu/r^2$.

This theorem allows us to start from a given surface of constant total curvature and generate another surface with the same constant total curvature. In other words, from a given solution to the SG-equation we can generate a new solution.

If $\varphi(u, t)$ describes a wave motion on a straight line u, and t is time, then the SG-equation has soliton solutions. The transformation described above will lead to new solutions, with an increased or decreased soliton number. In this way, one can obtain solutions to the SG-equation with an arbitrary soliton number.

A surface of constant negative total curvature generalizes in the case of higher dimensions to an n-dimensional constant curvature submanifold of the $(2n-1)$-dimensional Euclidean space E^{2n-1}. Such a submanifold is determined by a system of partial differential equations, which may be a higher-dimensional generalization of the SG-equation. Chuu Lian Terng and the Brazilian mathematician Keti Tenenblat have proved the higher-dimensional generalization of the Bäcklund theorem (Tenenblat and Terng 1980).

Surfaces with constant mean curvature, or minimal surfaces, find equally diversified applications in theoretical physics. If $f : X \longrightarrow Y$ is a mapping between two Riemannian manifolds, then we can define an energy functional $E(f)$. A critical mapping for this functional is called a harmonic mapping. This setup is a generalization of harmonic functions and minimal submanifolds. Harmonic mappings satisfy a system of elliptic second-order partial differential equations. When X is compact, harmonic mappings are relatively rare. Since the study of such mappings originates from variational principles, they are likely to find applications in physics.

From a geometrical perspective, for given manifolds X and Y with $\dim X <$ $\dim Y$, the problem of how to imbed or immerse X in Y as a minimal sub-manifold is an extremely interesting one. Even for the 2-sphere $X = S^2$, the problem is already non-trivial. Under this assumption for X, E. Calabi (Calabi 1967), S. S. Chern (Chern 1970), and L. Barbosa (Barbosa 1975) have stud-ied the case $Y = S^n$ (the n-sphere) many years ago. In 1980, the physicists A. M. Din and W. J. Zarkrzewski (Din and Zarkrzewski 1980) determined all the harmonic mappings $f : S^2 \longrightarrow P_n(\mathbb{C})$ (the n-dimensional complex pro-jective space) in the so-called σ-model. When Y is some other space, such as $SU(n)$, $Q_n(\mathbb{C})$ (the complex hyperquadric), or $G(n, k)$ (the Grassman mani-fold), knowledge of the identity of the minimal 2-sphere is eagerly sought after. This problem is as yet not completely solved[a].

There exist so-called strong "regularity" properties in the mathematical analysis of minimal surfaces. This means that under certain boundary condi-tions, there exist regular or smooth minimal surfaces. This important result has numerous applications in geometry. Within the context of the general theory of relativity, R. Schoen and S. T. Yau used it to prove the so-called "positive mass conjecture" (Schoen and Yau 1979).

§B–3   Gauge Field Theory

The mathematical basis of gauge field theory is the concept of vector bundles. The evolution of this concept is very natural within mathematics. The object of study in Newton's calculus is the function $y = f(x)$. We can generalize to the situation where the independent variables are the coordinates in an m-dimensional space to obtain a vector valued function of m variables. Usually we also represent this function by the mapping $f : X \longrightarrow Y$, where $X = \mathbb{R}^m$ and $Y = \mathbb{R}^n$. Such a mapping can be represented by a "graph" $F : X \longrightarrow X \times Y$, $F(x) = (x, f(x))$, $x \in X$. The range of the mapping F is the product of two topological spaces. Let $\pi : X \times Y \longrightarrow X$ be such that $\pi(x, y) = x$, $x \in X$, $y \in Y$, then F satisfies the condition $\pi \circ F(x) = x$.

The concept of the vector bundle is of critical importance in modern math-ematics. The crucial idea is to replace $X \times Y$ by a space E which is a product only locally. In other words, there is a space E and a mapping $\pi : E \longrightarrow X$ such that at every point $x \in X$ there is a neighborhood U satisfying the con-dition that $\pi^{-1}(U)$ is homeomorphic to $U \times Y$.

Does a space with local product structure necessarily also have a global product structure? Equivalently, is the space E introduced above necessarily

---

[a]This problem has recently been solved by Jon Wolfson. See Wolfson 1988.

homeomorphic to $X \times Y$? This problem is an extremely intriguing one in mathematics. Its solution entails the notion of the so-called characteristic classes. (The answer is that E is not necessarily $X \times Y$.)

Let a vector bundle be given by $\pi : E \longrightarrow X$. A mapping $F : X \longrightarrow E$ satisfying the condition $\pi \circ F(x) = x$, $x \in X$, is called a section. To carry out differentiation on a section, we need a connection. From the connection we obtain the curvature, which measures the non-commutativity of differentiations.

A gauge field is precisely a connection on a vector bundle. Physicists call it a gauge potential, and the curvature a field strength. This is a marvelous example of the synergistic relationship between differential geometry and theoretical physics.

My understanding is that all fundamental theories of physics have to ultimately undergo the process of "quantization." Mathematically, we will need to study infinite-dimensional spaces and the associated discrete phenomena.

§B–4   Conclusion

I will of course need to mention the relationship between general relativity and Riemannian geometry. Without the theory of relativity, Riemannian geometry would hardly have enjoyed the status it does among mathematicians.

Professor C. N. Yang has given a pictorial depiction of the relationship between mathematics and physics (Yang 1980). I offer another drawing (Figure 15) to conclude this essay.

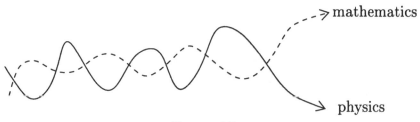

FIGURE 15.

References

L. Auslander and R. E. Mackenzie, *Introduction to Differential Manifolds* (Mc-Graw Hill, New York, 1963).

D. Bao and S. S. Chern, *On a Notable Connection in Finsler Geometry*, Houston Journal of Mathematics, Vol. **19**, No. 1 (1993), 135–180.

D. Bao, S. S. Chern and Z. Shen, ed., *Finsler Geometry (Proceedings of the Joint Summer Research Conference on Finsler Geometry, July 16–20, 1995, Seattle, Washington)*, Contemporary Mathematics, Vol. **196**, Amer. Math. Soc., Providence, RI (1996).

D. Bao, S. S. Chern and Z. Shen, *An Introduction to Riemann-Finsler Geometry* (Springer-Verlag, New York, to appear).

J. L. M. Barbosa, *On Minimal Immersions of S^2 into S^{2m}*, Trans. Amer. Math. Soc., Vol. **210**, (1975), 75–106

N. Berline, E. Getzler, and M. Vergne, *Heat Kernels and Dirac Oparators*, Vol. **298** of Grundlagen der Mathematischen Wissenschaften (Springer-Verlag, New York, 1991).

W. H. Boothby, *An Introduction to Differentiable Manifolds and Riemannian Geometry* (Academic Press, New York, 1975).

E. Calabi, *Minimal Immersions of Surfaces in Euclidean Spaces*, J. of Differential Geom., Vol. **1** (1967), 111–125.

E. Cartan, *La Théorie des Groupes Finis et Continus et la Géométrie Differentielle Traitées par la Méthode du Repére Mobile* (Gauthier-Villars, Paris, 1937).

E. Cartan, *Les Systémes Differentiels Extérieurs et leurs Applications Géométriques* (Hermann, Paris, 1945).

(a) S. S. Chern, *On a Theorem of Algebra and its Geometric Application*, J. Indian Math. Soc., Vol. **9** (1944), 29–36.

(b) S. S. Chern, *A simple Intrinsic Proof of the Gauss-Bonnet Formula for Closed Riemannian Manifolds*, Annals of Math., Vol. **45** (1944), 747–752. Reprinted in *Shiing-Shen Chern Selected Papers*(Springer-Verlag, New York, 1978), 83–88.

S. S. Chern, *Characteristic Classes of Hermitian Manifolds*, Annals of Math., Vol. **47**, No. 1 (1946), 85–121. Reprinted in *Shiing-Shen Chern Selected Papers* (Springer-Verlag, New York, 1978), 101–137.

S. S. Chern, *Local Equivalence and Euclidean Connections in Finsler Spaces*, Science Reports Tsing Hua University, Vol. **5** (1948), 95–121. Reprinted in *Shiing-Shen Chern Selected Papers, Vol. II* (Springer-Verlag, New York, 1989), 194–212.

S. S. Chern, *Differentiable Manifolds* (mimeographed), University of Chicago, (1953), 166pp.

S. S. Chern, *An Elementary Proof of the Existence of Isothermal Parameters on a Surface*, Proc. Amer. Math. Soc., Vol. **6** (1955), 771–782.

(a) S. S. Chern, *Curves and Surfaces in Euclidean Space*, in *Global Geometry and Analysis*, MAA Studies in Mathematics, Vol. **4**, ed. by S. S. Chern (The Mathematical Association of America, Prentice Hall, Inc., Inglewood Cliffs, 1967), 16–56.

(b) S. S. Chern, *Complex Manifolds Without Potential Theory* (D. Van Nostrand, Princeton, 1967); second edition, revised (Springer-Verlag, New York, 1979).

S. S. Chern, *On the Minimal Immersions of the 2-Sphere in a Space of Constant Curvature*, in *Problems in Analysis* (Princeton Univeristy Press, Princeton, 1970) 27–40. Reprinted in *Shiing-Shen Chern Selected Papers, Vol. III* (Springer-Verlag, New York, 1989), 141–154.

(a) S. S. Chern, *Finsler Geometry is just Riemannian Geometry without the Quadratic Restriction*, Notices of the AMS (Septemeber 1996) 959–963.

(b) S. S. Chern, *Riemannian Geometry as a Special Case of Finsler Geometry*, Contemporary Mathematics, Vol. **196** (1996), 51–58.

(c) S. S. Chern, *Remarks on Hilbert's 23rd Problem*, The Mathematical Intelligencer, Vol. **18**, No. 4 (1996), 7–8.

C. Chevalley, *Theory of Lie Groups* (Princeton University Press, Princeton, 1946).

G. de Rham, *Differentiable Manifolds*, Vol. **266** of Grundlagen der Mathematischen Wissenschaften (Springer-Verlag, New York, 1984).

A. M. Din and W. J. Zakrzewski, *General Classical Solutions in the CP^{n-1} Model*, Nuclear Physics **B174** (1980), 397–406.

M. P. do Carmo, *Riemannian Geometry*, translated by F. Flaherty (Birkhauser, Boston, 1992).

S. K. Donaldson and P. B. Kronheimer, *The Geometry of Four-Manifolds* (Clarendon Press, Oxford, New York, 1991).

J. Eells and J. H. Sampson, *Harmonic Mappings into Riemannian Manifolds*, Amer. J. Math., Vol. **86** (1964), 109–160.

A. Einstein, *Autobiographed Notes*, in *Albert Einstein: Philosopher-Scientist*, edited by P. A. Schilpp (Open Court, La Salle, Ill., 1949), Vol. **1**, p. 67.

P. A. Griffiths, *On Cartan's Mathod of Lie Groups and Moving Frames as applied to Uniqueness and Existence Questions in Differential Geometry*, Duke Math. J., Vol. **14** (1974), 775–814.

N. J. Hicks, *Notes on Differential Geometry* (D. Van Nostrand, Princeton, 1965).

W. Hurewicz, *Lectures in Ordinary Differential Equations* (M. I. T. Press, Cambridge, 1966).

S. Kobayashi and K. Nomizu, *Foundations of Differential Geometry*, Vols. I and II (Interscience, New York, 1963 and 1969).

J. Milnor, *On Manifolds Homeomorphic to the 7-Sphere*, Annals of Math., Vol. **64** (1956), 394–405.

J. Milnor, *Morse Theory* (Princeton University Press, Princeton, 1963).

J. R. Munkres, *Topology, A First Course* (Prentice-Hall, Inc., Inglewood Cliffs, 1975).

A. Newlander and L. Nirenberg, *Complex Analytic Coordinates in Almost Complex Manifolds*, Annals of Math., Vol. **65** (1957), 391–404.

K. Nomizu, *Lie Groups and Differential Geometry*, Publ. Math. Soc. of Japan, No. 2 (1956).

H. Rund, *The Differential Geometry of Finsler Spaces* (Springer-Verlag, Berlin, 1959).

R. Schoen and S. T. Yau, *On the Proof of the Positive Mass Conjecture in General Relativity*, Comm. Math. Phys., Vol. **65** (1969), 45–76.

I. M. Singer and J. A. Thorpe, *Lecture Notes on Elementary Topology and Geometry*, Undergraduate Texts in Mathematics (Springer-Verlag, New York, 1976).

M. Spivak, *A Comprehensive Introduction to Differential Geometry*, Vols. I–V (Publish or Perish, Inc., Boston, 1979).

N. Steenrod, *The Topology of Fiber Bundles* (Princeton University Press, Princeton, 1951).

K. Tenenblat and C. L. Terng, *Bäcklund's Theorem for n-Dimensional Submanifolds of $\mathbb{R}^{2n-1}$*, Annals of Math., Vol. **111** (1980), 477–490.

W. P. Thurston, *Three-Dimensional Geometry and Topology*, Vol. **1**, edited by Silvio Levy (Princeton University Press, Princeton, 1997).

J. G. Wolfson, *Harmonic Sequences and Harmonic Maps of Surfaces into Complex Grassmann Manifolds*, J. of Differential Geom., Vol. **27**, No. 1 (1988), 161–178.

C. N. Yang, *Fiber Bundles and the Physics of the Magnetic Monopole*, in *The Chern Symposium 1979* (Springer-Verlag, New York, 1980), 247–253.

Index